Mechanisms of Forest Response to Acidic Deposition

Alan A. Lucier Sharon G. Haines
Editors

Mechanisms of Forest Response to Acidic Deposition

With 12 illustrations

Springer-Verlag
New York Berlin Heidelberg
London Paris Tokyo Hong Kong

Alan A. Lucier
Research Forester
NCASI
New York, NY 10016, USA

Sharon G. Haines
Soils Research Section Leader
International Paper
Bainbridge, GA 31717, USA

Library of Congress Cataloging-in-Publication Data
Mechanisms of forest response to acidic deposition/Alan A. Lucier, Sharon G. Haines, editors.
p. cm.
Includes bibliographical references.
ISBN 0-387-97205-6 (alk. paper)
1. Trees—Effect of acid deposition on. 2. Forest ecology. 3. Trees—Nutrition. 4. Soil ecology. I. Lucier, Alan Alfred, 1984– . II. Haines, Sharon.
SB745.4.M43 1990
581.5'2642—dc20 89-48303

Printed on acid-free paper.

Media conversion by Impressions, Inc., Madison, Wisconsin.
Printed and bound by Edwards Brothers, Inc., Ann Arbor, Michigan.
Printed in the United States of America.

9 8 7 6 5 4 3 2 1

ISBN 0-387-97205-6 Springer-Verlag New York Berlin Heidelberg
ISBN 3-540-97205-6 Springer-Verlag Berlin Heidelberg New York

Preface

This volume addresses several key mechanisms by which forest ecosystems might respond to acidic deposition. These mechanisms have been identified in numerous research papers and general review articles during the past decade. The reviews presented here take the next step by analyzing the mechanisms in detail and developing specific recommendations for further research.

We have not attempted to provide a comprehensive treatment of acidic deposition and forest ecosystems. The strength of this volume lies in its in-depth treatment of selected technical issues, in the extensive lists of citations, and especially in the insights provided by the authors of the individual chapters. The book is intended for professionals and graduate students in soil science, forest ecology, and environmental science. It will also be of use to environmental policy makers and forest managers who need to understand technical issues affecting assessments of acidic deposition and its effects on natural resources.

A key mechanism not specifically addressed in this volume is accelerated leaching of basic nutrient cations from soils. This mechanism has been reviewed by J.O. Reuss and D.W. Johnson in their authoritative book *Acid Deposition and the Acidification of Soils and Waters* (Springer-Verlag 1986). Reuss and Johnson identified mineral weathering rate as an important uncertainty in their analysis. This uncertainty is analyzed in detail here by G.N. White, S.B. Feldman, and L.W. Zelazny in Chapter 4.

Funding for the reviews in this volume was provided by the National Council of the Paper Industry for Air and Stream Improvement (NCASI). The review topics were selected during initial planning of NCASI's Air Quality/Forest Health Program by a work group of soil scientists from U.S. forest products companies. We extend special thanks to Dr. Nicholas Berenyi of Westvaco Corp. and Mr. Anthony Filauro of Great Northern Paper Company for valuable contributions to this project from start to finish.

ALAN A. LUCIER
SHARON G. HAINES

Contents

Contributors

STEVEN B. FELDMAN, Research Associate, Department of Crop and Soil Environmental Sciences, Virginia Polytechnic Institute and State University, Blacksburg, Virginia 24061, USA.

SHARON G. HAINES, Leader of the Soils Research Section, International Paper, Bainbridge, Georgia 31717, USA.

ALAN A. LUCIER, Research Forester, National Council of the Paper Industry for Air and Stream Improvement (NCASI), New York, New York 10016, USA.

DAVID D. MYROLD, Associate Professor, Department of Soil Science, Oregon State University, Corvallis, Oregon 97331-2213, USA.

GEOFFREY G. PARKER, Forest Ecologist, Smithsonian Environmental Research Center, Edgewater, Maryland 21037-0028, USA (formerly with the New York Botanical Garden's Institute of Ecosystem Studies).

WILLIAM H. SMITH, Professor of Forest Biology, School of Forestry and Environmental Studies, Yale University, New Haven, Connecticut 06511, USA.

G. NORMAN WHITE, Research Associate, Department of Crop and Soil Environmental Sciences, Virginia Polytechnic Institute and State University, Blacksburg, Virginia 24061, USA.

JEFFREY D. WOLT, Project Leader, Environmental Soil Chemistry, Formulations and Environmental Chemistry Department, Dow Elanco, Midland, Michigan 48641-1706, USA.

LUCIAN W. ZELAZNY, Professor, Department of Crop and Soil Environmental Sciences, Virginia Polytechnic Institute and State University, Blacksburg, Virginia 24061, USA.

1
Overview and Synthesis

ALAN A. LUCIER and SHARON G. HAINES

I. Introduction

Acidification of forest ecosystems is a complex process controlled by numerous hydrologic, geochemical, and biological factors. In the context of soil genesis, acidification is a normal consequence of mineral weathering, biological activity, and base cation leaching in humid environments (precipitation exceeds evapotranspiration). Relevant time scales are decades to millenia. In the context of ecology and environmental science, acidification may result from biomass harvest, land use conversions, natural changes in vegetation, and intentional or unintentional inputs of acidifying compounds, especially those of sulfur and nitrogen. Relevant time scales range from days to centuries.

Atmospheric inputs of acidifying compounds derived from fossil fuel combustion are commonly known as acidic deposition or "acid rain." In many parts of the world, acidic deposition measurably alters the chemistry of solutions in forest ecosystems. Annual proton inputs by wet and dry acidic deposition exceed net proton production by natural processes at some sites (van Breemen et al. 1984). The capacity of forest ecosystems to buffer acid inputs is great, however, and the extent and magnitude of ecological effects are unknown. The numerous factors that control ecosystem response to acid inputs vary greatly in space and time and interact in ways which are not fully understood. Effects of acidic deposition are subtle, and methods for their measurement are still under development.

Numerous reviews and assessments of acidic deposition's potential impacts on forests have been published (e.g., Abrahamsen 1980, Kulp 1987, Morrison 1984, Rehfuess 1981). Available data suggest that ambient rainfall acidity does not visibly damage foliage or reduce tree growth. Questions remain about the potential for direct injury of tree tissues by other mechanisms. These include acid toxicity to reproductive structures and sensitive life stages (Cox 1983, Percy 1986); erosion of cuticle by long-term exposure to wet and dry acidic deposition (Huttunen and Laine 1983, Percy and Baker 1987); and foliar injury caused by highly acidic solutions deposited as liquid aerosols (e.g., fog) or formed on vegetation surfaces by dissolution of accumulated dry-deposited acids (Jacobson 1984). Well-designed experiments could determine the potential significance of these mechanisms within 5 to 10 years. Possible interactions of acidic deposition with other stresses pose a more formidable problem (Huttunen 1984, Laurence 1981).

Scientific opinion is extremely divided regarding potential impacts of acidic deposition on forest element cycles and tree nutrition. Theoretically sound mechanisms of impact have been described, but no consensus exists as to their importance as determinants of forest health and productivity. Frequently mentioned mechanisms include:

(a) Accelerated leaching of base cations from soils and foliage, leading to nutrient deficiency or imbalance.
(b) Increased mobilization of aluminum and other metals, leading to root damage, nutrient deficiency, and reduced drought tolerance.
(c) Inhibition of soil biological processes, leading to reduced organic matter decomposition, nutrient deficiency, damage to mycorrhizae, and altered host-pathogen relationships.
(d) Increased bioavailability of nitrogen, leading to accelerated organic matter decomposition and increased tree susceptibility to natural stresses. Increased tree susceptibility to stress may occur as a consequence of changes in growth rate, phenology, carbon allocation, nutrition, and water relationships.

The importance of these mechanisms is uncertain because knowledge of rates, interactions, and variability of biogeochemical and hydrologic processes is incomplete.

The five literature reviews in this volume explore several important uncertainties about the potential effects of acidic deposition on forest element cycles and tree nutrition. The following paragraphs present the rationale for topic selection and a brief summary of findings within each review.

A. Foliage Leaching

When rain comes in contact with the leaves and branches of a forest canopy, the chemistry of the rainwater changes; solutes pass from rain to plant tissues and vice versa. In addition, particles deposited on canopy surfaces before rainfall events may be dissolved or physically displaced. The "throughfall" solutions that ultimately reach the ground may bear little chemical resemblance to incoming rain.

Throughfall has been studied for many years and is recognized as an important pathway in forest nutrient cycles. More recently, throughfall has been studied in investigations of regionally distributed air pollutants and their possible effects on forest health. Some investigators are developing throughfall measurement techniques for quantifying dry acidic deposition to forest canopies. Other investigators are measuring throughfall chemical fluxes to determine pollutant effects on the leaching of nutrients and organic metabolites from foliage.

In Chapter 2, Dr. Geoffrey Parker reviews the extensive literature on characteristics of throughfall in forest ecosystems. He describes in detail the physical and biological processes that control throughfall chemistry and discusses hypothesized and observed effects of acidic deposition and other pollutants on throughfall composition. Available evidence suggests that pollutant effects on nutrient losses from foliage are small. Additional experimentation and methods development are required, however, to (a) determine possible long-term pollutant exposure effects on nutrient losses and (b) evaluate the utility of throughfall analyses for monitoring dry deposition and forest health.

B. Aluminum Mobilization

Although annual atmospheric inputs of protons are small compared to proton quantities in soils, acidic deposition does affect current soil chemistry (see Ruess and Johnson [1986] for an excellent review and synthesis). Deposition of sulfate and other ions can increase the ionic strength of soil solutions and thus increase soil solution acidity and aluminum concentrations through the "salt effect". In addition, "mobile anions" in acidic deposition (especially SO_4^{2-}) tend to move through soil profiles, thus promoting the co-leaching of cations. Mobile anion leaching may increase the net rate of base cation loss and reduce soil base saturation. Potential soil chemistry effects vary greatly among sites depending on rates of atmospheric deposition and numerous properties of soils and vegetation (Wiklander and Andersson 1971).

Increased bioavailability of soil aluminum and manganese is a potential consequence of acidic deposition's effects on soil chemistry. Aluminum and manganese are ubiquitous components of soils. In alkaline and circumneutral soils, aluminum and manganese exist primarily in solid forms. In acidic soils, however, substantial amounts of aluminum and manganese may be dissolved in soil solutions and available to plant roots. It is well established that high aluminum and manganese concentrations in soil solutions can damage sensitive crop plants. Farmers often apply lime to raise soil pH and thus reduce dissolved aluminum and manganese species to safe levels.

Although forest soils are often quite acidic, widespread aluminum and manganese toxicity to forest trees has generally been considered unlikely. The law of natural selection and the usual nonresponsiveness of tree growth to lime applications (Bengtson 1968) indicate that most forest tree species are well adapted to the naturally acidic soils they occupy. Recent concerns about acidic deposition and forest decline have challenged this traditional view, however. Some investigators have proposed that acidic deposition can increase soil acidity substantially and thus raise available aluminum and manganese to root-damaging levels.

In Chapter 3, Dr. Jeffrey Wolt reviews soil chemical and biological relationships relevant to possible effects of acidic deposition on forests through the mechanism of increased availability of aluminum and manganese. Aluminum and manganese toxicity are manifestations of "acid soil infertility", a complex phenomenon that may concurrently cause deficiencies in essential plant nutrients. Although forest trees appear to be tolerant to levels of aluminum and manganese found in most forest soil solutions, few studies have specifically addressed this concern. The most useful assessment criterion is likely to be the relationship of root elongation rate to activities of toxic chemical species in hydroponic or soil solutions. Future monitoring studies should better define spatial and temporal relationships between root function and concentrations of aluminum and manganese species in soil solutions. Future experimental work should determine how tree responses to aluminum and manganese are affected by ratios of elements in soil solutions and other environmental conditions.

C. Mineral Weathering

Implications of acidic deposition for forest health and productivity are uncertain because of inadequate knowledge of forest element cycles and tree nutrition. Investigations now underway at several sites are quantifying acidic deposition, nutrient uptake by vegetation, nutrient leaching from soils and foliage, and tree responses to changes in soil chemistry.

Although many information gaps are being filled, new methods are needed to quantify several processes by which forest ecosystems may resist accelerated acidification. Most notable of these processes is mineral weathering. Weathering reactions can buffer hydrogen ion inputs and replace base cations lost by leaching. Moreover, weathering reaction rates may increase in response to acid inputs and thus compensate for deposition-induced accelerations in acidification processes. Models that consider only cation exchange and aluminum hydrolysis as acid-buffering mechanisms may overstate potential adverse effects on sites where active weathering is occurring. Potential effects of acidic deposition on forest soil fertility and tree nutrition cannot be accurately assessed without better understanding of rates of nutrient release by weathering.

Dr. G. Norman White, Steven B. Feldman, and Professor Lucian W. Zelazny review weathering mechanisms and the factors that control weathering rates in Chapter 4. They identify many technical obstacles that have hindered field and laboratory investigations and conclude that a satisfactory approach to weathering rate estimation has not yet been developed. Development of a satisfactory approach requires basic research on weathering mechanisms, including mineral dissolution, interaction of minerals with organic matter, and hydrologic control of solute concentrations and fluxes.

D. Soil Biology

Numerous experiments have investigated hypothesized effects of acidic deposition on soil organisms and biological processes. Some of these experiments have revealed statistically significant responses to acidic deposition treatments. Unfortunately, the question of whether such treatments were realistic simulations of actual or potential acidic deposition rates has often received inadequate attention.

Dr. David Myrold presents a critical review and analysis of the literature on the responses of soil organisms to acidic deposition in Chapter 5. Dose-response tables summarizing published experimental results allow the reader to gauge the severity of acidic deposition treatments that have produced effects. Dr. Myrold suggests that acidic deposition could have effects on the relative population sizes of some species of soil organisms. However, significant near-term effects on organic matter decomposition and nutrient cycling appear unlikely. Possible long-term effects of chronic acidic deposition remain to be investigated.

In Chapter 6, Professor William H. Smith explores the idea that air quality impacts on soil biology are most likely to occur near root surfaces (i.e., in the "rhizosphere"). Professor Smith reviews the characteristics and functions of the rhizosphere and describes in detail how impacts on rhizosphere processes might influence tree health. He offers substantial indirect support for the hypothesis that air pollutant stress to aboveground tree tissues could alter rhizosphere processes through effects on the quality and quantity of metabolites transferred from shoots to roots and from roots to rhizosphere. The review is a useful and thought-provoking reference for soil and plant scientists involved in air quality-forest health research.

II. Future Directions

Research on acidic deposition effects in forest ecosystems is near the end of an initial exploratory stage. Substantial information on deposition rates and processes has been generated, potential effects have been better defined through experimentation, and necessary interdisciplinary collaborations have begun to develop. Additional progress in defining the extent and magnitude of effects is possible in the next decade.

Each of the literature reviews in this volume concludes with a research recommendations section. Several recurring themes are apparent.

A. Roles of Organic Matter

Many of the remaining uncertainties about effects of acidic deposition involve organic matter and its roles in element cycling and tree nutrition. Soil organic compounds influence weathering rates of minerals, activities of soil organisms, and the movement and bioavailability of water, nutrients, and toxic components. In addition, organic compounds from canopy sources constitute a substantial portion of the dissolved mass in throughfall. Basic understanding of the structure and function of organic matter in soils and throughfall is limited (Stevenson 1987, Parker, this volume), and interactions with acidic deposition cannot be predicted with confidence at this time.

Acidic deposition could cause many ecological changes by altering the solubility, movement, or decomposition rate of organic compounds. Anthropogenically induced humus disintegration is a central but often overlooked element of Professor Bernard Ulrich's theory of forest ecosystem destabilization (Ulrich 1981). Loss of humus from mineral horizons and development of an acid mor-type forest floor can radically alter soil fertility and site quality. Whether acidic deposition could be a significant factor in humus disintegration in comparison to past land use practices (especially agriculture) is debatable.

B. Need for More Precise Definition of Mechanisms

Acidic deposition clearly does alter ecosystem processes, but the effects are subtle at current and likely future deposition rates. Mechanistic hypotheses of acidic deposition effects have been refined substantially during the past decade, but many details remain unspecified. Chains of effects from deposition through element cycles to tree nutrition and vigor are difficult to trace amidst the extreme spatial and temporal variability of forest ecosystems. More precise definition of the links in these hypothetical chains would improve chances of detecting them in the forest if they exist.

Experimentation and forest health monitoring both can contribute to more precise definition of mechanistic hypotheses. For example, evolution of the magnesium depletion hypothesis from the more general nutrient depletion hypothesis was stimulated by Norwegian experiments in which simulated acid rain depleted soil and foliage magnesium reserves and by the occurrence of widespread magnesium deficiency in West German forests (Abrahamsen et al. 1987). The magnesium hypothesis is applicable to fewer sites than the more general nutrient depletion hypothesis but is perhaps more easily tested.

C. Need for Ecosystems Perspective

All the reviews in this volume reinforce the interdependency of the scientific disciplines investigating forest responses to acidic deposition. Large information gaps clearly remain to be filled by studies within disciplines, but interpretation of those studies will require frameworks of site-specific information on atmospheric and ecosystem processes. Data bases being developed by the multisite "Integrated Forest Study" (Pitelka et al. 1985) are excellent examples of the kinds of frameworks that are needed. These data bases contain information about air quality, acidic deposition, throughfall, soil solutions, and soil and biomass element pools.

The importance of an ecosystems approach in acidic deposition research is illustrated by the problem of interpreting measured changes in soil chemistry over time (Falkengren-Grerup 1987, Hallbacken and Tamm 1986). The contribution of acidic deposition to observed decreases in pH and base saturation cannot be estimated in the absence of time-series information on element fluxes associated with acidic deposition, biomass increase, harvest, erosion, and leaching. Estimates of past fluxes could perhaps be developed from historical deposition and rainfall data, analyses of tree rings and forest management records, and studies of current fluxes.

III. Conclusion

Acidic deposition does alter forest ecosystem processes but has not been shown to pose an immediate threat to the health and productivity of commercial forests in North America. Published research demonstrates that ambient levels of rainfall acidity in most forests do not cause acute injury to foliage. Effects on element cycles are subtle and unlikely to cause widespread and substantial adverse effects on soil fertility in the near future.

Current adverse impacts of acidic deposition are more plausible in high-elevation forests, especially those on peaks that are periodically immersed in clouds. Such forests may receive high rates of acidic deposition and be exposed to highly acidic cloudwater and fog. However, high-elevation forests are also subject to severe natural stresses and to stress from pollutants other than acidic deposition. Responses of high-elevation forests to acidic deposition and other pollutant and natural stresses are being investigated intensively (Hertel 1988).

The potential long-term effects of acidic deposition on forest ecosystems are not yet adequately defined. The numerous factors controlling ecosystem responses to acidic deposition and other acid-generating processes vary greatly in space and time and interact in ways that are not fully understood. A single, consistent relationship between deposition rate and forest response cannot exist because of the inherent variability among ecosystems. Regional impact assessment will require ecosystem classification and the development of frequency distributions of response (Goldstein et al. 1984).

References

Abrahamsen, G. 1980. Acid precipitation, plant nutrients, and forest growth. *In*: Proceedings, International Conference on the Ecological Impacts of Acid Precipitation, D. Drablos and A. Tollan (eds.), pp. 58–63. Ås, Norway: SNSF Project.

Abrahamsen, G., B. Tveite, and A.O. Stuanes. 1987. Wet acid deposition effects on soil properties in relation to forest growth—experimental results. Presented at the IUFRO conference, Woody Plant Growth in a Changing Physical and Chemical Environment, Vancouver, British Columbia, July 27–31, 1987.

Bengtson, G.W. 1968. Progress and needs in forest fertilization research in the South. *In*: Forest Fertilization—Theory and Practice. Muscle Shoals, Alabama: Tennessee Valley Authority.

Cox, R.M. 1983. Sensitivity of forest plant reproduction to long range transported air pollutants: *in vitro* sensitivity of pollen to simulated acid rain. New Phytol 95:269–276.

Falkengren-Grerup, U. 1987. Long-term changes in pH of forest soils in southern Sweden. Environ Pollut 43:79–90.

Goldstein, R.A., C.W. Chen, and S.A. Gherini. 1984. Integrated Lake Watershed Acidification Study: Summary. *In*: Final Report of the Integrated Lake Watershed Acidification Study, Vol. 4, Summary of Major Results, R.A. Goldstein and S.A. Gherini (eds.) EA-3221, Palo Alto, California: Electric Power Research Institute.

Hallbacken, L. and C.O. Tamm. 1986. Changes in soil acidity from 1927 to 1982–1984 in a forest area in South-West Sweden. Scand J For Res 1:219–232.

Hertel, G.D., tech. coord. 1988. Proceedings of the US/FRG research symposium: effects of atmospheric pollutants on the spruce-fir forests of the Eastern United States and the Federal Republic of Germany; 1987 October 19–23; Burlington, VT. Gen. Tech. Rep. NE-120. Radnor, PA: U.S. Department of Agriculture, Forest Service, Northeastern Forest Experiment Station. 543 p.

Huttunen, S. 1984. Interactions of disease and other stress factors with atmospheric pollution. *In*: Air Pollution and Plant Life, M. Treshow (ed.), pp. 321–356. New York: Wiley.

Huttunen, S. and K. Laine. 1983. Effects of air-borne pollutants on the surface wax structure of *Pinus sylvestris* needles. Ann Bot Fenn 20:79–86.

Jacobson, J.S. 1984. Effects of acidic aerosol, fog, mist and rain on crops and trees. Philos Trans R Soc Lond B 305:327–338.

Kulp, J.L. 1987. Effects on forests. *In*: Interim Assessment: The Causes and Effects of Acidic Deposition. Washington, D.C.: National Acid Precipitation Assessment Program.

Laurence, J.A. 1981. Effects of air pollutants on plant-pathogen interactions. J Plant Dis Prot 87:156–172.

Morrison, I.K. 1984. Acid rain: Review of literature on acid deposition effects in forest ecosystems. For Abstr 45:483–506.

Percy, K.E. 1986. The effects of simulated acid rain on germinative capacity, growth and morphology of forest tree seedlings. New Phytol 104:473–484.

Percy, K.E. and E.A. Baker. 1987. Effects of simulated acid rain on production, morphology and composition of epicuticular membrane development. New Phytol 107:577–589.

Pitelka, L.F., E.A. Bondietti, D.W. Johnson, and S.E. Lindberg. 1985. Project Summary: Integrated Forest Study of Effects of Atmospheric Deposition, An Oak Ridge National Laboratory Research Project Funded by the Electric Power Research Institute. Oak Ridge, Tennessee: Oak Ridge National Laboratory.

Rehfuess, K.E. 1981. On the impact of acid precipitation in forest ecosystems. Forstwiss Centralbl 6:363–381. (Oak Ridge National Laboratory Translation 4830.)

Ruess, J.O. and D.W. Johnson. 1986. Acid Deposition and the Acidification of Soils and Water. New York: Springer-Verlag.

Stevenson, F.J. 1987. Soil biochemistry: past accomplishments and recent developments. *In*: Future Developments in Soil Science Research, pp. 133–144. Madison, Wisconsin: Soil Science Society of America.

Ulrich, B. 1981. The destabilization of forest ecosystems by the accumulation of air contaminants. Forst Holzwirt 36:525–532. (EPA Translation TR-82-0257.)

van Breeman, N., C.T. Driscoll, and J. Mulder. 1984. Acidic deposition and internal proton sources in acidification of soils and waters. Nature (London) 307:599–604.

Wiklander, L. and A. Andersson. 1971. The replacing efficiency of hydrogen ion in relation to base saturation and pH. Geoderma 7:159–165.

2
Evaluation of Dry Deposition, Pollutant Damage, and Forest Health with Throughfall Studies

GEOFFREY G. PARKER

I. Introduction

A. Historical Perspective

This chapter summarizes some of the burgeoning literature on the alteration of precipitation chemistry by forest canopies, with emphasis on canopy buffering of acidic precipitation and the influence of dry deposited pollutants on throughfall chemistry. Field studies involving forest trees are emphasized, although reference is made to the literature on crop species, and that on laboratory experiments, where necessary.

The input of dry deposition to and the impacts of pollutants on forests are subjects of a large number of recent reviews. Most numerous are works on specific atmospheric pollutants (e.g., OZONE: Environmental Protection Agency 1984a, Skarby and Sellden 1984; NITROGEN OXIDES: National Research Council 1976; SULFUR DIOXIDE: Knabe 1976, Shreffler 1978; STRONG ACIDS: Abrahamsen 1980, Cowling and Linthurst 1981, Environmental Protection Agency 1984b, Evans 1984, Fuhrer and Fuhrer-Fries 1982, Hosker and Lindberg 1982, Linthurst 1984, Overrein et al. 1980, Society of American Foresters 1984, Tamm 1976, Tamm and Cowling 1977, Ulrich 1981, 1982). General treatments of air pollutant effects on forest vegetation may be found in McLaughlin (1985) and Smith (1981), and on forest ecosystems in Bormann (1982) and Woodwell (1970).

Much of the recent increase in the air pollution/forest health literature has been motivated by the potential contribution of air pollutants to forest declines observed in central Europe (Hileman 1984, McLaughlin 1985, Rehfuess 1981, Tomlinson 1983) and eastern North America (Carrier and Gagnon 1985, Friedland and Johnson 1985, Johnson and Siccama 1983, 1984). Many recent discussions have emphasized the potential effects of acidic deposition on nutrient cycling, that is, changes in soil nutrient availability, increased concentration of toxic elements in the soil solution, and decreased solute uptake caused by damage to fine roots (e.g., Ulrich 1981, 1982). The potential of atmospheric pollutants to affect the canopy-mediated portion of the nutrient cycle is also recognized (Johnson et al. 1982, Lepp and Fairfax 1976, Miller 1984, Rehfuess 1981, Tukey 1980) but has not been systematically investigated.

B. Scope of this Review

This review is motivated by questions relating to three issues concerning solute interactions in the forest canopy:

1. To what extent does atmospheric dry deposition contribute to the solutes detected in throughfall and stemflow? Are there reliable methods that can partition the solute deposition in throughfall and stemflow into those portions deriving directly from atmospheric deposition (inputs) and those originating within the ecosystem (recycled materials)?
2. Does atmospheric deposition promote the leaching of substances from forest canopies? Is throughfall more highly concentrated in materials removed from plants after episodes of high dosages of phytotoxic substances than during cleaner periods? Alternatively, are forests in polluted regions losing foliar materials at higher rates than those in unpolluted zones? If so, what might be the consequences to the forest nutrient cycle and to forest health?
3. Might the overall health of the forest be indicated by the presence or concentrations of some inorganic or organic component (or some combination of components) of throughfall? Is the occurrence of specific or general stresses to the canopy indicated by specific compounds detectable in precipitation collected beneath plant canopies? If so, could the amount of such an indicator substance reflect the degree of the stress?

These issues are, by necessity, arranged hierarchically: the evaluation of leaching responses to atmospheric deposition (issue 2) presupposes that the leachate fraction of throughfall may be unambiguously identified (issue 1). Additionally, the assessment of general forest health by throughfall analysis (issue 3) would be aided by identification of specific effects from air pollutants alone (issue 2). For each of these issues I shall evaluate existing data and methods, indicate what critical information is lacking, and suggest approaches to obtain that information.

Before examining these specific issues, I shall first review the nature, extent, and influences of solute interactions in forest canopies. This treatment will require introductions to some aspects of forest hydrology, atmospheric deposition to forests, and the potential effects of various species in the pollutant complex on solute exchange in canopy tissues.

C. Definition of Terms

Liquid or frozen water that contacts the canopy is termed *incident precipitation*. If it falls down more or less vertically, as in rain, snow, sleet, or hail, it is called *direct precipitation*, but if it impinges onto the canopy more or less laterally, as in fog, cloudwater, mist, or rime ice, it is called *indirect precipitation.* Some events combine both forms of incident precipitation. That fraction which passes through the canopy and drips to the ground is called *throughfall* and may exceed incident precipitation in quantity at some locations in the forest (Banaszak 1975, Prebble and Stirk 1980, Zinke 1962). *Stemflow* is the portion that reaches the ground at the base of the tree by draining down the branches and trunk. Incident precipitation that does not arrive at the forest floor by either of these routes is termed *interception loss* (Kittredge 1948). These principal compartments of the canopy hydrologic balance can be expressed in the following equation:

$$P = I + T + S$$

where P, I, T, and S are incident precipitation (direct and indirect), interception loss, throughfall, and stemflow, respectively (Helvey and Patric 1965). The literature on forest hydrology contains numerous reports on the partitioning of these categories in many forest systems (e.g., Mahendrappa and Kingston 1982, Zinke 1966). For direct precipitation, throughfall amounts to between 60% and 95% of incident precipitation, stemflow from 0 to 35%, and interception loss from 5% to 35% depending on climate, the nature of the precipitation event, the geometry of the tree, and the structure of the stand. The sum of throughfall and stemflow may exceed the direct precipitation input in periods with indirect precipitation also.

The flux of material per unit ground area per unit time in precipitation pathways is termed *deposition.* Such fluxes are obtained by multiplying the estimated flux of water by the estimated mean concentration of solutes in that water. Incident precipitation is called *bulk precipitation* if the sampler is continuously left uncovered, but *wet precipitation* if it is closed between events (Galloway and Parker 1980). The simple difference between bulk and wet deposition from such collectors is *not* equivalent to dry deposition to canopies (Lindberg et al. 1986). The gross deposition in throughfall and stemflow includes the component due to incident precipitation. The net effect of the canopy in altering solute fluxes is obtained by subtracting the incident precipitation deposition from the throughfall and stemflow depositions to give net throughfall and net stemflow, either of which can be a negative quantity.

Materials contributing to the net solute load in throughfall and stemflow derive from two broad classes of sources. Those materials originating within the forest system itself (from the canopy surfaces or from the activities of organisms inhabiting those surfaces) are often referred to as "leachates." Those materials originating outside the forest ecosystem, primarily material of atmospheric origin, are called "washoff" or "interception deposition" (Ulrich 1984). The identification of their relative contributions to throughfall and stemflow has been a major concern of the recent literature.

II. Uptake and Release of Solutes by Forest Canopies

A. Chemical Interactions in the Canopy

Figure 2.1 indicates the complex of physical, microbiological, and chemical processes interacting between the ambient atmosphere, the solute pools in various canopy tissues, and the epibiota that inhabit the surfaces. Gaseous, particulate, and dissolved matter is potentially released from, taken up by, and transformed at, all canopy surfaces. The relative importance of each of the pathways diagramed will depend on the species of trees (and the epibiota supported); ambient air quality; the type, amount, and timing of precipitation events; the quality of the substrate (i.e., soil) solute pool; the geometry of the tree and the stand; and the

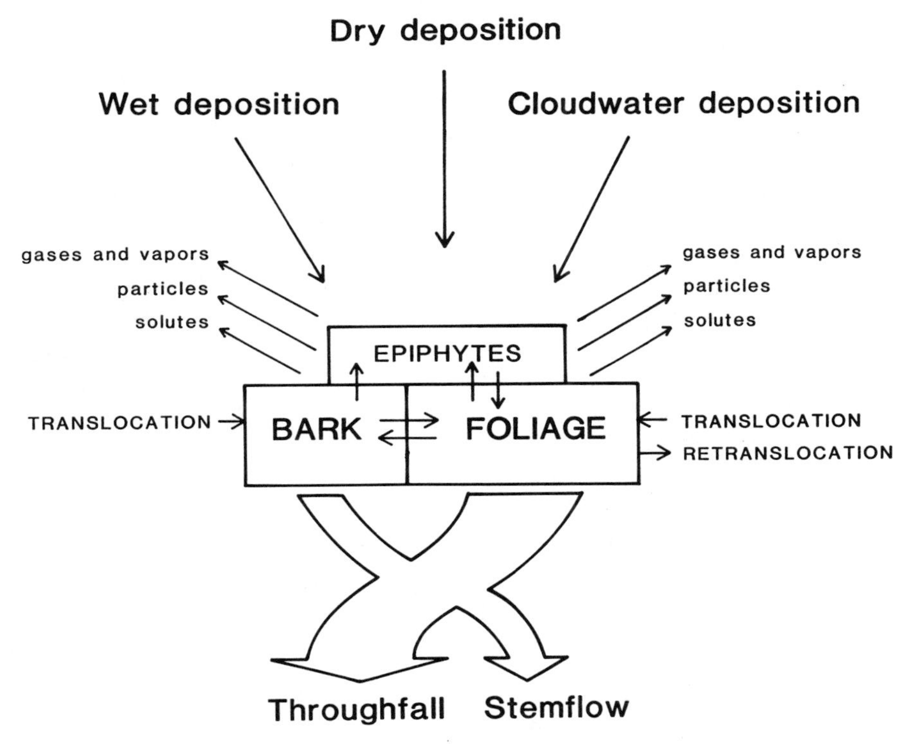

FIGURE 2.1. Conceptualization of complexity of chemical interactions potentially operating in forest canopy that may affect quality and quantity of throughfall and stemflow.

position of the surface within the canopy (e.g., upper layers may behave differently than do lower ones: Schaefer et al. 1985). Solute interactions in forest canopies involve three phases of matter and at least three biological kingdoms interacting with extremely large spatiotemporal variation on a substrate of great physical and chemical complexity.

A variety of atmospheric materials may accumulate on plant surfaces. These include gases, large particles (sea salts, dusts), droplets (cloud, fog, and mist), and aerosols (both wet and dry). Much of the accumulated material will be released from the canopy or remain tightly bound to the canopy surface. The remainder will be taken up into canopy tissues (Bentley and Carpenter 1984). Sources of solutes for canopy uptake include incident precipitation and sprays, particles dissolved in canopy water films, and epibiotically fixed dinitrogen gas (Waughman et al. 1981).

While accumulating atmospheric materials, plant surfaces are simultaneously supplied with soluble materials from internal sources. For leaves, the ultimate source is the translocation stream, but the proximate pool may be the "inner space" or "free space" (Epstein 1972), part of the apoplast. Rates of supply via these avenues may vary in time. The leaf surface, which may be relatively depleted of free materials following heavy storms, may require several days to replenish solutes (Lausberg 1935, Reiners and Olson 1984). Also, in some forests there may be a seasonal movement of solutes from foliar pools to other tissues (retranslocation), such as in the pre-abscission period (Ryan and Bormann 1982).

Material released from forest canopies includes not only soluble and particulate matter in throughfall but also nonthroughfall gases and particulates. The predominant gas-releasing processes are photosynthesis (O_2), respiration (CO_2), and evapotranspiration (H_2O). Other processes release ammonia (Farquhar et al. 1979), various reduced sulfur compounds, and hydrocarbons (Rasmussen 1972, Went 1960). Particulate matter released from the canopy may derive from impacted material that becomes resuspended (Gregory 1961); from plant parts that have been abscised, eroded, or decomposed (Baker and Hunt 1986, Beauford et al. 1977, Schnell and Vali 1972, 1973); or from microbial or animal sources, such as feces, exuviae, and frass (Carlisle et al. 1966).

The complexity of canopy processes that contribute to altering aqueous chemistry observed in throughfall and stemflow has only recently become appreciated. The emphasis of the study of throughfall and stemflow has shifted from the descriptive (typical of many earlier nutrient cycling studies) to the mechanistic. Studies now focus on the importance of specific processes outlined here, on factors that may modify the relative importance of these processes under given conditions, and on the development of procedures for their routine estimation. Currently, only a few of the fluxes indicated by arrows in Figure 2.1 can be confidently estimated for any forest canopy, and there is no system in which attempts have been made to estimate them all. The study of processes contributing to solute exchanges in plant canopies is and will remain an area of extremely active research.

B. Observation of Solute Interactions in Forest Canopies

The uptake and release of solutes by forest canopies is commonly studied by comparing the chemical composition of throughfall and stemflow with that of incident bulk precipitation. The principal motivation for the majority of such efforts is the quantification of the contribution of material released from forest canopies to the biological portion of the forest nutrient cycle (e.g., Attiwill 1980, Brown 1974, Corlin 1971). Currently, the estimation of throughfall deposition is a routine part of nutrient cycling studies. The estimation of stemflow, however, is often ignored. Thus, the bulk of the literature tends to emphasize foliar-based interactions over those that occur at bark tissues. Nutrient cycling studies usually focus on macronutrients, most commonly the elements N, P, K, Ca, and Mg; only rarely are attempts made to account for all the species contributing to the charge balance. Furthermore, observations of throughfall and stemflow chemistry are rarely conducted by or within the precipitation event. Thus, biological and chemical processes occurring during the aging of a sample in its collector over hours and days are superimposed on the effects of processes that take place in the canopy in only minutes.

Understanding of within-storm processes of solute exchange from multievent, whole-canopy observations of macronutrients is difficult. Nonetheless, information at such scales forms the bulk of our knowledge about canopy interactions and may provide some understanding about general patterns. The following general conclusions about the quality of throughfall and stemflow are condensed from observations taken principally for the study of the forest nutrient cycle. A more detailed treatment of these phenomena can be found in Parker (1983).

C. Inorganic Constituents of Throughfall and Stemflow

It is universally observed that the chemical composition of precipitation changes dramatically after interacting with the canopies of live or dead plants (Cole and Rapp 1981, Parker 1983, Rodin and Basilevich 1967, Stenlid 1958). This chemical alteration occurs in all habitats observed, in both pristine (Basilevich and Rodin 1966, Brinkmann 1983, Brinkmann and dos Santos 1973, Edwards 1982, Grimm and Fassbender 1981, Hesse 1957, Johnson 1981, Kazimirov and Morozova 1973, Turvey 1974) and polluted (Baker et al. 1977, Haughbotn 1973, Ulrich 1984, Zielinski 1984) environments; in laboratories and greenhouses (Tukey 1970a) and in the field (LeClerc and Breazeale 1908, Parker 1983); and with every species of plant studied.

Generally, concentrations and fluxes of both organic and inorganic compounds are greater in throughfall than in incident precipitation despite interception losses in the canopy. Reductions in concentration and flux, however, are fairly common for hydrogen ion and inorganic nitrogen species. Throughfall enrichment factors (throughfall flux/incident precipitation flux) for frequently measured solutes are shown in Figure 2.2 for numerous forests throughout the world. Figure 2.2 shows that mean enrichment factors vary greatly among solutes, ranging from 1.27 for nitrate-nitrogen to 11.2 for potassium. The canopy may either increase or decrease the acidity of incident precipitation. The factors influencing changes in pH are discussed in Section IV.B.

The flux of soluble matter in throughfall and stemflow is a normal part of the forest nutrient cycle. The solutes removed from plant canopies become available to be taken up by the plant from the soil and litter (Tukey 1966, Tukey et al. 1958) and subsequently released again. For some elements, such as S, K, Na, Cl, and Mg, the solution pathway of nutrient return is often larger than that in litterfall (Gosz et al. 1976, Parker 1983, Will 1955, 1959, 1968). Smaller, but still significant, contributions occur for N (0–15% of total nutrient return in net throughfall),

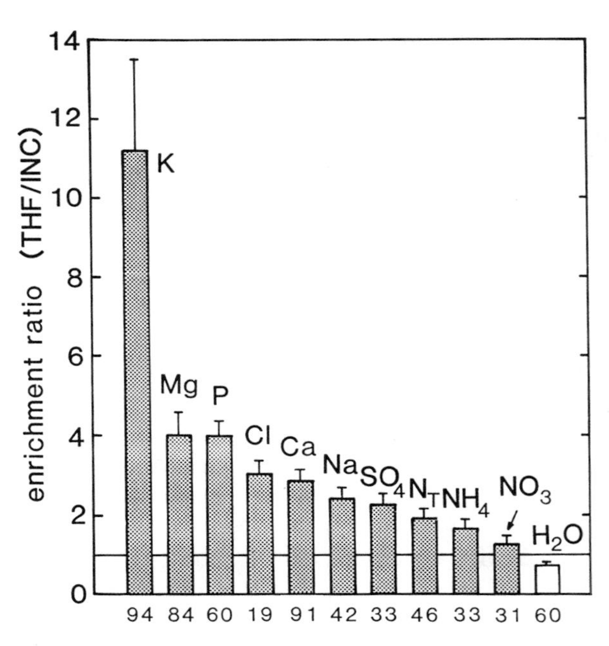

FIGURE 2.2. Mean enrichment ratios [throughfall deposition/incident deposition (THF/INC)] for major solutes in throughfall. N_T is total nigrogen. Error bars are one standard error. Number of values used to calculate mean is given on x-axis (adapted from Parker 1983).

P (10–20%), Ca (10–20%), Mn (≈ 20%), and carbon (≈ 5%) on an annual basis. In certain seasons of the year, solutes in throughfall and stemflow may be more available than those from decomposing litter (Carlisle et al. 1967, Eaton et al. 1973, Henderson et al. 1977, Miller 1963).

Stemflow ion concentrations are typically much higher than those in throughfall (Iwatsubo and Tsutsumi 1967, Mahendrappa 1974, 1983, Mina 1967). The stemflow contribution to the total deposition in throughfall and stemflow averages 12% (range 0–55%), but can be relatively important for some solutes (H, Ca, S, Mg, and organic matter) and some canopies (e.g., smooth-barked trees with large crowns and ascending branches). While not a dominant transfer to the forest floor, stemflow is very important to the trunk epiflora (Nieboer et al. 1978, Nye 1961, Pike 1978, Rasmussen and Johnsen 1976). In addition, stemflow dominates all other routes of solute transfer to the small area at the base of trees, where it may influence the distribution of understory shrubs (Cloutier 1985) and the physical, chemical, and microbiological properties of soils (Bollen et al. 1968, Gersper and Hollowaychuck 1970, 1971, Kaul and Billings 1965, Lunt 1934, Mina 1965, Patterson 1975, Skeffington 1983).

D. Organic Constituents of Throughfall and Stemflow

Carbon is a relatively minor component of incident precipitation. Total organic carbon (TOC) concentrations are on the order of 1–3 mg C/liter (Carlisle et al. 1966, Likens et al. 1983) with the dissolved form contributing more than three-quarters of the total on an annual basis. Some of the nitrogen and much of the phosphorus in incident precipitation is also contained in the organic (usually Kjeldahl-digestible) fraction.

Carbon is a major constituent in precipitation that has passed through the forest canopy, with concentrations ranging as high as 20 mg C/l (De Boois and Jansen 1975, Eaton et al. 1973, Horntvedt 1979, McDowell and Likens 1988). It is often the element with the greatest mass flux, with depositions as high as several hundred kilograms per hectare annually (Carlisle et al. 1966). Stemflow often appears distinctly amber- or tea-colored (Yadav and Mishra 1980) and commonly has total carbon concentration from 30 to more than 200 mg C/liter (Carlisle et al. 1966, Mahendrappa 1974). Most of the nitrogen, and presumably much of the phosphorus, in forest solutions is associated with the organic carbon fraction (Stachurski and Zimka 1984).

The organic chemistry of throughfall is poorly known, beyond a few general characterizations. Wet precipitation organics in rural New York and New Hampshire were found to be dominated by low molecular weight carboxylic acids, aldehydes, carbohydrates, and tannin/lignin (Galloway et al. 1976, Likens et al. 1983). Numerous other compounds may be found in incident precipitation though their concentrations usually make insignificant contributions to the dissolved organic carbon (DOC). McDowell

(1982) reported that the dominant molecular size class increased in throughfall relative to incident precipitation. Carbohydrates (both monomeric and polymeric) and phenolics dominated throughfall organics. Hoffman et al. (1980b), using gas chromatography and mass spectrometry, found throughfall under *Quercus prinus* in eastern Tennessee to be primarily "aliphatic saturates and neutral esters." Organic acid anions may also be an important throughfall component. Cronan and Schofield (1979) estimated the concentration of organic anions to rise nearly threefold between precipitation and throughfall (from 12 to 33 μeq/l).

Specific compounds may occasionally dominate the organics in throughfall. Carlisle (1965) found that the carbohydrate fraction (78% of net throughfall DOC under sessile oak) was dominated by the trisaccharide melezitose, a major component of aphid honeydew.

A wide variety of organic compounds have been detected in leachates including carbohydrates, amino acids, vitamins, hormones, and alkaloids (Lee and Tukey 1972, Morgan and Tukey 1964). In fact, Tukey (1970a) wrote that "failure of certain substances to appear in the list of leached substances may be due to failure to search for them or to insufficient sensitivity of detection methods."

E. Factors Influencing Throughfall and Stemflow

Factors that influence the composition of throughfall and stemflow may be grouped into three classes: (1) those concerned with the species of trees making up the forest, (2) those concerned with the location and trophic status of the site, and (3) those concerned with the atmosphere and climate. Of course, these factors are not independent, but the descriptive literature on throughfall solutes has not focused on interactions among these factors.

The composition of the forest has a strong effect on the quality of both throughfall and stemflow (Denaeyer-DeSmet 1966, Hart and Parent 1974, Kaul and Billings 1965, Madgwick and Ovington 1959, Mahendrappa 1974, 1983, Patterson 1975, Tamm 1951, Voigt 1960). The presence of the forest may even affect the quality of incident precipitation for some elements, for example, elevated potassium concentrations in forested regions (Gosz 1980, Lindberg et al. 1986).

Numerous studies comparing adjacent forest types confirm that annual throughfall and/or stemflow depositions under hardwoods generally exceed those under conifer stands (Best and Monk 1975, Cronan and Reiners 1983, Henderson et al. 1977, Rapp 1969, Reiners 1972, Tarrant et al. 1968, Verry and Timmons 1977, Wells et al. 1972). Forests of *Picea abies* may be an exception to this rule (Heinrichs and Mayer 1977, Mina 1965, Nihlgard 1970), but the reason is not understood.

The relatively high throughfall and stemflow depositions of hardwoods reinforce the idea that rates of biological cycling are higher in hardwood ecosystems than in coniferous ecosystems (Bazilevich and Rodin 1966, Cole and Rapp 1981). However, the relatively high depositions of hardwoods are somewhat surprising when one considers that (1) most temperate zone hardwoods are leafless for much of the year and (2) needle-leaved canopies seem to filter air more effectively than broad-leaved canopies (Belot and Gauthier 1975, Chamberlain 1975a,b, Droppo 1976, 1980, Graustein and Armstrong 1978, 1983, Sehmel 1980, Slinn 1977, Wedding et al. 1976). A comparison of the seasonal behaviors of throughfall and stemflow of deciduous conifers (such as *Larix* or *Taxodium* spp.) with those of evergreen conifers in the same area would be of interest in this regard.

It is unknown whether each tree species has a characteristic chemical signal in throughfall, although the finding that different forest types have more or less predictable effects on solute fluxes is suggestive. Strong support for such species effects could involve demonstrating that forests of conspecifics in widely separated locations have more similar throughfall chemistry than unlike, although adjacent, forests.

Throughfall and stemflow fluxes appear to increase with increasing stand age (Binkley et al. 1982, Lemee 1974, Miller 1983, 1985, Miller et al. 1985, O'Connell 1985). The enhanced solute concentrations with the aging of the canopy might be caused by increased influence of bark and bark epibiota on an ever-increasing nonleafy surface area. Foliage may have relatively little effect on this process because foliage area stabilizes after canopy closure and because the age distribution of leaves is similar on young and old trees.

Comparisons of forests on rich and poor soils indicate that throughfall and stemflow solute concentrations increase with the nutritional status of the site (Astrup and Bülow-Olsen 1979, Leininger and Winner 1985, Petersen 1985, Tarrant et al. 1968, Tsutsumi and Yoshimitsu 1984). Robitaille (personal communication) has observed extremely low potassium concentrations in throughfall under sugar maple on K-deficient soils in Quebec. In forests on sites classified as oligotrophic (low nutrient status), such as two forests in the northern Amazon, net throughfall depositions of nutrients can be very low or even negative (Jordan 1978, Jordan et al. 1979). Marked solute uptake by such canopies has been interpreted as a nutrient retention mechanism (Herrera et al. 1978). Negative net throughfall nitrogen depositions have been observed in many forest stands, especially for ammonium nitrogen (Carlisle et al. 1966, Evans et al. 1986, Feller 1977, Foster 1974, Foster and Gessel 1972, Horntvedt and Joranger 1974, Miller 1963, Miller et al. 1976, Päiväinen 1974, Parker 1981, Richter and Granat 1978, Rolfe et al. 1978, Stachurski and Zimka 1984, Verry and Timmons 1977, Wells and Jorgenson 1974, Wells et al. 1975). Throughfall nitrogen in stands with trees associated with nitrogen-fixing bacteria is elevated relative to that in stands without such additional inputs (Binkley et al. 1982, Bollen et al. 1968, Tarrant et al. 1968, van Miegroet and Cole 1984).

Throughfall quality within a stand may change in response to changes in stand nutrient status. Effects of fertilizer amendments to stands often appear as elevated concentrations in throughfall soon after application. The rise and decline of canopy-leached solutes generally parallels changes in nutrient availability in the soil, although some variability in responses is observed (Khanna and Ulrich 1981, Mahendrappa and Ogden 1973, Matzner et al. 1983, Miller et al. 1976, Päiväinen 1974, Wells et al. 1975, Yawney et al., 1978). The converse experiment, on the response of throughfall solutes to a decrease in nutrient availability, has not been conducted.

Latitude, which is generally negatively correlated with site quality, influences the nutrient cycle through changes in the length of the growing season, temperature, the amount of insolation, and often, the amount of precipitation. Throughfall and stemflow solute depositions are generally inversely related to the latitude of the site, as are nutrient uptake, and other releases such as litterfall (Cole and Rapp 1981, Rodin and Bazilevich 1967).

The phase, amount, frequency, and duration of precipitation all have potential influences on the amount of solutes released from plant canopies. Liquid water is much better at mobilizing canopy solutes than is frozen water; rainwater is the best solvent. However, even snow (Bockheim et al. 1983, Fahey 1979, Pastor and Bockheim 1984) and possibly rime ice (Falconer and Falconer 1979, Lovett 1984, Lovett et al. 1982, Schlesinger and Reiners 1974) may also remove material from canopies. Indirect precipitation in cloud droplets (and probably in fog and mists as well) has a great potential for removing canopy materials. Such precipitation often has higher concentrations of strong acids and heavy metals than direct precipitation (Deal 1983, Fuhrer 1985, Lovett et al. 1982, Scherbatskoy and Bliss 1983, Waldman et al. 1982, 1985, Weathers et al. 1986) and may interact over extended periods with both the upper and lower surfaces of leaves. Dews and frosts that condense on canopy surfaces may also remove plant solutes if they drip off (Jacobsen 1984, Wisniewski 1982).

The amount of precipitation appears to be the single most important aspect of atmospheric precipitation contributing to the loss of material from plant canopies. Throughfall deposition (net or gross) is strongly correlated to the amount of rain per collection period (Attiwill 1966, Berhard-Renversat 1975, Lindberg and Harriss 1981, Madgwick and Ovington 1959, Parker et al. 1980, Reiners 1972, Szabo 1977, Verry and Timmons 1977). The correlation holds even during periods with frequent rainfall (Bache 1977, Parker 1981).

Throughfall concentrations decrease quickly with increasing precipitation amount but depositions increase as precipitation increases. These patterns are most evident when data are taken by event (Parker et al. 1980). The timing of precipitation events (length of preceding dry interval) may be correlated with throughfall sulfate deposition (McColl and Bush 1978) and may be a predictor of throughfall pH (Moore 1983).

The chemistry of wet precipitation may affect the subsequent release of canopy materials, from either surface deposits or from internal pools. Tukey (1970a) reports that some substances increase leaching (e.g., Na and K salts) while others decrease leaching (salts of Ca). The potential of atmospheric pollutants to influence solute losses is treated in Section IV.

The quality of throughfall is directly affected by materials deposited to the canopy in the dry interval between storms. The importance of solutes originating from external sources depends on the strength of the source and its proximity to the forest stand. Throughfall under numerous stands has been identified as containing dry deposits of natural origins, including saline aerosols and soil dust (Attiwill 1966, Eriksson 1955, Hart and Parent 1974, Ingham 1950a,b, Potts 1978, Roose and LeLong 1981, Tamm and Troedsson 1955, Westman 1978). A variety of anthropogenic materials may also contribute to interception deposition, including heavy metals (Friedland and Johnson 1985, Hughes et al. 1980, Lindberg et al. 1982); deicing salt (Krause 1977); pesticides (McDowell et al. 1984); SO_x and NO_x gases and vapors (Lindberg and Lovett 1985, Lindberg et al. 1986); and fertilizers (van Breeman et al. 1982). The degree to which these materials influence solute fluxes depends on (1) the total amount deposited, (2) surface retention (not all deposits are retained on canopy surfaces), and (3) mobility in precipitation (not every deposit may be washed off, although the soluble ones are rapidly washed). Precipitation quality itself may influence the mobility of surface deposits. For example, some deposited metals may be made more soluble by the acidity of precipitation.

F. Leaching and Foliar Absorption

Leaching is "the removal of plant parts by external solutions" (Tukey, 1970a). Leaching often results in the temporary reduction of element concentrations in the affected tissues (primarily leaves). The material released through leaching is ultimately derived from the translocation stream, but proximally from either the surfaces of plants or from the "intercellular free space" of the tissues in contact with externally applied water (the leachant). Leaching includes a number of processes that are conceptually distinct, although operationally difficult to distinguish (Lovett et al. 1985, Schaefer et al. 1985). The leaching of particulate material may only involve dislocation through the rinsing action of precipitation. Salts and weak acids may diffuse from highly concentrated internal pools through cuticular layers into the relatively dilute solution of the leachant.

Ions in the external solution may exchange with ions either weakly bound on leaf and bark surfaces or with ions that have diffused into the external water film (Lovett and Lindberg 1984). Evaporites from previous precipitation events or solutes exuded to the leaf exterior between storms (Reiners et al. 1986) may also be dissolved in precipitation and contribute to the leachate.

Foliar materials are most easily lost to the external solution during the initial stages of wetting. Additional applications of water yield progressively lower concentrations of leached materials. This decline in leachate concentrations with increasing water added has been observed both in the laboratory (Clements et al. 1972, Mecklenberg et al. 1966, Reiners and Olson 1984, Tukey et al. 1958) and in the field, where throughfall has been sequentially sampled (Cole and Johnson 1977, Olson et al. 1985, Parker 1981, Richter and Granat 1978, Sollins and Drewry 1970, Yawney and Leaf 1971, Schaefer et al. 1988). The solutes lost from the free space are probably replaced within 3–4 days after a large event (Lausberg 1935), depending on the season, the type of tissue (Reiners and Olson 1984), the solute species (Olson et al. 1981), and the size of the storm. Substances leached in greatest quantities include organic matter, bicarbonate, potassium, calcium, magnesium, and manganese. Relative to total foliar contents, Na and Mn are most leachable; Ca, Mg, K, and Sr are intermediate; and Fe, Zn, P and Cl are lowest (Tukey et al. 1958).

The susceptibility to leaching increases with leaf age and may be related to changes in leaf wettability (the tendency of water droplets not to bead up (Fogg 1947, Martin and Juniper 1970). Younger leaves, with intact cuticle and low surface wettability, are difficult to leach (Tukey 1970a). The leachability of newly emerged leaves, with immature cuticles, has not been studied. Leaching of bark tissues presumably shows similar age-dependence, but has also not been studied.

Rapid applications of a given volume of leachant release solutes at lower rates than do slower applications of the same volume (Tukey 1970a). Difference in residence time for solutes in water droplets to react with leaf-bound materials are likely to explain much of this effect. Thus, the greatest loss of solutes through leaching is expected to occur from old (presenescent) leaves in extended, low intensity storms.

Solutes may be removed from external solutions through foliar absorption (Boynton 1954, Franke 1967, Witter and Bukovac 1969, Wittwer and Teubner 1959, Wittwer et al. 1965). On a canopy basis, uptake by leaves and bark is difficult to distinguish from uptake by phyllosphere and bark epibiota. Indeed, canopy organisms without access to the soil solution, such as lichens (Crittenden 1983, Denison et al. 1977, Lang et al. 1976, Nieboer et al. 1978, Pike 1978, Puckett et al. 1973), mosses (Rasmussen and Johnsen 1976, Tamm 1950), algae (Witkamp 1970), cyanobacteria (Jones 1970,1976), bromeliads (Schlesinger and Marks 1977, Tukey 1970b), and orchids (Benzing 1973) likely require some degree of nonroot uptake for survival. The efficacy of foliar uptake appears to follow the same ion order as for leaching. Foliar uptake is undoubtedly part of the process of solution exchange in canopies, because throughfall in many forests is less enriched in some solutes than in incident precipitation. The degree of foliar uptake has been estimated (e.g., for nitrogen, approximately 80% of incident deposition: Stachurski and Zimka 1984), but never directly measured for a forest canopy.

Many of the solute exchanges in the canopy may involve both leaching (L) and uptake (U) simultaneously. Although radiotracer studies could provide estimates of foliar uptake and leaching, it is currently difficult to judge their relative contributions. As the leaching process probably dominates in most cases, foliar uptake is usually ignored, or subsumed into the term "net leachate" (L − U).

G. Dry Deposition

The term dry deposition combines all the processes that deliver gases, vapors, large particles, and aerosols to the canopy in the interval between storms. The rate of dry deposition per unit land area is widely assumed to be greater to tree canopies than to less complex surfaces such as grasses, soil, and water, and far greater than to the buckets employed to monitor dry deposition. The major contributing processes are sedimentation, gaseous absorption, and gas and aerosol impaction.

Large particles such as sea salts, soil dust, smoke and haze particles, and spray droplets settle gravitationally (sediment) onto canopy surfaces, principally the upper layers. The rate of large particle deposition depends less on canopy factors than on features of the atmosphere (particle concentration and size distribution; wind speed and turbulence). Sedimentation is an important component of dry deposition in the vicinity of large particle sources such as *the ocean* (Art et al. 1974, Attiwill 1966, Carlisle et al. 1966, Clements and Colon 1975, Golley et al. 1975, Ingham 1950a, Miller 1963, 1979, Miller and Miller 1980, Miller et al. 1976, Nihlgard 1970, Potts 1978, Rapp 1969, Westman 1978), *roads* (Krause 1977, Smith and Staskwicz 1977, Tamm and Troedsson 1955), *and recently plowed or fertilized fields* (van Breeman et al. 1982). Sedimentation probably contributes little in large forest tracts removed from such sources.

Aerosol particles (less than 2 μm in diameter) and vapors are captured by plant canopies by inertial impact on small foliar elements (Hallam and Juniper 1971, Holloway 1971). The ability of the canopy to retain the impacted material depends on its surface physical and chemical characteristics ("stickiness" or "hairiness"), surface wetness, and the solubility of the deposited material. Some of the larger impacted particles may be subsequently lost through "bounce off" (immediate resuspension) or "blow off" (delayed resuspension) (Gregory 1971). Deposition rates of particles greater than 0.2 μm in diameter increase with higher particle density and diameter, with increased canopy roughness, and with higher wind speed and atmospheric turbulence (Sehmel 1980). Impaction is most efficient when the ratio of particle size to obstacle size is large (Chamberlain 1975a). Thus particles are better captured by canopies with small-sized obstacles (e.g., needle-leaved forests) than by canopies with large-sized obstacles (broad-leaved forests).

Cloudwater deposition to forest canopies is distinct from wet or dry deposition, since it occurs during immersion in nonprecipitating clouds or fogs. Both sedimentation and impaction contribute to water capture by canopy surfaces by this pathway. The rate of water and solute capture depends on the liquid water content of the air, its droplet size distribution, the wind speed, turbulence, and the geometry of the canopy (Lovett et al. 1982).

Absorption through leaf stomates is the pathway for the acquisition of not only CO_2, but of other gaseous species as well (O_3, SO_2, NO_x, HF (hydrogen fluoride), and PAN (peroxyacetylnitrate). Some of these gases may sorb to cuticular surfaces (e.g., SO_2: Raybould et al. 1977). The deposition of these compounds is greatly affected by their water solubility (e.g., Hill 1971), leaf wetness, ambient relative humidity, and leaf stomatal behavior (Reich and Amundson 1985).

Dry deposition and cloudwater deposition are commonly estimated empirically and modeled with parameters of bulk collection efficiency (fraction of material removed from the atmosphere). The most common such parameter is the deposition velocity, which is equivalent to the speed at which the airborne mass at some reference height (commonly 1 m above) is transferred to the surface (Chamberlain 1975a,b, Sehmel 1980). The input rate is then obtained by multiplying the ambient concentration by the appropriate deposition velocity for that canopy, compound, and period. Typical values range from 0.04 to 7.5 cm sec^{-1} (for SO_2 gas), 0.002 to 2.0 cm sec^{-1} (for O_3), negative to 0.5 cm sec^{-1} (for NO_x), 0.0001 to 1.0 cm sec^{-1} for small particles (<2 μm in diameter) and greater than 0.01 cm sec^{-1} for large particles, depending on diameter and density (Sehmel 1980). Cloudwater deposition velocities may range from 0.1 to as high as 100 cm sec^{-1}. Much effort has gone into the estimation of such velocities, but they have proved to be extremely variable and difficult to measure (Droppo 1976, 1980, Sehmel 1980, White and Turner 1970). The appeal of the deposition velocity approach lies in the simplicity with which fluxes may be calculated. However, the technique is not exact, and small differences in deposition velocity may lead to large errors in deposition estimates.

III. Use of Throughfall to Estimate Dry Deposition

The processes that alter solution chemistry in the canopy may be summarized in the following mass balance equations:

$$NTS = (THF + STF) - INC$$
$$NTS = (L - U) + D$$

where NTS, THF, STF, and INC are the depositions in net (throughfall + stemflow), gross throughfall, gross stemflow, and incident precipitation, and L, U, and D are the contributions from leaching, foliar uptake of materials in INC, and dry deposition, respectively; (L − U) is the net leachate. The distinction between materials released from within the canopy and from dry deposition is not always as straightforward as these formulae suggest, because (1) some canopy-derived materials (e.g., particulate litter) may also wash off from canopy surfaces, and (2) some portions of dry deposited material may be retained in the canopy and not appear in throughfall. The degree to which the formulae depict actual relationships among processes (i.e., the amount of overlap in Figure 2.3) is likely to vary from one canopy to another.

The contributions of dry deposition washoff and leaching in net throughfall deposition have been estimated with a variety of methods. Most methods have focused on the estimation of the washoff component of dry deposition; few attempts have been made to quantify the leaching contribution directly. Unfortunately, there is no standard method for measuring dry deposition against which other approaches might be calibrated or evaluated. Without such a basis, it is necessary to discuss the array of existing methods by comparison and by examination of the assumptions in the approaches.

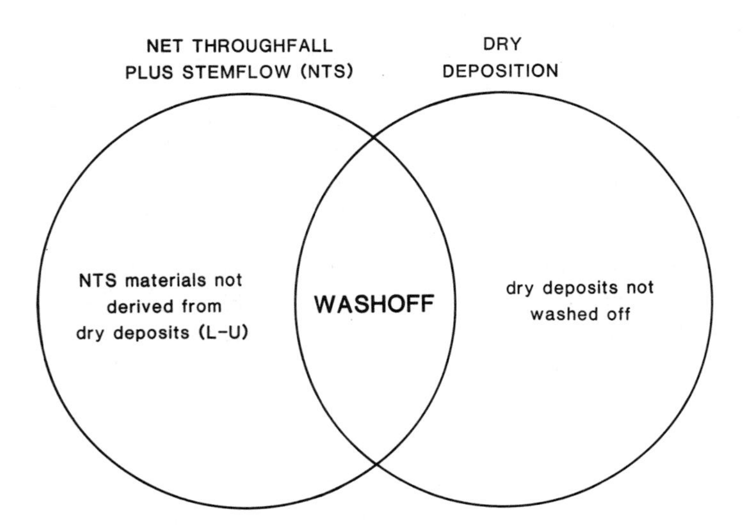

FIGURE 2.3. Washoff is portion of dry deposition that may be mobilized by precipitation.

A. Direct Approaches

The most direct approach to estimating dry deposition contribution to net throughfall would be to compare throughfall fluxes from canopies exposed to dry deposition with those from adjacent ones protected from such inputs. Banaszak (1975) found that net throughfall quality from young loblolly pines in Mississippi exposed to ambient deposition was similar to that under canopies covered by plastic tents between storms, suggesting a limited contribution from dry deposition. Logistical difficulties will probably prevent use of this approach in forest trees. Additionally, plants protected from exposure may differ sufficiently from open-grown plants (in leaf-surface characteristics and leaching characteristics) to complicate comparisons between their responses (Keever and Jacobsen 1983, Shriner 1977).

Another direct approach is to examine net throughfall deposition as a function of the strength of the dry deposition, where the degree of input is indicated by the distance from a suspected source. Such an effect has been observed in the case of marine aerosols. For example, Rapp (1969) found a strong gradient in net sodium deposition under four different forests with increasing distance from the Mediterranean Sea. Similarly, Potts (1978) reported a marked decrease in sodium, and, to a lesser extent, magnesium, depositions in throughfall collectors at increasing distance from the seaward edge of a Sitka spruce stand. For dry deposition of anthropogenic origin, Baker et al. (1977) found net throughfall sulfate deposition under spruces downwind of a power plant in Alberta to be 12.3 times higher than under spruces upwind of the plant. A similar result was observed by Haughbotn (1973) under conifers at three locations in southern Norway. In that study, net throughfall deposition of sulfate sulfur at a location next to a power plant was estimated at 78.9 kg S ha^{-1} yr^{-1}, the highest value ever reported.

Care must be taken in gradient studies to avoid problems that may interfere with ascribing net throughfall deposition differences among locations to concurrent dry deposition only. First, the observations must be carefully controlled by the matching of canopy species and site quality, because these factors may affect throughfall quality. Such confounding influences may have contributed to the lack of distance effect on throughfall sulfate deposition found by Parker et al. (1980). Second, the reference site should not to be within a region of high background deposition. Kelly (1984) found no statistical differences in throughfall depositions between carefully matched sites that differed in distance from a power plant; he hypothesized that the lack of difference was from regionally distributed inputs which were more important than those of local origin. Finally, higher net throughfall depositions in areas close to sources could include elevated leaching of existing materials resulting from past deposition, rather than to only the washoff of materials deposited in the current dry period.

B. Indirect Approaches

Numerous attempts have been made to estimate dry deposition using artificial collectors intended to simulate canopy surfaces. One approach commonly used is to collect throughfall under chemically inert artificial canopies (Art et al. 1974, Etherington 1967, Hart and Parent 1974, Iwatsubo and Tsutsumi 1967, Juang and Johnson 1967, Lakhani and Miller 1980, Lindberg and Lovett 1985, Lindberg et al. 1982, Miller and Miller 1980, Nihlgard 1970, Schlesinger and Reiners 1974, White and Turner 1970). Alternatively, small surrogate surfaces have been exposed in actual canopies, both with and without protection from precipitation (Lindberg and Lovett 1985, Vandenberg and Knoerr 1985). Dry deposition rates to such surrogate collectors are estimated as the mass of solutes released from the surfaces by washing, divided by the time of exposure and the area of the collector. The deposition rates from artificial surfaces are then extrapolated to estimate dry deposition to whole canopies. The use of substitute surfaces to estimate deposition has a number of drawbacks, all related to unknown fidelity with which they mimic actual canopy behaviors. The gas and particle trapping efficiency are related to the material of which the surrogate collectors are constructed (Vandenberg and Knoerr 1985). Artificial surfaces do not absorb deposited gases as real canopies do, nor can they leach or take up solutes from incident precipitation.

The problems of simulating the behavior of actual canopies may be circumvented somewhat by controlled washing of actual canopy tissues (Lindberg et al. 1982, Reiners and Olson 1984, White and Turner 1970, author's unpublished data). Canopy components are periodically sampled and washed in solutions of known concentrations. Differences in washwater concentrations among periods are used to estimate dry deposition. Such techniques have the drawback that the leaching rate is not necessarily constant. Leaching rate may be influenced by (1) the interval since the last storm (i.e., recovery of the pool of leachable solutes), (2) differences in leachability between tree species, tissue ages, and canopy positions, and (3) experimental conditions such as the foliage mass/leachant volume, the method of leaching, and the concentration of leachant used.

Fluxes of leachates and washoff are both highest in the initial stages of wetting, declining with cumulative precipitation. The surface load of dry deposits is finite and easily dislodged and thus the washoff contribution is rapidly exhausted. The supply of leachable substances is not similarly limited because it is continuously supplied from foliar tissues. Thus, the importance of leaching relative to washoff is likely to be greater in large storms than in small storms.

Where dry deposition has a characteristic chemical signal distinct from that of plant leachates, it may be estimated by a ratio approach. This has been most commonly done where dry deposition is known to have a strong sea salt component (Art et al. 1974, Miller 1963, Miller and Miller 1980,

Potts 1978). Typically, the investigator assumes that all of one component in throughfall (such as sodium or chloride) originates from outside the system and calculates the dry deposition contribution of other species assuming constant ratios. However, the element ratios in sea salt aerosols are neither always the same as for seawater nor are they always constant. There can be considerable changes in ratios during aging of sea salts (Clayton 1972, Miller and Miller 1980).

Graustein and Armstrong (1983) and Gosz et al. (1983) extended the ratio approach in estimating dry deposition inputs to high elevation forests in New Mexico, using strontium isotopes. They assumed that the ^{87}Sr:^{86}Sr ratio found in net throughfall was a combination of the ratios found in (1) the sapwood of tree boles (taken as the leaching fraction), (2) the underlying rocks (taken as the dust washoff fraction), and (3) bulk precipitation. The percentage of net throughfall strontium from washed-off dust input was calculated to be 66% and 25% for a spruce-fir and aspen canopy, respectively. The ratio approach can occasionally predict negative depositions for some elements. Also, the commonly used assumption that leachates contain none of the tracer material is probably not realistic in systems having a history of deposition; in such cases, material from earlier input may currently be circulating in the system.

Dry deposition to canopies is rarely estimated by attempting to account for the leaching rate, but in theory there is no reason why tracers could not be introduced into trees to follow the course of leaching. Tracers have been employed in the estimation of other materials released from plants, for example, transpired water using tritium (Kline et al. 1970). In nutrient redistribution studies, a variety of tracers have been used, including dyes (Thomas 1967, 1969), radioisotopes (Dayton 1970, Kimmins 1972, Riekerk 1978, Waller and Olsen 1967, Witherspoon 1964, Witkamp and Frank 1964), and element ratios (Stone 1981, Stone and Kszystyniak 1977). The difference between measured net throughfall and leaching of a tracer can underestimate dry deposition unless foliar uptake is accounted for. This approach has been utilized in the study of canopy solute exchange (Garten 1988, Garten and Lindberg 1988) in eastern Tennessee, where very little of net throughfall sulfur ($<15\%$) was attributable to leaching.

Some authors have suggested that intrinsic mobility (leachability) of an ionic species might be useful in estimating the relative importance of leaching in throughfall (Eaton et al. 1973, Henderson et al. 1977). An indication of an ion's leachability might be provided by controlled leaching studies (e.g., Gosz et al. 1975, Tukey et al. 1958) or by the ratio of net throughfall flux to the maximum canopy content of each element. Such an index might indicate the minimum contribution of leaching to net throughfall for some solutes.

Mayer and Ulrich (1972) computed annual dry deposition to a beech canopy by assuming that the rate of dry deposition is equal to the flux of net throughfall during the dormant period. However, dry deposition is a function of both canopy surface area and ambient gas and aerosol loads, both of which may be greater in the growing season (e.g., Galloway and Parker 1979). Thus, dormant-season throughfall could underestimate growing-season dry deposition. On the other hand, throughfall in dormant periods is not solely dry deposition because nonfoliar tissues can be leached (Reiners and Olson 1984, Tukey 1970a). Thus, dry deposition during the dormant season may be overestimated with this approach. The degree to which dry deposition overestimation in the dormant season balances underestimation in the growing season is not known.

C. Statistical and Simulation Approaches

Miller et al. (1976) estimate the leaching fraction of net throughfall from the positive y-intercept of linear regressions of throughfall deposition on incident wet deposition. This suggests that throughfall in small storms consists of leachates, whereas washoff studies (Lindberg et al. 1982, Little 1973, 1977) have shown that water-soluble dry deposits are dislodged in the initial stages of wetting. Also, the relationship between net throughfall and incident deposition is commonly curvilinear (Attiwill 1966, Bernhard-Renversat 1975, Madgwick and Ovington 1959, Parker 1981). A refinement of the regression approach was suggested by Lakhani and Miller (1980), who assumed that dry deposition to canopies would be proportional to dry deposition to bulk precipitation collectors covered with inert screening.

Stachurski and Zimka (1984) compared literature reports of the net throughfall deposition of nitrogen and employed a multiple regression approach to separate the portions due to (1) foliar uptake (proportional to the incident deposition of nitrogen), (2) leaching induced by phytophagous insects (proportional to the nitrogen in particulate litter filtered from throughfall samples in their own studies), and (3) dry deposition (from the positive y-intercept of the regression of the deposition of net throughfall on incident deposition). Dry deposition estimated by this method ($\approx$1.9 kg N ha^{-1} yr^{-1}) is likely to be very sensitive to factors influencing the magnitude of foliar uptake (e.g., site trophic status, presence of epibiota) and canopy herbivory.

Lovett and Lindberg (1984) estimated both dry deposition and canopy exchange of solutes in individual storms in two deciduous forests. They assumed (1) leaching is proportional to the amount of precipitation (P) and (2) dry deposition is proportional to the length of the dry period preceding the storm (A). They then estimated the relative contributions of leaching and dry deposition to net throughfall of various solutes by examining the fitted parameters of the following multiple regression equation:

$$\begin{aligned} NTF &= (L-U) + D \\ &= b_1{*}A + b_2{*}P \end{aligned}$$

where NTF, L, U, and D are rates of net throughfall, leaching, uptake, and dry deposition, A and P are the length of the dry period and the amount of precipitation, and b_1 and b_2 are fitted parameters assumed to relate to the mean rates of dry deposition and canopy exchange.

Recently there have been attempts at comprehensive models of the solute interactions in forest canopies that contribute to throughfall deposition (Chen et al. 1983, Vasudevan 1982). The model of Chen et al. (1983) requires solute- and canopy-specific rates of dry deposition and foliar exudation as inputs to the calculation of throughfall concentrations. Their model is therefore not suitable for inferring the relative importance of these contributions to net throughfall. Their model estimates the leaching of foliar solutes by a constant fraction of the foliar content (ignoring difference in leachability between elements). Also, ammonium lost in the canopy is considered to have been converted to nitrate through nitrification, a process not known to occur in forest canopies. Nonetheless, the model predictions agreed reasonably well with observed monthly throughfall depositions under conifer and hardwood canopies in the Woods Lake watershed in the Adirondacks. The generality of this model is unknown because the critical test of applying it to other forest systems has not been attempted yet. Mathematical models of solute interactions in canopies that consider all major processes with realistic rates have much potential but remain relatively unexplored.

D. Summary of Available Results

Table 2.1 summarizes estimates of dry deposition contributions to net throughfall for the basic cations Ca, Mg, K, and Na. Despite differences in methods employed and in forests studied, there is modest agreement among studies for this group of elements. Dry deposition appears to supply 10–60% of potassium, 30–50% of magnesium, 20–80% of calcium, and 50–60% of sodium in net throughfall, except in maritime forests, where sodium and magnesium in net throughfall may be predominantly caused by atmospheric input. These ranges are quite broad and may not be useful as predictors of dry deposition in net throughfall for a given stand.

A similar comparison of studies reporting on the dry deposition contribution to net throughfall sulfate (often the major anion) shows even less pattern. The dry deposition contribution ranges from 8% to nearly 100% of net throughfall and exhibits no identifiable trend (Table 2.2). Of particular interest are the different estimates made for dry deposition in throughfall of chestnut oak at the Walker Branch watershed in eastern Tennessee. Lindberg et al. (1979) calculated the dry deposition contribution of net throughfall sulfur to be 26.3% by extrapolating data on dry deposition to inert plates, using scaling factors that depended on canopy phenology. Subsequently, the dry deposition component of throughfall

TABLE 2.1. Estimates computed from the literature for the percentage of net throughfall from dry deposition for basic cations.[a]

Forest system	Element				Source
	Na	Ca	Mg	K	
Holly forest, Long Island, New York	100[b]	46.4	98.1	18.6	Art et al. 1974
Corsican pine Scotland	54.5	70.0	41.2	43.3	Miller et al. 1976
Beech forest, central Germany	60.7	35.0	46.1	25.7	Mayer and Ulrich 1972
Hard beech, maritime New Zealand	100[b]	50.0	100[b]	15.6	Miller 1963
Spruce-fir,[c] high elevation New Mexico	100[b] (150)	51.6 (79.1)	34.3 (51.4)	17.2 (27)	Graustein 1980 Graustein and Armstrong 1978
Chestnut oak	—	26–47	—	58–63	Lovett and Lindberg 1984
White oak, Tennessee	—	18	—	48–62	

[a]Adapted from Parker 1983, with permission of Academic Press.
[b]Defined as 100% in the method employed.
[c]Using sodium-ratio method (parenthetical values are strontium isotope-ratio method).

sulfate at the same site was estimated to be 38% (growing season) and 46% (dormant season) by Lovett and Lindberg (1984), who separated leaching and dry deposition with a multiple regression analysis of throughfall from individual events. Finally, Lindberg et al. (1986) used a combined approach to estimate dry deposition of large particles (by scaling up deposition from inert plates) and small particles, gases, and vapors (by multiplying ambient air concentrations by estimated deposition velocities). They suggested that dry deposition could account for nearly 100% of net throughfall and stemflow sulfate (though the uncertainty in the vapor and fine particle deposition was estimated at 50–70%). These estimates made at the same site demonstrate the strong influence of methodology on estimates of dry deposition contribution to net throughfall.

A reasonable estimate of the dry deposition component of net throughfall can be obtained in certain fortuitous circumstances, where the dry deposition is distinct, dominant, and relatively invariant. However, dry deposition is commonly a mixture of components, with time-varying proportions, and therefore information in addition to throughfall chemistry, such as atmospheric concentrations and foliar surface chemistry, would be useful for refining dry deposition estimates. Ideally, one would attempt to estimate dry deposition using micrometeorological and/or surrogate surface methods simultaneously with the throughfall observations.

TABLE 2.2. Incident precipitation (INC), throughfall (THF), and net throughfall (NTF) sulfate-sulfur depositions; and the percentage of net throughfall from dry deposition in systems where attempts were made to partition net throughfall sources.[a]

References	Sampling period	Deposition (kg S ha^{-1})			NTF from dry deposition (%)
		INC	THF	NTF	
Raybould et al. 1977	66 days[b]	2.06	4.42	2.36	13
Lovett and Lindberg, 1984	2 years[c]	13.0	22.6	9.8	38–46
	1 year[d]	13.0	38.5	25.5	8–19
Eaton et al., 1978	INC many years, THF 5 months[e]	12.7	33.7	21.0	21.9
Lindberg et al. 1979	2 years[c]	13.0	32.0	19.0	26.3
Nicholson et al. 1980	1 year	11.8	37	25.2	68
Bache 1977	153 days[f]	5.8	19.6	13.8	89.1
	127 days[g]	—	—	4.5	14.8–44.4
Ulrich et al. 1978	8 years[h]	23.8 ±2.7[j]	53.1[k] ±11.2	29.3[k] ±9.4	88.9 ±12.0
Miller 1963	1 year[i]	8.4	10.4	2.0	100
Lindberg et al. 1986	2 years[c]	11.2	25.7[k]	14.2[k]	100

[a]Extended from Parker et al. 1980.
[b]Wheat canopy.
[c]Chestnut oak.
[d]White oak.
[e]Northern hardwoods.
[f]Corsican and Scots pine.
[g]Pine forest, using data from Bjor et al. (1974, in Bache 1977).
[h]European beech.
[i]Hardwood.
[j]Standard deviation.
[k]Includes stemflow.

IV. Effects of Pollutants on Canopy Solute Interactions

A. Overview

Although much is known about the effects of specific pollutants on various aspects of plant metabolism, little is understood about their potential to increase leaching in forest canopies. Several pollutant-dependent mechanisms that could promote leaching have been described. For example, an increase in the permeability of the plasma membrane is likely with damage caused by ozone (Evans and Ting 1973) or strong acidity (Chia et al. 1984), but not with damage from SO_2, NO_x, or HF (Mudd et al. 1984). Also, particular pollutants may increase concentrations of certain soluble and potentially leachable metabolites, for example, carbohydrates (Koziol 1984) and amino acids (Darral and Jager 1984). Enhanced activities of several enzymes, such as peroxidases, may be relatively general responses to stress (Mejnartowicz 1984), but such changes are likely to be modified by both the species of tree and the particular pollutant. Separate effects of individual pollutants may be difficult to distinguish in field situations because one pollutant compound is usually found in combination with others, resulting in an often poorly defined "pollutant complex."

B. Canopy Processing of Acidic Deposition

An assessment of pollutant effects on foliar leaching may be handled less speculatively in the case of acidic deposition. This is because (1) the processes of hydrogen ion exchange (Mecklenberg et al. 1966) and of weak base exchange (Lovett et al. 1985) provide reasonable mechanisms for potential leaching increases caused by acidic deposition, and (2) field data are available that may help assess the importance of such process.

The pH of precipitation measures only the acids in the dissociated form, that is, the "free" acidity. The total acidity of the solution, including the undissociated acids, must be quantified by other means, such as titration. It is common to distinguish between "strong" acids (e.g., H_2SO_4, HNO_3, and HCl, which dissociate easily at pH levels typical of rainfall and throughfall) and "weak" acids (e.g., various organic acids, which dissociate less readily). Strong acids may dominate the free acidity in precipitation and throughfall, but partially dissociated weak acids also contribute (Lindberg and Coe 1984, Tyree 1985). The distinction between acid strengths is of potential use in understanding the changes in acidity induced by forest canopies, because strong acids are likely of anthropogenic origin while weak acids are usually naturally derived (Galloway et al. 1976, 1984).

Forest canopies commonly change the pH of precipitation reaching the forest floor (Figure 2.4). Because pH is a log scale, the magnitude of the free acidity change indicated by a given change in pH increases greatly as initial pH declines.

Hardwood canopies generally increase the pH of precipitation. There are, however, a number of exceptions to this rule (e.g, Künstle et al. 1981, Skeffington 1983). For coniferous stands, reports of increased pH (e.g., Abrahamsen et al. 1976, Mahendrappa 1983, Miller 1984) are about as common as reports of decreased pH. Miller (1984) suggested that stand age and nutrient deficiency have an important influence on pH changes

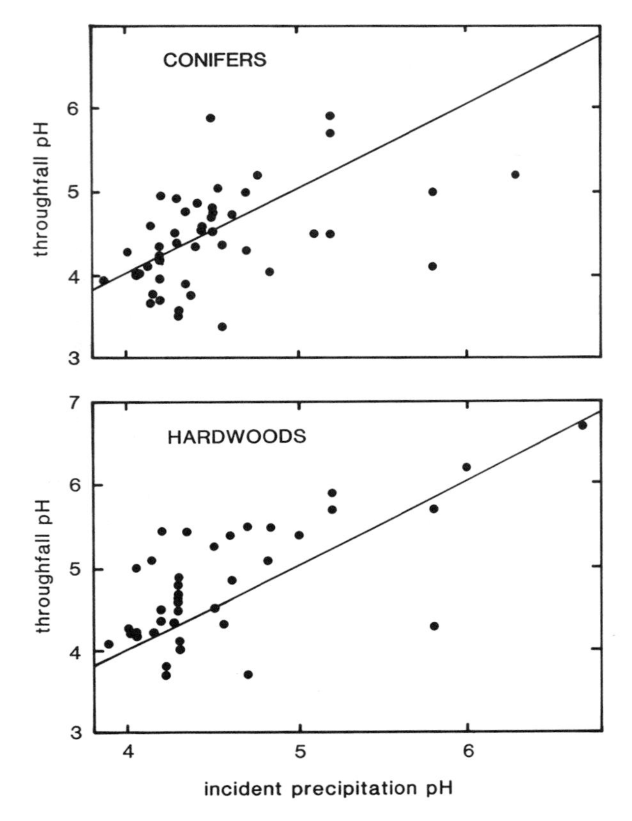

FIGURE 2.4. Forest canopy-induced alterations of pH of precipitation under coniferous (upper panel) and hardwood (lower panel) stands (from literature reports). Oblique line indicates no net effect on precipitation pH. Points plotted here are taken or calculated from Abrahamsen et al. 1976, Alcock and Morton 1981, Binkley et al. 1982, Bockheim et al. 1983, Bringmark 1980, Brinson et al. 1980, Chen et al. 1983, Cole and Johnson 1977, Cronan and Reiners 1983, Dethier and Jones 1985, Eaton et al. 1973, Ferres et al. 1984, Foster 1985, Horntvedt 1979, Johnson et al. 1985, Künstle et al. 1981, Langkamp et al. 1982, Mahendrappa 1983, McColl and Bush 1978, Miller 1985, Mollitor and Raynal 1982, Moore 1983, Nicholson et al. 1980, Nihlgard 1970, Parker 1981, Pastor and Bockheim 1984, Potts 1978, Raynal et al. 1983, Richter and Granat 1978, Richter et al. 1983, Roose and Lelong 1981, Shearer et al. 1977, Singer et al. 1978, Skeffington 1983, Sollins et al. 1980, Swank and Swank 1983, Ugolini et al. 1977, van Breeman et al. 1982, van Miegroet and Cole 1984, Westman 1978.

for conifers. Young forests (<60 years) often elevate pH while adjacent older stands (>60 years) tend to depress it (Binkley et al. 1982, Miller 1984). Stands on rich or recently fertilized sites appear to have lower hydrogen ion depositions in net throughfall than nearby stands on poor sites (Khanna and Ulrich 1981, Mahendrappa 1983, Miller et al. 1976, Richter et al. 1983).

Three processes are involved in the alteration of precipitation pH. First, interception deposition may contribute either basic materials (e.g., soil dust; Hart and Parent 1974), or acid-forming materials (e.g., SO_2 and NO_x gases and sulfate aerosols; Ulrich et al. 1978). Second, the canopy may acidify precipitation through the release of weak organic acids (Hoffman et al. 1980a). Finally, canopy surfaces may increase pH by releasing basic cations in exchange for hydrogen ions in the incident precipitation. Clearly, change in pH alone is not a good indicator of the relative importance of these processes, which could conceivably interact to produce small changes in free acidity. An understanding of the causes and significance of canopy-mediated changes in pH must wait until more is learned about the nature and extent of these processes.

It would be particularly useful to investigate the nature of acidity and acid buffering on bark and leaf surfaces. Cellular pH is known to be closely regulated to near neutrality even when the pH of the external solution varies (Nieboer et al. 1984, Wind 1979). The buffering capacity (the change in acid or base input required to produce a given change in pH) of leaf homogenates is affected by tree species, time of year, and pollutant dosage, and has been suggested as an index of foliage sensitivity to acidification (Pylypec and Redman 1984, Scholz and Stephan 1974, Sidhu and Zakrevsky 1982). The acidity of pulverized bark in solution (which may have a pH as low as 3) and its buffering capacity are also affected by the tree species and proximity to pollution sources (Grodzinska 1971, Staxäng 1969).

The regulation of acidity at the surfaces of intact leaves and bark is also not well known. The acid content of single water droplets applied to leaves decreases relative to droplets on inert surfaces (Adams and Hutchinson 1984, Evans et al. 1985, Oertli et al. 1977). The decrease in acidity of the external solution may be related to buffering by the bicarbonate system (Tukey 1970a), but full chemical characterization of leaf droplets has not been reported. The pH of droplets much more acidic than precipitation (e.g., pH 2.5–3.1) may decrease during foliar contact, although the acid content of the evaporating droplet actually declines (Adams and Hutchinson 1984, Evans et al. 1985). The fact that droplets residing a long time on dry leaf surfaces can decline in acidity may not pertain to leaf buffering capacities in actual storms. The capacity of canopy surfaces to alter droplet pH may be quickly exhausted and important only in the initial fractions of the storm.

The hypothesis that solutes released from forest canopies may be related to the hydrogen ions retained by the canopy (hydrogen ion exchange hypothesis) may be examined by comparing data from throughfall studies. Figure 2.5 shows the release of the major basic cations (sum of Ca, Mg, K, and Na) in net throughfall as a function of the amount of hydrogen

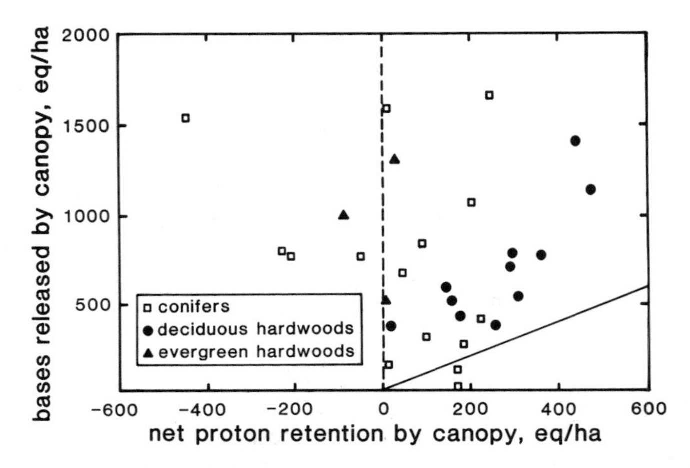

FIGURE 2.5. Relationship between release of basic cation (sum of Ca, Mg, K, and Na) equivalents (eq) to retention of hydrogen ion equivalents in forest canopies (from literature reports). Oblique line indicates hypothetical case of perfect hydrogen-cation exchange. Plotted points calculated from data in Binkley et al. 1982, Bockheim et al. 1983, Bringmark 1980, Cole and Johnson 1977, Cronan and Reiners 1983, Eaton et al. 1973, Ferres et al. 1984, Foster 1985, McColl and Bush 1978, Miller 1985, Nihlgard 1970, Parker 1981, Pastor and Bockheim 1984, Potts 1978, Raynal et al. 1983, Richter and Granat 1978, Roose and Lelong 1981, Sollins et al. 1980, Ugolini et al. 1977, van Miegroet and Cole 1984. Data from Ulrich 1981 and Westman 1978 were far out of range of this graph and were not plotted.

ions retained or released by the canopy. The units of both axes are charge deposition (equivalents per hectare [eq ha^{-1}]). The diagonal line indicates the hypothetical case of pure hydrogen-cation exchange, where all the bases released may be accounted for by hydrogen ions retained. Points above this line suggest that processes other than hydrogen ion exchange contribute to cation release, and points below the line imply that hydrogen ion uptake exceeds cation loss. Points on the left side of the graph suggest that the canopy releases both hydrogen ions and basic cations.

The scatter in the points may be interpreted somewhat better if different classes of tree canopies are considered separately. The evergreen hardwood canopies (eucalypts, evergreen oaks, and tropical hardwoods, shown as triangles) exhibit little net retention of hydrogen ions in wet precipitation, but much variation in the amount of cationic charges released. The coniferous canopies show very little uniformity: they may either take up or release hydrogen ions, and show much variation in the release of cationic charge.

In this data set, all the deciduous hardwood canopies (most of them in the north temperate zone) release cations and retain some hydrogen ions. Despite the great range in amounts of precipitation and high variation in dry deposition inputs to these forests, there appears to be some proportionality between hydrogen ion retention and total cation release. The mean value of the ratio (H retention/base release) was 0.39 ± 0.17 (range 0.05-0.67). Thus, roughly 40% of cations released from these deciduous hardwood canopies might be attributed to hydrogen ion exchange. This value is within the range (40–60%) calculated by Lovett et al. (1985) for three Tennessee canopies.

However, the interpretation of the correlation between base release and hydrogen ion retention in temperate hardwoods as cross-system evidence for the hydrogen exchange mechanism is hampered by several uncertainties. First, neither the total hydrogen ion input to the canopy nor the leaching fraction of net throughfall was specified in most studies. The studies in which the dry deposition contribution has been estimated are few and, considering the differences in methods employed, probably not comparable. Many reports of mean precipitation or throughfall pH do not indicate whether the value is volume weighted. Also, both the total hydrogen ion inputs and cation releases from the canopy are strongly dependent on the amount of precipitation (i.e., a wet site with low precipitation acidity might have the same hydrogen ion input as a drier site with highly acidic rain). The deposition in stemflow (always more acidic than in throughfall) is usually not reported, and therefore hydrogen ion retention from bulk precipitation is likely to have been overestimated. The failures to account for additional hydrogen ion inputs in dry deposition (which leads to underestimating H retention) and stemflow (which may cause overestimates of H retention) might cancel out, but compensation by these opposing errors will likely be system dependent.

Some of the problems in comparing the cation-hydrogen ion balances of different forests may be alleviated by studying the hydrogen ion-cation exchange in many events under one canopy. Figure 2.6 presents the same chemical interactions as in Figure 2.5 but for separate summertime rainstorms under both white pine and mixed oaks in western Virginia (author's data). These data show that there is considerable storm-to-storm variation in the cation-hydrogen ion behavior. Often, more proton charges were retained in the canopy than cationic charges released. On the whole, there is little or no relationship that might support the importance of hydrogen ion exchange on an event basis. Ultimately, the interpretation of these data suffers from some of the problems mentioned above for cross-system comparisons; for example, neither dry deposition inputs of hydrogen ions nor stemflow releases of hydrogens and cations were accounted for.

C. Evidence for Effects of Pollutants on Leaching

Numerous laboratory and greenhouse trials have investigated the effect of altering leachant pH on the leaching of solutes. Fairfax and Lepp (1974) found Ca in leachates from tobacco leaves increased 3.7-fold when pH

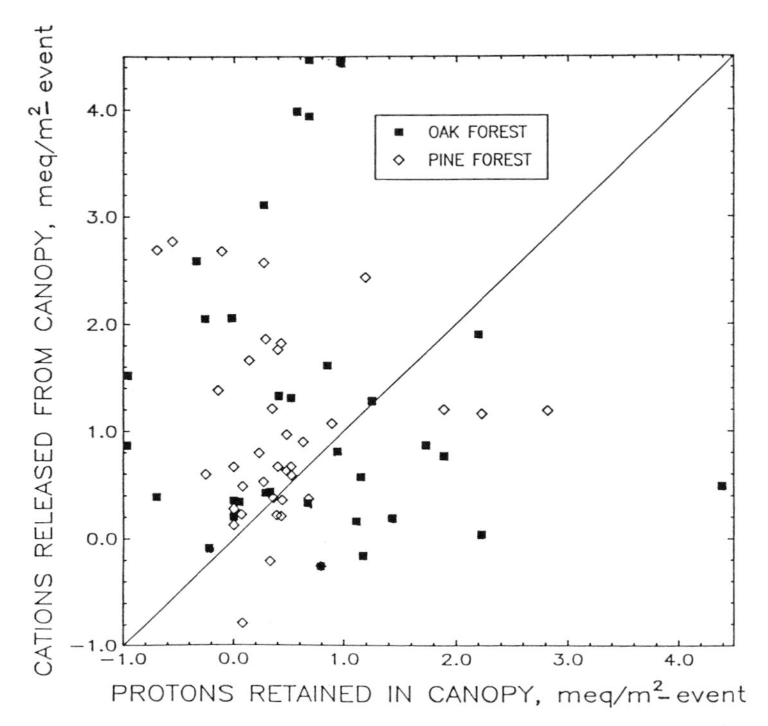

FIGURE 2.6. Relation between cation equivalents released and hydrogen ion equivalents retained in single summertime rainstorms under white pine and mixed oak forests in western Virginia (authors' data). Oblique line gives hypothetical case of perfect hydrogen ion-cation exchange.

was reduced from 6.7 to 3.0; however, K decreased with increasing acidity and magnesium showed no effect. Kratky et al. (1974) reported that Ca and Mg concentrations in tomato leaf disk leachates increased twofold (and potassium threefold) when the pH was reduced from 5.6 to 4.0. Wood and Bormann (1975) found a marked pH effect on Ca, K, and Mg leaching from *Phaseolus vulgaris* (pH 3.0 vs. 5.0) and *Acer saccharum* (pH 2.3 vs. 5.0), but no effect for Na. Evans et al. (1981) also worked with *Phaseolus vulgaris* and found significant pH effects (2.7 vs. 5.7) for Ca, K, NO_3, and SO_4 but not for NH_4, Mg, Cl, and Zn. In that study, the degree of leaching was not always proportional to hydrogen ion concentration; for example, K showed an inverse effect. Lee and Weber (1982), in a 3-year study with whole plants of *Alnus rubra* and *Acer saccharum*, found no effect of pH (range 3.0–5.7) on throughfall concentrations of numerous ions. Keever and Jacobsen (1983) found a pH effect on the leaching of ^{86}Rb (72% increase) from *Glycine max* leaves, but only between the extreme treatments (pH 2.7 and 5.6). Scherbatskoy and Klein

(1983) reported that amino acids increased and soluble proteins decreased in leachates from *Betula alleghaniensis* and *Picea glauca* with decreasing leachant pH (5.7 to 2.8); in birch leachates, K and Ca also increased with acidity, while carbohydrates, nitrate, and phosphate showed no effect. Skeffington and Roberts (1985) found some pH effect on leachates from *Pinus sylvestris* for Mg and Al, but not for Ca, P, K, or Mn. Haines et al. (1985) found a strong species effect but no pH effect (5.5 to 2.5) for leachate concentrations from whole plants of nine species.

Effects of leachant pH on throughfall quality have also been studied in a field experiment (Abrahamsen et al. 1976). Acidified water (pH 6.0–2.5) was applied above the canopy (4–5 m high) in plantation plots of *Picea abies* in southern Norway. Increases in throughfall concentrations were observed for organic carbon (19%), Mg (67%), K (94%), Ca (120%), and S (15.4-fold increase). No effects were seen for nitrate- or ammonium-nitrogen (Horntvedt 1979). In a related study, acidified water was added to the forest floor below the canopy. Throughfall acidity increased but other solutes showed no response to treatment. These results suggest that hydrogen ion leaching is affected by the acidity of the substrate but that the release of other solutes reacts to the quality of incident precipitation.

Recent studies of gaseous pollutant effects on foliar leaching have been restricted to ozone, always in combination with acidified water. Krause et al. (1984) exposed *Picea abies* seedlings to ozone (0, 200, and 600 $\mu g\ m^{-3}$) in combination with pH 3.5 fog. They reported increased leaching of nitrate, Mg, Ca, K, and, to some extent, sulfate. Because only one level of acidity was employed, interaction effects between acidity and ozone could not be assessed. Skeffington and Roberts (1985) conducted a factorial experiment on *Pinus sylvestris* involving acid mist (distilled water vs. pH 2.99 solution) and ozone (<1, 94, 203, and 304 $\mu g\ O_3\ m^{-3}$). They found no ozone effect on foliage solute exchange (except some inhibition of nitrate uptake) and no acidity-ozone interaction.

Investigations of pollutant effects on leaching have employed a great variety of procedures. Differences in experimental protocols (whole plant, single leaf, leaf disk), plant species (crops and trees), ages (seedlings to mature trees), and methods of reporting results (leachate mass per unit leaf area or per plant, throughfall concentrations) may account for some of the variation in reported responses under controlled conditions. Nonetheless, an increase in foliar leaching in response to increasing hydrogen ion concentrations, and to some extent for increasing ozone doses, has been frequently shown, especially for the cations Ca and Mg, and often for K as well.

In many of the studies reviewed, however, the maximum or "control" pH was $\geq$5.6. Such pH levels probably exceed those of "pristine" rain (Galloway et al. 1984). The minimum leachant pH levels employed (2.3–3.0) represent hydrogen ion concentrations far higher than encountered in bulk precipitation in industrialized regions (Jacobson 1984). Considering the range of leachant acidities used, the degree of enhanced leaching is relatively small. Typically, observed changes in leachate concentrations are more closely related to the change in pH than to the change in absolute hydrogen ion concentration. For example, Fairfax and Lepp (1974) found a 3.7-fold increase in calcium leaching with a reduction of pH from 6.7 to 3.0. This suggests that the mechanism of hydrogen ion-cation exchange is relatively less important at low pH than at high pH. Moreover, there is no evidence of a threshold pH below which solute losses are dramatically increased. Thus, effects of spatial or temporal variation in ambient precipitation acidity on foliage leaching may be difficult to detect in field situations.

The implications of pollutant-enhanced foliar leaching for the forest nutrient cycle depends on the ability of the canopy to replace the ions lost (Johnson et al. 1985, Lovett et al. 1985, Nilsson et al. 1982). If replacement of lost ions is not limited by nutrient availability, root uptake, or translocation, then the nutrient cycle may simply accelerate. In this case, soil solution losses might then increase (if immobilization is slow), and the increases in ion uptake might increase the excretion of hydrogen ions into the soil, enhancing acidification of the soil around the fine feeder roots. If, on the other hand, the canopy cannot be supplied with solutes to compensate for pollutant-enhanced leaching, then the foliar tissues will likely suffer mineral deficiencies, especially of relatively immobile elements such as Ca (Krause et al. 1984, Prinz et al. 1985).

V. Throughfall and Forest Health

The potential use of throughfall to indicate the overall health of forest canopies (and of forests themselves) has intuitive appeal but little direct support. The indirect support for this proposition derives from reports concerning the short-term response of leachate chemistry to a number of factors. Here I explore some of the known responses of leachate or throughfall chemistry to various beneficial and harmful agents.

As previously discussed, throughfall and stemflow composition are affected by site quality: (1) rich sites have higher concentrations than poor sites, (2) trees with nitrogen-fixing bacterial associates have higher nitrogen depositions in throughfall than those without them, and (3) throughfall responds to stand fertilizations with increased solute concentrations, apparently in parallel with soil nutrient availability.

Throughfall quality can respond to abiotic stress, for example, leaf abrasion, drought, and extremes of temperature (Tukey and Morgan 1963). It can also respond to biotic stresses: leachate solute concentrations are often elevated following insect activities in the canopy (Carlisle et al. 1966, 1967, Crossley and Seastedt 1981, Kimmins 1972, Krause 1977, Nilsson 1978, Seastedt et al. 1983). Insect-enhanced leaching is caused primarily by leakage of cellular contents from damaged leaves (Kimmins 1972, Tiedemann et al. 1980), although solutes from insect frass (Kimmins 1972), feces (Stachurski and Zimka 1984), exuviae (Kimmins 1972), and exudates such as honeydew (Carlisle 1965, Carlisle et al. 1966, 1967) may also contribute. Defoliation of *Abies grandis* by Paraquat (60% of needles lost) did not increase solute loss relative to a control plot (Tiedemann et al. 1980).

Experiments involving atmospheric pollutants (see Section IV) provide some support to the hypothesis that ozone and/or acidic rain may accelerate foliage leaching and thus contribute to forest health problems. Additional support for the hypothesis comes from field observations of foliar nutrient deficiencies in West German trees exhibiting the "new type forest damage" (Prinz et al. 1984, 1985). I am not aware that any measurements of throughfall chemistry have been made in the affected stands to confirm the leaching possibility.

No data relate the occurrence or concentrations of specific inorganic or organic species in throughfall to the action of specific or combined stressing agents. The lack of stress indicators in throughfall may only reflect a failure to seek them. The main impediment to identifying solute responses to stresses is the complex nature of the interacting process even in the absence of apparent stress. Effects of an agent of stress would add yet another level of complexity to an already difficult problem.

Potential indicator organic compounds for canopy stress include (1) pigments, such as chlorophyll or its degradation products, (2) soluble reducing sugars, (3) low molecular weight organic acids, especially those from the Kreb's cycle, (4) free amino acids, (5) products of peroxidase activity (Mejnartowicz 1984), and (6) cuticle waxes (Baker and Hunt 1986). Inorganic compounds potentially responding to stress could include those of magnesium, manganese, nitrogen, and phosphorus. Even the turbidity of the throughfall solution might be an indicator of damage (Darrall and Jager 1984).

VI. Conclusions and Research Needs

The release of inorganic and organic ions from plant canopies is a normal part of the forest nutrient cycle. It has been studied in numerous forests under a variety of edaphic, climatological, and anthropogenic conditions. Despite some general understanding of throughfall and stemflow chemistry (e.g., its importance relative to other nutrient cycling pathways), there is little understanding of the mechanisms and rates of processes contributing to the alteration of solute chemistry by forest canopies.

Throughfall and stemflow quality may have species-specific components. Comparative studies of throughfall quality for several species at many sites could test this possibility. Norway spruce, black spruce, Scots pine, white pine, Douglas fir, and European beech are potential candidate species because they are widely distributed and frequent subjects of ecophysiological research. More information on the composition of stemflow (especially its acidity), including the influences of epiphytes and tree age, could provide an understanding of the role of tree bark in solute alterations. Studies on the seasonal behavior of deciduous conifers and evergreen hardwoods species could provide insight into the role of bark-derived acids in the hydrogen ion balance of forest canopies. Finally, a characterization of the specific organic compounds commonly found in throughfall and stemflow, especially those contributing to the undissociated acidity (particularly in conifer stands), could advance the understanding of the sources of acidity and exchange in forest canopies.

A major area of uncertainty is the amount of dry deposition input and the degree to which the deposited materials may be washed off. Dry deposition probably makes some contribution to net throughfall in almost every case. It is important to know the dry deposition contribution to net throughfall in order to quantify the nutrient uptake by forest vegetation, to balance the nutrient budgets of forest ecosystems, to assess the removal of materials from the lower atmosphere, and to identify the responses of leachates to various factors.

A review of attempts to estimate dry deposition from analyses of throughfall chemistry alone suggests that there is no simple, reliable, and generally applicable method for determining sources of solutes in throughfall. Many of the methods employed depend on unique features of the system investigated and may therefore not be useful at most locations. Dry deposition alone is extremely difficult to estimate; its contribution to solute fluxes in throughfall and stemflow appears even more intractable. It is doubtful that examination of throughfall fluxes only can provide reliable estimates of dry deposition to plant canopies.

However, the analysis of throughfall might be employed in setting bounds on estimates of dry deposition to forest canopies, especially if used in conjunction with other techniques, such as surrogate surfaces, tracers, eddy-correlation, and modeling. An intensive study using multiple approaches could aid in the development of standard methods that might be applied in other locations.

Air pollutants can influence the quantity and quality of leachates lost from plant canopies, but several methodological problems interfere with the discrimination of such effects in field studies. Chief among these is the current inability to distinguish the plant-derived contributions from the atmospheric contributions to net throughfall. Another problem is to distinguish effects of individual air pollutants from those of other natural and pollutant stresses. Controlling for additional variation in throughfall and stemflow chemistry resulting from such factors as the amount, intensity, and frequency of precipitation, is yet another obstacle.

Controlled laboratory and greenhouse studies remain a good source of information for assessing the significance of accelerated leaching caused by pollutants, despite the difficulty of relating their results to field situations. This area of research could benefit from (1) standardization of experimental protocols, (2) longer term experiments with a greater variety of forest tree species, (3) the use of combinations of both pollutants (e.g., ozone, sulfur dioxide, acidified water) and other stresses (e.g., drought, frost, insect infestation), and, (4) the use of stress levels that might reasonably be encountered in the field.

Specific substances in throughfall and stemflow (or combinations of compounds) could serve as indicators of canopy ill-health or well-being. That there is no evidence to substantiate the existence of such a substance may only reflect the failure to seek it. There are several desirable characteristics of a compound or group of compounds to be used as an indicator of forest health. The substance should be plant derived, without a significant atmospheric source. Thus, polyaromatic hydrocarbons (PAH) detected by Matzner (1984) in both throughfall and incident precipitation are not suitable. Also, the substance should appear in elevated concentrations in throughfall before the appearance of foliar injury or reductions in growth. Finally, the material should be detectable in a wide variety of forest types by using a relatively simple and inexpensive method.

Acknowledgment. This is a contribution from the program of the Institute of Ecosystem Studies, The New York Botanical Garden. Financial support was provided by NCASI and the Mary Flagler Cary Charitable Trust. The author thanks Catherine Parker for her assistance in organizing this work and numerous colleagues for sharing their thoughts, comments, and recent results.

References

Abrahamsen, G. 1980. Impact of atmospheric sulfur deposition on forest ecosystems. *In*: Atmospheric Sulfur Deposition, D.S. Shriner, C.R. Richmond, and S.E. Lindberg (eds.), pp. 397–415. Ann Arbor, Michigan: Ann Arbor Science.

Abrahamsen, G, K. Bjor, R. Horntvedt, and B. Tveite. 1976. Effects of acid precipitation on coniferous forests. *In*: Proceedings of the First International Symposium on Acid Precipitation and the Forest Ecosystem, L.S. Dochinger and T.A. Seliga (eds.), pp. 991–1009. USDA Forest Service Gen Tech Rep NE-23.

Adams, C.M. and T.C. Hutchinson. 1984. A comparison of the ability of leaf surfaces of three species to neutralize acid rain drops. New Phytol 97:463–478.

Alcock, M.R. and A.J. Morton. 1981. The sulphur content and pH of rainfall and of throughfalls under pine and birch. J Appl Ecol 18:835–839.

Art, H.W., F.H. Bormann, G.K. Voigt, and G.M. Woodwell. 1974. Barrier island forest ecosystem: role of meteorologic nutrient inputs. Science 184:60–62.

Astrup, M. and M. Bülow-Olsen. 1979. Nutrient cycling in two Danish beech (*Fagus sylvatica*) forests growing on different soil types. Holarct Ecol 2:125–129.

Attiwill, P.M. 1966. The chemical composition of rainwater in relation to cycling of nutrients in mature eucalyptus forest. Plant Soil 24:390–408.

Attiwill, P.M. 1980. Nutrient cycling in a *Eucalyptus obliqua* (L'Herit.) forest. II. Nutrient uptake and nutrient return. Aust J Bot 28:199–222.

Bache, D.H. 1977. Sulphur dioxide uptake and the leaching of sulfates from a pine forest. J Appl Ecol 14:881–895.

Baker, E.A. and Hunt, G.M. 1986. Erosion of waxes from leaf surfaces by simulated rain. New Phytol 102:161–173.

Baker, J., D. Hocking, and M. Nyborg. 1977. Acidity of open and intercepted precipitation in forests and effects on forest soils in Alberta, Canada. Water Air Soil Pollut 7:449–460.

Banaszak, K.T. 1975. Relative throughfall enrichment by biological- and aerosol-derived materials in loblolly pines. PB-245-258/9ST. Mississippi Water Research Institute, Mississippi State University. Springfield, Virginia: National Technical Information Center.

Basilevich, N.I. and Rodin, L.E. 1966. The biological cycle of nitrogen and ash elements in plant communities of the tropical and subtropical zones. For Abstr 27:357–368. (Transl. from Botaniceskig Zurnal 49:185–209.)

Beauford, W., J. Barber, and A.R. Barringer. 1977. Release of particles containing metals from vegetation into the atmosphere. Science 195:571–573.

Belot, Y. and D. Gauthier. 1975. Transport of micronic particles from atmosphere to foliar surfaces. *In*: Heat and Mass Transfer in the Biosphere, Part I, Transfer Processes in the Plant Environment, D.A. deVries and N.H. Afgan (eds.), pp. 583–591. New York: Wiley.

Bentley, B.L. and E.J. Carpenter. 1984. Direct transfer of newly-fixed nitrogen from free-living epiphyllous microorganisms to their host plant. Oecologia 63:52–56.

Benzing, D.H. 1973. The monocotyledons: Their evolution and comparative biology—mineral nutrition and related phenomena in Bromeliaceae and Orchidaceae. Q Rev Biol 48:277–290.

Bernhard-Renversat, F. 1975. Nutrients in throughfall and their quantitative importance in rain forest mineral cycles. *In*: Tropical Ecological Systems—Trends in Terrestrial and Aquatic Research, E. Medina and F.B. Golley (eds.), pp. 153–154. New York: Springer-Verlag.

Best, G.R. and C.D. Monk. 1975. Cation flux in hardwood and white pine watersheds. *In*: Mineral Cycling in Southeastern Ecosystems F.G. Howell, J.B. Gentry, and M.H. Smith (eds.), pp. 847–861. Springfield, Virginia: National Technical Information Center.

Binkley, D., J.P. Kimmins, and M.C. Feller. 1982. Water chemistry profiles in an early successional and a mid-successional forest in coastal British-Columbia Canada. Can J For Res 12:240–248.

Bjor, K., R. Horntvedt, and E. Joranger. 1974. Distribution and chemical composition of precipitation in a southern Norway forest stand (in Norwegian). Report 1/74, Norwegian Council for Scientific and Industrial Research, Oslo-Ås (referenced in Bache 1977).

Bockheim, J.G., S.W. Lee, and J.E. Leide. 1983. Distribution and cycling of elements in a *Pinus resinosa* plantation ecosystem, Wisconsin, USA. Can J For Res 13:609–619.

Bollen, W.B., C.S. Chen, K.C. Lu, and R.F. Tarrant. 1968. Effect of stemflow precipitation on chemical and microbiological soil properties beneath a single alder tree. *In*: Biology of Alder, J.M. Trapp, J.F. Franklin, R.F. Tarrant, and G.M. Hansen (eds.), pp. 148–156. PNW Forest and Range Experiment Station, USDA Forest Service, Portland, Oregon.

Bormann, F.H. 1982. The effects of air pollution on the New England landscape. Ambio 11:338–346.

Boynton, D. 1954. Nutrition by foliar application. Annu Rev Plant Physiol 5:31–54.

Bringmark, L. 1980. Ion leaching through a podsol in a Scots pine stand. *In*: Structure and Function of a Northern Coniferous Forest—An Ecosystem Study, T. Persson (ed.). Ecol Bull (Stockholm) 32:341–361.

Brinkmann, W.L.F. 1983. Nutrient balance of a central Amazonian rainforest: Comparison of natural and man-managed systems. *In*: Hydrology of Humid Tropical Regions with Particular Reference to the Hydrological Effects of Agriculture and Forestry Practice, R. Keller, (ed.), pp. 153–163. International Association of Hydrological Sciences Publ. 140.

Brinkmann, W.L.F. and A. dos Santos. 1973. Natural waters in Amazonia. VI. Soluble calcium properties. Acta Amazonica 3:33–40.

Brinson, M.M., H.D. Bradshaw, R.N. Holmes, and J.B. Elkins, Jr. 1980. Litterfall, stemflow, and throughfall nutrient fluxes in an alluvial swamp forest. Ecology 61:827–835.

Brown, A.H.F. 1974. Nutrient cycles in oakwood ecosystems in northwest England. *In*: The British Oak—Its History and Natural History, M.G. Morris and F.H. Perring (eds.) Cambridge, United Kingdom: Pendragon Press.

Carlisle, A. 1965. Carbohydrates in the precipitation beneath a Sessile Oak *Quercus petraea* (Mattushka) Liebl. canopy. Plant Soil 24:399–400.

Carlisle, A., A.H.F. Brown, and E.J. White. 1966. The organic matter and nutrient elements in the precipitation beneath a Sessile Oak (*Quercus petraea*) canopy. J Ecol 54:87–98.

Carlisle, A., A.H.F. Brown, and E.J. White. 1967. The nutrient content of tree stemflow and ground flora litter and leachates in a Sessile Oak (*Quercus petraea*) woodland. J Ecol 55:615–627.

Carrier, L. and G. Gagnon. 1985. Maple dieback in Quebec. Report for Ministère de l'Energie et des Ressources. Service de la Recherche Appliquée, Sainte-Foy, Québec.

Chamberlain, A.C. 1975a. The movement of particles in plant communities. *In*: Vegetation and the Atmosphere, Vol. 1, Principles, J.L. Monteith (ed.), pp. 155–203. New York: Academic Press.

Chamberlain, A.C. 1975b. Pollution in plant canopies. *In*: Heat and Mass Transfer in the Biosphere, Part I, Transfer Processes in the Plant Environment, D.A. deVries and H.H. Afgan (eds.), pp. 561–582. New York: Wiley.

Chen, C.W., R.J.M. Hudson, S.A. Gherini, J.D. Dean, and R.A. Goldstein. 1983. Acid rain model: Canopy module. J Environ Engineering 109:585–603.

Chia, L.S., C.I. Mayfield, and J.E. Thompson. 1984. Simulated acid rain induces lipid peroxidation and membrane damage in foliage. Plant Cell Environ 7:333–338.

Clayton, J.L. 1972. Salt spray and mineral cycling in two California coastal ecosystems. Ecology 53:74–81.

Clements, R.G. and J.A. Colon. 1975. The rainfall interception process and mineral cycling in a Montane Forest in eastern Puerto Rico. *In*: Mineral Cycling in Southeastern Ecosystems, F.G. Howell, J.B. Gentry, and M.H. Smith (eds.), pp. 813–823. Springfield, Virginia: National Technical Information Center.

Clements, C.R., L.P.H. Jones, and M.J. Hopper. 1972. The leaching of some elements from herbage plants by simulated rain. J Appl Ecol 9:249–260.

Cloutier, A. 1985. Microdistribution des espèces végétales au pied des troncs d'*Acer saccharum* dans une érablière du sud du Québec. Can J Bot 63:274–276.

Cole, D.W. and D.W. Johnson. 1977. Atmospheric sulfate additions and cation leaching in a Douglas Fir ecosystem. Water Resour Res 13:313–317.

Cole, D.W. and M. Rapp. 1981. Elemental cycling in forest ecosystems. *In*: Dynamic Properties of Forest Ecosystems, D.E. Reichle (ed.), pp. 341–409. New York: Cambridge University Press.

Corlin, J.W. 1971. Nutrient cycling as a factor in site productivity and forest fertilization. *In*: Tree Growth and Forest Soils, C.T. Youngberg and C.B. Davey (eds.), pp. 313–325. Corvallis, Oregon: Oregon State University Press.

Cowling, E.B. and R.A. Linthurst. 1981. The acid precipitation phenomenon and its ecological consequences. BioScience 31:649–654.

Crittenden, P.D. 1983. The role of lichens in the nitrogen economy of subarctic woodlands: nitrogen loss from the nitrogen-fixing lichen *Stereocaulon paschal* during rainfall. *In*: Nitrogen as an Ecological Factor, J.A. Lee, S. McNeill, and I.H. Rorison (eds.), pp. 43–68. Oxford: Blackwell Scientific.

Cronan, C.S. and W.A. Reiners. 1983. Canopy processing of acidic precipitation by coniferous and hardwood forests in New England. Oecologia (Berlin) 59:216–223.

Cronan, C.S. and C.L. Schofield. 1979. Aluminum leaching response to acid precipitation: Effects on high elevation watersheds in the northeast. Science 204:304–306.

Crossley, D.A., Jr. and T.R. Seastedt. 1981. Effects of canopy arthropod consumption and leaf biomass on throughfall chemistry of a successional forest in the southern Appalachians (abstr.). Bull Ecol Soc Am 62:106.

Darral, N.M. and H.J. Jager. 1984. Biochemical diagnostic tests for the effect of air pollution on plants. *In*: Gaseous Air Pollutants and Plant Metabolism, M.J. Koziol and F.R. Whatley (eds.), pp. 333–349. London: Butterworths.

Dayton, B.R. 1970. Slow accumulation and transfer of radio-strontium by young loblolly trees (*Pinus taeda* L.). Ecology 51:204–216.

Deal, W.J. 1983. The quantity of acid in acid fog. J Air Pollut Control Assoc 33:691.

De Boois, H.M. and E. Jansen. 1975. Effects of nutrients in throughfall water and of litterfall upon fungal growth in a forest soil layer. Pedobiologia 16:161–166.

Denaeyer-DeSmet, S. 1966. Bilan annuel des apports d'éléments minéraux par les eaux de précipitation sous couvert forestier dans la forêt mélangée caducifoliée de Blaimont. Bull Soc R Bot Belg 99:345–375.

Denison, R., B. Caldwell, B. Bormann, L. Eldred, C. Swanberg, and S. Anderson. 1977. The effects of acid rain on nitrogen fixation in western Washington coniferous forests. Water Air Soil Pollut 8:21–34.

Dethier, D.P. and S.B. Jones. 1985. Atmospheric and weathering contributions to stream chemistry, northwestern Massachusetts. Northeastern Environ Sci 4:8–17.

Droppo, J.G. 1976. Dry removal of air pollutants by vegetation canopies. *In*: Proceedings of 4th National Conference on Fire and Forest Meteorology, B.H. Baker and M.A. Basberg (eds.), pp. 200–208. USDA Gen. Tech. Rep.

Droppo, J.G. 1980. Experimental techniques for dry-deposition measurements. *In*: Atmospheric Sulfur Deposition, D.S. Shriner, C.R. Richmond, and S.E. Lindberg (eds.), pp. 209–221. Ann Arbor, Michigan: Ann Arbor Science.

Eaton, J.S., G.E. Likens, and F.H. Bormann. 1973. Throughfall and stemflow chemistry in a northern hardwood forest. J Ecol 61:495–508.

Eaton, J.S., G.E. Likens, and F.H. Bormann. 1978. The input of gaseous and particulate sulfur to a forest ecosystem. Tellus 30:546–551.

Edwards, P.J. 1982. Studies of mineral cycling in a montane forest in New Guinea. V. Rates of cycling in throughfall and litterfall. J Ecol 70:807–827.

Environmental Protection Agency. 1984a. Air quality criteria for ozone and other photochemical oxidants. EPA-600/8-84-020A, External review draft. Washington, D.C.: U.S. Environmental Protection Agency.

Environmental Protection Agency. 1984b. The acidic deposition phenomenon and its effects: critical assessment review papers. EPA-600/8-83-016A. Washington, D.C.: U.S. Environmental Protection Agency.

Epstein, E. 1972. *Mineral Nutrition of Plants: Principles and Perspectives.* New York: Wiley.

Eriksson, E. 1955. Airborne salts and the chemical composition of river water. Tellus 7:243–250.

Etherington, J.H. 1967. Studies of nutrient cycling and productivity in oligotrophic systems. I. Soil potassium and windblown sea spray in South Wales dune grassland. J Ecol 55:743–752.

Evans, L.S. 1984. Botanical aspects of acidic precipitation. Bot Rev 50:449–490.

Evans, L.S. and I.P. Ting. 1973. Ozone-induced membrane permeability changes. Am J Bot 60:155–162.

Evans, L.S., D.C. Canada, and K.A. Santucci. 1986. Foliar uptake of 15N from rain. Environ Exp Bot 26:143–146.

Evans, L.S., T.M. Curry, and K.F. Lewin. 1981. Responses of leaves of *Phaseolus vulgaris* L. to simulated acid rain. New Phytol 88:403–420.

Evans, L.S., K.A. Santucci, and M.J. Patti. 1985. Interactions of simulated rain solutions and leaves of *Phaseolus vulgaris* L. Environ Exp Bot 25:31–40.

Fahey, T.J. 1979. Changes in nutrient content of snow water during outflow from Rocky Mountain coniferous forest. Oikos 32:422–428.

Fairfax, J.A.W. and N.W. Lepp. 1974. Effect of simulated "acid rain" on cation loss from leaves. Nature 255:324–325.

Falconer, R.E. and P.D. Falconer. 1979. Determination of cloudwater acidity at a mountain observatory in the Adirondack mountains of New York State. Atmospheric Sciences Research Center, Publ. No 741. Albany: State University of New York.

Farquhar, G.D., R. Wetselaar, and P.M. Firth. 1979. Ammonia volatilization from senescing leaves of maize. Science 203:1257–1258.

Feller, M.C. 1977. Nutrient movement through western hemlock—western red cedar ecosystems in southwestern British Columbia. Ecology 58:1269–1283.

Ferres, U., F. Roda, A.M.C. Verdes, and J. Terradias. 1984. Circulation de nutrientes en algunos ecosistemas forestales del Montseny (Barcelona). Mediterr Ser Biol 7:139–166.

Fogg, G.F. 1947. Quantitative studies on the wetting of leaves by water. Proc R Soc London Biol 134:503–522.

Foster, N.W. 1974. Annual macroelement transfer from *Pinus banksiana* Lamb. forest to soil. Can J For Res 4:470–476.

Foster, N.W. 1985. Acid precipitation and soil solution chemistry within a maple-birch forest in Canada. For Ecol Manage 12:215–231.

Foster, N.W. and S.P. Gessel. 1972. The natural addition of nitrogen, potassium, and calcium to a *Pinus banksiana* Lamb. forest floor. Can J For Res 2:448–455.

Franke, W. 1967. Mechanisms of foliar penetration of solutions. Annu Rev Plant Physiol 18:281–300.

Friedland, A.J. and A.H. Johnson. 1985. Lead distribution and fluxes in a high-elevation forest in northern Vermont. J Environ Qual 14:332–336.

Fuhrer, J. 1985. Formation of secondary air pollutants and their occurrence in Europe. Experientia 41:286–301.

Fuhrer, J. and C. Fuhrer-Fries. 1982. Interactions between acidic deposition and forest ecosystem processes. Eur J For Pathol 12:377–390.

Galloway, J.N. and G.G. Parker. 1979. Sulfur deposition in the eastern United States. *In*: MAP3S Update: Progress Report for FY 1977 and FY 1978, M.C. McCracken (ed.), pp. 124–134. Springfield, Virginia: National Technical Information Center.

Galloway, J.N. and G.G. Parker. 1980. Difficulties in measuring wet and dry deposition on forest canopies and soil surfaces. *In*: Effects of Acid Precipitation on Terrestrial Ecosystems, T.C. Hutchinson and M. Havas (eds.), pp. 57–68. New York: Plenum Press.

Galloway, J.N., G.E. Likens, and E.S. Edgerton. 1976. Hydrogen ion speciation in the acid precipitation of the northeastern United States. Water Air Soil Pollut 6:423–433.

Galloway, J.N., G.E. Likens, and M.E. Hawley. 1984. Acid rain: Natural versus anthropogenic components. Science 225:829–831.

Garten, C.T., Jr. 1988. Fate and distribution of sulfur-35 in yellow poplar and red maple trees. Oecologia 76:43–50.

Gersper, D.L. and N. Hollowaychuck. 1970. Effects of stemflow water on a Miami soil under a beech tree. I. Morphological and physical properties. Soil Sci Soc Am Proc 34:779–786.

Gersper, D.L. and N. Hollowaychuck. 1971. Some effects of stemflow from forest canopy trees on chemical properties of soils. Ecology 52:691–702.

Golley, F.B., J.T. McGinnis, R.G. Clements, G.I. Child, and M.J. Duever. 1975. Mineral Cycling in a Tropical Moist Ecosystem. Athens, Georgia: University of Georgia Press.

Gosz, J.R. 1980. Nutrient budget studies for forests along an elevational gradient in New Mexico. Ecology 61:515–521.

Gosz, J.R., D.G. Brookins, and D.I. Moore. 1983. Using strontium isotope ratios to estimate inputs to ecosystems. BioScience 33:23–30.

Gosz, J.R., G.E. Likens, and F.H. Bormann. 1976. Organic matter and nutrient dynamics of the forest and forest floor in the Hubbard Brook Forest. Oecologia (Berlin) 22:305–320.

Gosz, J.R., G.E. Likens, J.S. Eaton, and F.H. Bormann. 1975. Leaching of nutrients from leaves of selected tree species of New Hampshire. *In*: Mineral

Cycling in Southeastern Ecosystems, F.G. Howell, J.B. Gentry, and M.H. Smith (eds.), pp. 638–641. Springfield, Virginia: National Technical Information Center.

Graustein, W.C. 1980. The effects of forest vegetation on chemical weathering and solute acquisition: A study of the Tesuque Watersheds near Santa Fe, New Mexico. Ph.D. Dissertation, Department of Geology and Geophysics, Yale University.

Graustein, W.C. and R.L. Armstrong. 1978. Measurement of dust input to a forested watershed using strontium-87/strontium-86 ratios. Geol Soc Am Abstracts with Programs 10:411.

Graustein, W.C. and R.L. Armstrong. 1983. The use of strontium-87/strontium-86 ratios to measure atmospheric transport into forested watersheds. Science 219:289–292.

Gregory, P.H. 1961. *The Microbiology of the Atmosphere.* London: Leonard Hill.

Gregory, P.H. 1971. The leaf as a spore trap. *In*: Ecology of Leaf Surface Microorganisms, T.F. Preece and C.H. Dickinson (eds.), pp. 239–243. London: Academic Press.

Grimm, U. and H.W. Fassbender. 1981. Ciclos biogeoquimicos en un ecosistema forestal de los Andes Occidentales de Venezuela. III. Ciclo hidrologica y translocacion de elementos quimicos con el agua. Turrialba 31:89–99.

Grodzinska, K. 1971. Acidification of tree bark as a measure of air pollution in southern Poland. Bull Acad Pol Sci, Ser Sci Biol 14:189–195.

Haines, B., J. Chapman, and C.D. Monk. 1985. Rates of mineral element leaching from leaves of nine plant species from a southern Appalachian forest succession subjected to simulated acid rain. Bull Torrey Bot Club 112:258–264.

Hallam, N.D. and B.E. Juniper. 1971. The anatomy of the leaf surface. *In*: Ecology of Leaf Surface Micro-organisms, T.F. Preece and C.H. Dickinson (eds.), pp. 3–37. London: Academic Press.

Hart, G.S. and D.R. Parent. 1974. Chemistry of a throughfall under Douglas Fir and Rocky Mountain Juniper. Am Midl Nat 92:191–201.

Haughbotn, O. 1973. Nedbørundersøkelsor i Sarpsborgdistriktet og undersøkelser over virkninger av forsurende nedfall på jordas kjemiske egenskaper, Ås-NLH 1973 (referenced in Horntvedt 1975).

Heinrichs, H. and R. Mayer. 1977. Distribution and cycling of major and trace elements in two Central European forest ecosystems. J Environ Qual 6:402–407.

Helvey, J.D. and J.H. Patric. 1965. Canopy and litter interception of rainfall by hardwoods in the eastern United States. Water Resour Res 1:193–206.

Henderson, G.S., W.F. Harris, D.E. Todd, Jr., and T. Grizzard. 1977. Quality and chemistry of throughfall as influenced by forest type and season. J Ecol 65:365–374.

Herrera, R., C.F. Jordan, H. Klinge, and E. Medina. 1978. Amazon ecosystems, their structure and functioning with particular emphasis on nutrients. Interciencia 3:223–232.

Hesse, P.R. 1957. Sulfur and nitrogen changes in forest soils of East Africa. Plant Soil 9:86–96.

Hileman, B. 1984. Forest decline from air pollution. Environ Sci Technol 18:A8–A9.

Hill, A.C. 1971. Vegetation: a sink for atmospheric pollutants. J Air Pollut Control Assoc 21:341–346.

Hoffman, W.A., Jr., S.E. Lindberg, and R.R. Turner. 1980a. Precipitation acidity: The role of the forest in acid exchange. J Environ Qual 9:95–100.
Hoffman, W.A., Jr., S.E. Lindberg, and R.R. Turner. 1980b. Some observations of organic constituents in rain above and below a forest canopy. Environ Sci Technol 14:999–1002.
Holloway, P.J. 1971. The chemical and physical characteristics of leaf surfaces. *In*: Ecology of Leaf Surface Micro-organisms, T.F. Preece and C.H.Dickinson (eds.), pp. 41–53. London: Academic Press.
Horntvedt, R. 1975. Kjemisk innhold i nedbør under trær. Et litteratursammendrag. Teknisk notat TN 18/75. SNSF Prosjektet. Oslo-Ås, Norway.
Horntvedt, R. 1979. Leaching of chemical substances from tree crowns by artificial rain. Contribution FA 33/78, SNSF Project, P.O. Box 333, Blindern, Oslo 3, Norway.
Horntvedt, R. and E. Joranger. 1974. Nedbørens fordeling og kjemiske innhold under traer: Juli-november 1973. Teknisk notat TN 3/74. SNSF-Prosjektet, Oslo-As, Norway.
Hosker, R.P. and S.E. Lindberg. 1982. Review: Atmospheric deposition and plant assimilation of gases and particles. Atmos Environ 16:889–910.
Hughes, M.K., N.W. Lepp and D.A. Phipps. 1980. Aerial heavy metal pollution and terrestrial ecosystems. Adv Ecol Res 11:217–327.
Ingham, G. 1950a. The mineral content of air and rain and its importance to agriculture. J Agric Sci 40:55–61.
Ingham, G. 1950b. Effect of materials absorbed from the atmosphere in maintaining soil fertility. Soil Sci 70:205–212.
Iwatsubo, G. and T. Tsutsumi. 1967. On the amount of plant nutrients supplied to the ground by rainwater in adjacent open plot and forest. Kyoto Univ For Bull 39:110–120.
Jacobson, J.S. 1984. Effects of acidic aerosol, fog, mist, and rain on crops and trees. Philos Trans R Soc Lond B 305:327–338.
Johnson, A.H. and T.G. Siccama. 1983. Acid deposition and forest decline. Environ Sci Technol 17:2442–3052.
Johnson, A.H. and T.G. Siccama. 1984. Decline of red spruce in the northern Appalachians: Assessing the possible role of acid deposition. Tappi J 67:68–72.
Johnson, D.W. 1981. The natural acidity of some unpolluted waters in southeastern Alaska and potential impacts of acid rain. Water Air Soil Pollut 16:243–252.
Johnson, D.W., J. Turner, and J.M. Kelly. 1982. The effects of acid rain on forest nutrient status. Water Resour Res 18:449–461.
Johnson, D.W., D.D. Richter, G.M. Lovett, and S.E. Lindberg. 1985. The effects of atmospheric deposition on potassium, calcium, and magnesium cycling in two deciduous forests. Can J For Res 15:773–782.
Jones, K. 1970. Nitrogen fixation in the phyllosphere of the Douglas fir, *Pseudotsuga douglasii*. Ann Bot 34:239–244.
Jones, K. 1976. Nitrogen fixing bacteria in the canopy of conifers in a temperate forest. *In*: Microbiology of Aerial Plant Surfaces, C.H. Dickinson and T.F. Preece (eds.), pp. 451–463. London: Academic Press.
Jordan, C.F. 1978. Stemflow and nutrient transfer in a tropical rain forest. Oikos 31:257–263.
Jordan, C., F. Golley, T. Hall, and J. Hall. 1979. Nutrient scavenging of rainfall by the canopy of an Amazonian rainforest. Biotropica 12:61–66.

Juang, F.H.T. and N.M. Johnson. 1967. Cycling of chlorine through a forested watershed in New England. J Geophys Res 72:5641–5647.

Kaul, O.N. and W.D. Billings. 1965. Cation content of stemflow in some forest trees in North Carolina. Indian For 91:367–370.

Kazimirov, N.I. and R.N. Morozova. 1973. Biologicheskii krugovorot veshchestv v elnikakh karelii (Biological Cycling of Matter in Spruce Forests of Karelia). Leningrad: Nauka Publishing House.

Keever, G.J. and J.S. Jacobsen. 1983. Response of *Glycine max* L. Merril to simulated acid rain: Environmental and morphological influence on the foliar leaching of 86-Rb. Fl Crops Res 6:241–250.

Kelly, J.M. 1984. Power plant influences on bulk precipitation throughfall and stemflow nutrient inputs. J Environ Qual 13:405–409.

Khanna, P.K. and B. Ulrich. 1981. Changes in the chemistry of throughfall under stands of beech (*Fagus sylvatica*) and spruce (*Picea abies*) following the addition of fertilizers. Acta Oecol Plant 2:155–164.

Kimmins, J.P. 1972. Relative contributions of leaching, litterfall and defoliation by *Neodiprion sertifer* (Hymenoptera) to the removal of cesium-134 from red pine. Oikos 23:226–234.

Kittredge, J. 1948. *Forest Influences.* New York: McGraw-Hill.

Kline, J.R., J.R. Martin, C.F. Jordan, and J.J. Koranda. 1970. Measurement of transpiration in tropical trees with tritiated water. Ecology 51:1068–1073.

Knabe, W. 1976. Effects of sulfur dioxide on terrestrial vegetation. Ambio 5:213–218.

Koziol, M.J. 1984. Interactions of gaseous pollutants with carbohydrate metabolism. *In*: Gaseous Air Pollutants and Plant Metabolism, M.J. Koziol and F.R. Whatley (eds.), pp. 251–273. London: Butterworths.

Kratky, B.A., E.T. Fukunaga, J.W. Hylin, and R.T. Nakano. 1974. Volcanic air pollution: Deleterious effects on tomatoes. J Environ Qual 3:138–140.

Krause, G.H., B. Prinz, and K.D. Jung. 1984. Forest effects in West Germany. *In*: Air Pollution and the Productivity of the Forest. Arlington, Virginia: Izaak Walton League of America.

Krause, P.D. 1977. Mineral cycling in aspen: the interactions of throughfall, litterfall, road salt and insect defoliation. Ph.D. Dissertation. Department of Biology, University of New Mexico, Albuquerque, New Mexico.

Künstle, E. von, G. Mitscherlich, and G. Ronicke. 1981. Untersuchungen über die Konzentration und den Gehalt an Schwefel, Chlorid, Kalium und Calcium sowie den pH-Wert im Freilandniederschlag und Kronendurchlass von Nadel- und Laubholzbeständen bei Freiburg. Allg Forst Jagdztg 152:147–165.

Lakhani, K.H. and H.G. Miller. 1980. Assessing the contribution of crown leaching to the element content of rain water beneath trees. *In*: Effects of Acid Precipitation on Terrestrial Ecosystems, T.C. Hutchinson and M. Havas (eds.), pp. 151–172. New York: Plenum Press.

Lang, G.E., W.H. Reiners, and R.K. Heier. 1976. Potential alteration of precipitation chemistry by epiphytic lichens. Oecologia 25:229–241.

Langkamp, P.J., G.F. Farnell, and M.J. Dalling. 1982. Nutrient cycling in a stand of *Acacia holosericea*. A. Lunn. ex Lt. Don. I. Measurements of precipitation interception, seasonal acetylene reduction and nitrogen requirement. Aust J Bot 30:87–106.

Lausberg, T. 1935. Quantitative Untersuchungen über die kutikulare Exkretion des Laubblattes. Jahrb Wiss Bot 81:769–806.

LeClerc, J.A. and J.F. Breazeale. 1908. Plant food removed from growing plants by rain or dew. USDA Yearbook of Agriculture 1908:389–402.

Lee, C. and H.B. Tukey, Jr. 1972. Effect of intermittant mist on the development of fall color in foliage of *Euonymous alata* Sieb. 'Compactus'. J Am Soc Hortic Sci 97:97–101.

Lee, J.J. and D.E. Weber. 1982. Effects of sulphuric acid rain on major cation and sulfate concentrations of water percolating through two model hardwood forests. J Environ Qual 11:57–64.

Leininger, T. and W. Winner. 1985. A comparison of rainfall and throughfall chemistry on two forested sites differing markedly in soil fertility (Abstr.) Phytopathology 75:626–627.

Lemee, G. 1974. Recherches sur les écosystèmes des réserves biologiques de la forêt de Fontainebleau IV. Entrées d'éléments minéraux par les précipitations et transfert au sol par le pluviolessivage. Oecol Plant 9:187–200.

Lepp, N.W. and J.A.W. Fairfax. 1976. The role of acid rain as a regulator of foliar nutrient uptake and loss. *In*: Microbiology of Aerial Plant Surfaces, C.H. Dickinson and T.F. Preece (eds.), pp. 107–118. London: Academic Press.

Likens, G.E., E.S. Edgerton, and J.N. Galloway. 1983. The composition and deposition of organic carbon in precipitation. Tellus 35B:16–24.

Lindberg, S.E. and J.M. Coe. 1984. Dissociation of weak acids during Gran plot free acidity titrations. Tellus 36B:186–191.

Lindberg, S.E. and C.T. Garten, Jr. 1988. Sources of sulfur in forest canopy throughfall. Nature 336:148–151.

Lindberg, S.E. and R.C. Harriss. 1981. The role of atmospheric deposition in an eastern deciduous forest. Water Air Soil Pollut 16:13–31.

Lindberg, S.E. and G.M. Lovett. 1985. Field measurements of particle dry deposition rates to foliage and inert surfaces in a forest canopy. Environ Sci Technol 19:238–244.

Lindberg, S.E., R.C. Harriss, and R.R. Turner. 1982. Atmospheric deposition of metals to forest vegetation. Science 215:1609–1611.

Lindberg, S.E., G.M. Lovett, D.D. Richter, and D.W. Johnson. 1986. Atmospheric deposition and canopy interactions of major ions in a forest. Science 231:141–144.

Lindberg, S.E., R.C. Harriss, R.R. Turner, D.S. Shriner, and D.D. Huff. 1979. Mechanisms and rates of deposition of selected trace elements and sulfate to a deciduous forest watershed. ORNL/TM-6674, Oak Ridge National Laboratory, Oak Ridge, Tennessee.

Linthurst, R.A. (ed.). 1984. Direct and Indirect Effects of Acidic Deposition on Vegetation. Boston: Butterworth.

Little, P. 1973. A study of heavy metal contamination of leaf surfaces. Environ Pollut 5:159–172.

Little, P. 1977. Deposition of 2.75, 5.0 and 8.5 μm particles to plant and soil surfaces. Environ Pollut 12:293–305.

Lovett, G.M. 1984. Atmospheric deposition to forests. *In*: Forest Responses to Acidic Deposition, C.S. Cronan (ed.), pp. 7–18. Orono, Maine: Land and Water Resources Center, University of Maine.

Lovett, G.M. and S.E. Lindberg. 1984. Dry deposition and canopy exchange in a mixed oak forest as determined by analysis of throughfall. J Appl Ecol 21:1013–1027.

Lovett, G.M., W.A. Reiners, and R.K. Olson. 1982. Cloud droplet deposition in subalpine balsam fir forests: Hydrological and chemical inputs. Science 218:1303–1304.

Lovett, G.M., S.E. Lindberg, D.D. Richter, and D.W. Johnson. 1985. The effects of acidic deposition on cation leaching from three deciduous forest canopies. Can J For Res 15:773–782.

Lunt, H.A. 1934. Distribution of soil moisture under isolated forest trees. J Agric Res 49:695–703.

Madgwick, H.A.I. and J.D. Ovington. 1959. The chemical composition of precipitation in adjacent forest and open plots. Forestry 32:14–22.

Mahendrappa, M.K. 1974. Chemical composition of stemflow from some Eastern Canadian tree species. Can J For Res 4:1–7.

Mahendrappa, M.K. 1983. Chemical characteristics of precipitation and hydrogen input in throughfall and stemflow under some eastern Canadian forest stands. Can J For Res 13:948–955.

Mahendrappa, M.K. and D.G.O. Kingston. 1982. Prediction of throughfall quantities under different forest stands. Can J For Res 13:474–481.

Mahendrappa, M.K. and E.O. Ogden. 1973. Effects of fertilization of a black spruce stand on nitrogen contents of stemflow, throughfall and litterfall. Can J For Res 3:54–60.

Martin, J.T. and B.F. Juniper. 1970. The cuticles of plants. New York: St. Martin's Press.

Matzner, E. 1984. Annual rates of deposition of polycyclic aromatic hydrocarbons in different forest ecosystems. Water Air Soil Pollut 21:425–434.

Matzner, E., P.K. Khanna, K.J. Meiwes, and B. Ulrich. 1983. Effects of fertilization on the fluxes of chemical elements through different forest ecosystems. Plant Soil 74:343–358.

Mayer, R. and B. Ulrich. 1972. Conclusions on the filtering action of forests from ecosystems analysis. Oecol Plant 9:157–168.

McColl, J.G. and P.S. Bush. 1978. Precipitation and throughfall chemistry in the San Francisco Bay area. J Environ Qual 7:352–357.

McDowell, L.L, G.H. Willis, L.M. Southwick, and S. Smith. 1984. Methyl parathion and EPN wash-off from cotton *Gossypium hirsutum* plants by simulated rainfall. Env Sci Technol 18:423–427.

McDowell, W.H. 1982. Mechanisms controlling the organic chemistry of Bear Brook, New Hampshire. Ph.D. Thesis, Cornell University, Ithaca, New York.

McDowell, W.H. and G.E. Likens. 1988. Origin, composition, and flux of dissolved organic carbon in the Hubbard Brook Valley. Ecol Monogr 58:177–195.

McLaughlin, S.B. 1985. Effects of air pollution on forests . . . a critical review. J Air Pollut Control Assoc 35:512–534.

Mecklenburg, H.A., H.B. Tukey, Jr., and J.V. Morgan. 1966. A mechanism for leaching of calcium from foliage. Plant Physiol 41:610–613.

Mejnartowicz, L.E. 1984. Enzymatic investigations on tolerance in forest trees. *In*: Gaseous Air Pollutants and Plant Metabolism, M.J. Koziol and F.R.Whatley (eds.), pp. 381–398. London: Butterworths.

Miller, H.G. 1979. The nutrient budgets of even-aged forests. *In*: The Ecology of Even-Aged Forest Plantations, E.D. Ford, D.C. Malcolm, and J. Atterson (eds.), pp. 221–256. Cambridge, United Kingdom: Institute of Terrestrial Ecology.

Miller, H.G. 1983. Studies of proton flux in forests and heaths in Scotland. *In*: Effects of Accumulation of Air Pollutants in Forest Ecosystems, B. Ulrich and J. Pankrath (eds.), pp. 183–193. Dordrecht, Holland: D. Reidel.
Miller, H.G. 1984. Deposition-plant-soil interactions. Philos Trans R Soc Lond Biol 305:339–352.
Miller, H.G. 1985. Acid flux and the influence of vegetation. *In*: Symposium on the Effects of Air Pollution on Forest and Water Ecosystems, pp. 37–46. Helsinki, Finland.
Miller, H.G. and J.D. Miller. 1980. Collection and retention of atmospheric pollutants by vegetation. *In*: Ecological Impact of Acid Precipitation, D. Drabløs and A. Tollan (eds.), pp. 33–40. SNSF Project, Oslo-Ås.
Miller, H.G., J.M. Cooper, and J.D. Miller. 1976. Effect of nitrogen supply on nutrients in litterfall and crown leaching in a stand of Corsican pine. J Appl Ecol 13:233–248.
Miller, H.G., J.D. Miller, and J.M. Cooper. 1985. Transformations in rainwater chemistry on passing through forested ecosystems. *In*: British Ecology Society/Industrial Ecology Group Symposium, "Pollution Transport and Fate in Ecosystems" (in press).
Miller, R.B. 1963. Plant nutrients in hard beech. III: The cycle of nutrients. New Zealand J Sci 6:388–413.
Mina, V.N. 1965. Leaching of certain substances from woody plants and its importance in the biological cycle. Sov Soil Sci 1965:605–617.
Mina, V.N. 1967. Influence of stemflow on soil. Sov Soil Sci 1967:1321–1329.
Mollitor, A.V. and D.J. Raynal. 1982. Acid precipitation and ionic movements in Adirondack forest soils. Soil Sci Soc Am J 46:137–141.
Moore, I.D. 1983. Throughfall pH: The effect of precipitation timing and amount. Water Resour Bull 19:961–965.
Morgan, J.V. and H.B. Tukey, Jr. 1964. Characterization of leachate from plant foliage. Plant Physiol 39:590–593.
Mudd, J.B., S.K. Banerjee, M.M. Dooley, and K.L. Knight. 1984. Pollutants and plant cells: effects on membranes. *In*: Gaseous Air Pollutants and Plant Metabolism, M.J. Koziol and F.R. Whatley (eds.), pp. 105–116. London: Butterworths.
National Research Council. 1976. Nitrogen Oxides. Washington, D.C.: National Academy of Sciences.
Nicholson, I.A., N. Cape, D. Fowler, J.W. Kinnaird, and I.S. Paterson. 1980. Effects of a Scots pine (*Pinus sylvestris* L.) canopy on the chemical composition and deposition pattern of precipitation. *In*: Ecological Effects of Acid Precipitation, D. Drabløs and A. Tollen (eds.), pp. 148–149. SNSF, Oslo-Ås, Norway.
Nieboer, E., J.D. MacFarlane, and D.H.S. Richardson. 1984. Modification of plant cell buffering capacities by gaseous air pollutants. *In*: Gaseous Air Pollutants and Plant Metabolism, M.J. Koziol and F.R. Whatley (eds.), pp. 313–330. London: Butterworths.
Nieboer, E., D.H.S. Richardson, and F.D. Tomassini. 1978. Mineral uptake and release by lichens: an overview. Bryologist 81:226–246.
Nihlgard, B. 1970. Precipitation, its chemical composition and effect on soil water in a beech and spruce forest in South Sweden. Oikos 21:208–217.
Nilsson, I. 1978. The influence of *Dasychira pudibunda* (Lepidoptera) on plant nutrient transports and tree growth in a beech *Fagus sylvatica* forest in southern Sweden. Oikos 30:133–148.

Nilsson, S.I., H.G. Miller, and J.D. Miller. 1982. Forest growth as a possible cause of soil and water acidification: an examination of the concepts. Oikos 39:40–49.

Nye, P.H. 1961. Organic matter and nutrient cycles under moist tropical forest. Plant Soil 13:333–346.

O'Connell, A.M. 1985. Nutrient accessions to the forest floor in Karri (*Eucalyptus diversicolor*) forests of varying age. For Ecol Manage 10:283–296.

Oertli, J.J., J. Harr, and R. Guggenheim. 1977. The pH value as an indicator for the leaf surface microenvironment. Z Pflanzenkr Pflanzenschutz 84:729–737.

Olson, R.K., W.A. Reiners, C.S. Cronan, and G.E. Lang. 1981. The chemistry and flux of throughfall and stemflow in subalpine fir forests. Holarct Ecol 4:291–300.

Olson, R.K., W.A. Reiners, and G.M. Lovett. 1985. Trajectory analysis of forest canopy effects on chemical flux in throughfall. Biogeochemistry 1:361–373.

Overrein, L.N., H.M. Seip, and A. Tollan. 1980. Acid precipitation—effects on forest and fish. Final report of the SNSF project 1972–1980. SNSF Project, Norway.

Päiväinen, J. 1974. Nutrient removal from Scots pine on drained peatland by rain. Acta For Fenn 61:5–19.

Parker, G.G. 1981. Quality, variability and sources of summertime throughfall in two Virginia Piedmont forests. Ph.D Thesis, University of Virginia, Charlottesville, Virginia.

Parker, G.G. 1983. Throughfall and stemflow in the forest nutrient cycle. Adv Ecol Res 13:57–133.

Parker, G.G., S.E. Lindberg, and J.M. Kelly. 1980. Canopy-atmosphere interactions of sulfur in the southeastern United States. *In*: Atmospheric Sulfur Deposition, D.S. Shriner, C.R. Richmond, and S.E. Lindberg (eds.), pp. 477–493. Ann Arbor, Michigan: Ann Arbor Science.

Pastor, J. and J.G. Bockheim. 1984. Distribution and cycling of nutrients in an aspen-mixed-hardwood-spodosol ecosystem in northern Wisconsin. Ecology 65:339–353.

Patterson, D.T. 1975. Nutrient return in the stemflow and throughfall of individual trees in the Piedmont. *In*: Mineral Cycling in Southeastern Ecosystems, F.G. Howell, J.B. Gentry, and M.H. Smith (eds.), pp. 800–812. Springfield, Virginia: National Technical Information Center.

Petersen, C. 1985. The effects of sugar maple canopies on the chemical composition of precipitation. MS Thesis, University of Massachusetts, Amherst.

Pike, L.H. 1978. The importance of epiphytic lichens in mineral cycling. Bryologist 81:247–257.

Potts, M.J. 1978. The pattern of deposition of air-borne salt of marine origin under a forest canopy. Plant Soil 50:233–236.

Prebble, R.E. and G.B. Stirk. 1980. Throughfall and stemflow on silver leaf ironbark (*Eucalyptus melanophloia*) trees. Aust J Ecol 5:419–427.

Prinz, B., G.H.M. Krause, and K.D. Jung. 1984. Neuere Untersuchungen der LIS zu den neuartigen Waldschäden. Düsseldorfer Geobot Kolloq 1:25–36.

Prinz, B., G.H.M. Krause, and K.D. Jung. 1985. Responses of German Forests in Recent Years: Cause for Concern Elsewhere? *In*: Effects of Atmospheric Pollutants on Forests, Wetlands, and Agricultural Ecosystems, T.C. Hutchinson and K.M. Meema (eds). New York: Springer-Verlag.

Puckett, K.J., E. Nieboer, M.J. Gorzynski, and D.H.S. Richardson. 1973. The uptake of metal ions by lichens: A modified ion-exchange process. New Phytol 72:329–342.

Pylypec, B. and R.E. Redman. 1984. Acid-buffering capacity of foliage from boreal forest species. Can J Bot 62:2650–2653.

Rapp, M. 1969. Apport d'éléments minéraux au sol par les eaux de pluviolessivage sous des peuplements de *Quercus ilex* L., *Quercus lanuginosa* Lamk. et *Pinus halapensis* Mill. Oecologia 4:71–92.

Rasmussen, L. and I. Johnsen. 1976. Uptake of minerals, particularly metals, by epiphytic *Hypnum cupressiforme.* Oikos 27:483–487.

Rasmussen, R.A. 1972. What do hydrocarbons from trees contribute to air pollution. J Air Pollut Control Assoc 22:537–543.

Raybould, C.C., M.H. Unsworth, and P.J. Gregory. 1977. Sources of sulphur in rain collected below a wheat canopy. Nature 267:14–16.

Raynal, D.J., F.S. Raleigh, and A.V. Molliter. 1983. Characterization of atmospheric deposition and toxic input at Huntington Forest. Rep. ESF 83-003. College of Environmental Science and Forestry, State University of New York, Syracuse.

Rehfuess, K.E. 1981. Über die Wirkungen der sauren Niederschläge in Waldökosystemen. Forstwiss Centralbl 100:363–381.

Reich, P.B. and R.G. Amundson. 1985. Ambient levels of ozone reduce net photosynthesis in tree and crop species. Science 230:566–570.

Reiners, W.A. 1972. Nutrient content of canopy throughfall in three Minnesota forests. Oikos 27:483–487.

Reiners, W.A. and R.K. Olson. 1984. Effects of canopy components on throughfall chemistry: An experimental analysis. Oecologia 63:320–330.

Reiners, W.A., R.K. Olson, L. Howard, and D.A. Shaeffer. 1986. Ion migration from interiors to outer surfaces of balsam fir needles during dry, interstorm periods. J Environ Exp Bot 227–231.

Richter, A. and L. Granat. 1978. Pine forest canopy throughfall measurements. Report AL–43, Department of Meteorology, University of Stockholm, Sweden.

Richter, D.D., D.W. Johnson, and D.E. Todd. 1983. Atmospheric sulfur deposition, neutralization, and ion leaching in two deciduous forest ecosystems. J Environ Qual 12:263–270.

Riekerk, H. 1978. Mineral cycling in a young Douglas fir forest stand. *In*: Environmental Chemistry and Cycling Processes, D.C. Adriano and I.L. Brisbin, Jr. (eds.), pp. 601–616. Springfield, Virginia: National Technical Information Center.

Rodin, L.E. and N.I. Bazilevich. 1967. Production and Mineral Cycling in Terrestrial Vegetation (Transl. ed. by G.E. Goff), Edinburgh: Oliver and Boyd.

Rolfe, G.L., M.A. Akhtar, and L.E. Arnold. 1978. Nutrient distribution and flux in a mature oak-hickory forest. For Sci 24:122–130.

Roose, E.J. and F. Lelong. 1981. Factors of the chemical composition of seepage and groundwaters in the Intertropical Zone (West Africa). J Hydrol 54:1–22.

Ryan, D.F. and F.H. Bormann. 1982. Nutrient resorption in northern hardwood forests. BioScience 32:29–32.

Schaefer, D.A., R.K. Olson, and W.A. Reiners. 1985. Vertical patterns of canopy effects on stemflow and throughfall chemistry (Abstr.). Bull Ecol Soc Am 66:263.

Schaefer, D.A., W.A. Reiners, and R.K. Olson. 1988. Factors controlling the chemical alteration of throughfall in a subalpine Balsam fir canopy. Environ Expt Bot 28:175–189.

Scherbatskoy, T. and M. Bliss. 1983. Occurrence of acidic rain and cloud water in high elevation ecosystems in the Green Mountains of Vermont. Proceedings of APCA Specialty Meeting on the Meteorology of Acidic Deposition. Pittsburgh, Pennsylvania: Air Pollution Control Association.

Scherbatskoy, T. and R.M. Klein. 1983. Response of spruce and birch foliage to leaching by acidic mists. J Environ Qual 12:189–195.

Schlesinger, W.H. and P.L. Marks. 1977. Mineral cycling and the niche of Spanish moss. Am J Bot 64:1254–1262.

Schlesinger, W.H. and W.A. Reiners. 1974. Deposition of water and cations on artificial foliar collectors in fir krumm-holtz of New England mountains. Ecology 55:378–386.

Schnell, R.C. and G. Vali. 1972. Atmospheric ice nuclei from decomposing vegetation. Nature 236:163–165.

Schnell, R.C. and G. Vali. 1973. Worldwide sources of leaf-derived freezing nuclei. Nature 246:212–213.

Scholz, F. and B.R. Stephan. 1974. Physiologische Untersuchungen über die unterschiedliche Resistenz von *Pinus sylvestris* gegen *Lophodermium pinastri*. I. Die Pufferkapazitat in Nadeln. Eur J For Pathol 4:118–126.

Seastedt, T.R., D.A. Crossley, Jr., and W.W. Hargrove. 1983. The effects of low level consumption by canopy arthropods on the growth and nutrient dynamics of black locust and red maple trees in the southern Appalachians, USA. Ecology 64:1040–1048.

Sehmel, G.A. 1980. Model predictions and summary of dry deposition velocity rates. *In*: Atmospheric Sulfur Deposition, D.S. Shriner, C.R. Richmond, and S.E. Lindberg (eds.), pp. 223–235. Ann Arbor, Michigan: Ann Arbor Science.

Shearer, M.Y., G.B. Coltharp, and R.F. Wittwer. 1977. The quality of water received as precipitation on a forest watershed in eastern Kentucky. Trans Ky Acad Sci 38:111–115.

Shreffler, J.H. 1978. Factors affecting dry deposition of sulfur dioxide on forests and grasslands. Atmos Environ 12:1497–1504.

Shriner, D.S. 1977. Effects of simulated rain acidified with sulfuric acid on host-parasite interactions. Water Air Soil Pollut 8:9–14.

Sidhu, S.S. and J.G. Zakrevsky. 1982. A standardized method for determining buffering capacity of plant foliage. Plant Soil 66:173–179.

Singer, M., F.C. Ugolini, and J. Zachara. 1978. In situ study of podzolization on tephra and bedrock. Soil Sci Soc Am J 42:105–111.

Skarby, L. and G. Sellden. 1984. The effects of ozone on crops and forests. Ambio 13:68–72.

Skeffington, R.A. 1983. Soil properties under three species of tree in southern England in relation to acid deposition in throughfall. *In*: Effects of Accumulation of Air Pollutants in Forest Ecosystems, B. Ulrich and J. Pankrath, (eds.), pp. 219–231. Dordrecht, Holland: D. Reidel.

Skeffington, R.A. and T.M. Roberts. 1985. The effects of ozone and acid mist on Scots pine saplings. Oecologia (Berlin) 65:201–206.

Slinn, W.G.N. 1977. Some approximations for the wet and dry removal of particles and gases from the atmosphere. Water Air Soil Pollut 7:513–543.

Smith, W.H. 1981. Air Pollution and Forests. New York: Springer-Verlag.
Smith, W.H. and B.J. Staskawicz. 1977. Removal of atmospheric particles by leaves and twigs of urban trees: Some preliminary observations and assessment of research needs. Environ Manage 1:317–330.
Society of American Foresters. 1984. Acidic deposition and forests. Society of American Foresters Pub. 84–12. Bethesda, Maryland.
Sollins, P. and G. Drewry. 1970. Electrical conductivity and flow rate of water through the forest canopy. *In*: A Tropical Rain Forest, H.T. Odum and R.F. Pigeon (eds.), pp. H137-H153. Washington, D.C.: USAEC.
Sollins, P., C.C. Grier, F.M. McCorison, K. Cromack, Jr., R. Fogel, and R.L. Frederiksen. 1980. The internal nutrient cycles of an old growth Douglas-fir stand in western Oregon. Ecol Monogr 50:261–285.
Stachurski, A. and J.R. Zimka. 1984. The budgets of nitrogen dissolved in rainfall during its passage through the crown canopy in forested ecosystems. Ekol Pol 32:191–218.
Staxäng, B. 1969. Acidification of bark of some deciduous trees. Oikos 20:224–230.
Stenlid, G. 1958. Salt losses and redistribution of salts in higher plants. *In*: Encyclopedia of Plant Physiology, Vol IV, Mineral Nutrition of Plants, W. Ruhland (ed.), pp. 615–637. Berlin: Springer-Verlag.
Stone, E.L. 1981. Persistence of potassium in forest systems: further studies with the rubidium/potassium technique. Soil Sci Soc Am J 45:1215–1218.
Stone, E.L. and R. Kszystyniak. 1977. Conservation of potassium in the *Pinus resinosa* ecosystem. Science 198:192–194.
Swank, W.T. and W.T.S. Swank. 1983. Dynamics of water chemistry in hardwood and pine ecosystems. *In*: Catchment Experiments in Fluvial Geomorphology, T.P. Burt and F.N.D. Walling (eds.), pp. 335–346. Norwich, United Kingdom: Geo Books.
Szabo, M. 1977. Nutrient content of throughfall and stemflow water in an oak-forest (*Quercetum petraea-Cerris*) ecosystem. Acta Agron Acad Sci Hung 26:241–258.
Tamm, C.O. 1950. Growth and plant nutrient concentration in *Hylocomium proliferum* (L.) Lindb. in relation to tree canopy. Oikos 2:60–64.
Tamm, C.O. 1951. Removal of plant nutrients from tree crowns by rain. Physiol Plant 4:184–188.
Tamm, C.O. 1976. Acid precipitation: Biological effects in soil and on forest vegetation. Ambio 5:235–239.
Tamm, C.O. and E.B. Cowling. 1977. Acidic precipitation and forest vegetation. Water Air Soil Pollut 7:503–511.
Tamm, C.O. and T. Troedsson. 1955. An example of the amounts of plant nutrients supplied to the ground in road dust. Oikos 6:61–70.
Tarrant, R.F., K.C. Lu, W.B. Bollen, and C.S. Chen. 1968. Nutrient cycling by throughfall and stemflow precipitation in three coastal Oregon forest types. Res Paper PNW-54, USDA Forest Service, Pacific Northwest Forest and Range Experiment Station, Portland, Oregon.
Thomas, W.A. 1967. Dye and calcium ascent in dogwood trees. Plant Physiol 42:1800–1802.
Thomas, W.A. 1969. Accumulation and cycling of calcium by dogwood trees. Ecol Monogr 39:101–120.

Tiedemann, A.R., J.D. Helvey, and T.D. Anderson. 1980. Effects of chemical defoliation of an *Abies grandis* habitat on amounts and chemistry of throughfall and stem flow. J Environ Qual 9:320–328.

Tomlinson, G.H. 1983. Air pollutants and forest decline. Environ Sci Technol 17:246a–256a.

Tsutsumi, T. and N. Yoshimitu. 1984. On the effects of soil fertility on the throughfall chemicals in a forest. Jpn J Ecol 34:321–330.

Tukey, H.B., Jr. 1966. Leaching of metabolites from above-ground plant parts and its implications. Bull Torrey Bot Club 93:385–401.

Tukey, H.B., Jr. 1970a. The leaching of substances from plants. Annu Rev Plant Physiol 21:305–329.

Tukey, H.B., Jr. 1970b. Leaching of metabolites from foliage and its implication in the tropical rain forest. *In*: A Tropical Rain Forest, H.T. Odum and R.F. Pigeon (eds.), pp. H155-H160. Washington, D.C.: AEC.

Tukey, H.B., Jr. 1980. Some effects of rain and mist on plants, with implications for acid precipitation. *In*: Effects of Acid Precipitation on Terrestrial Ecosystems, T.C. Hutchinson and M. Havas (eds.), pp. 141–150. New York: Plenum Press.

Tukey, H.B., Jr. and J.V. Morgan. 1963. Injury to foliage and its effect upon the leaching of nutrients from above-ground plant parts. Physiol Plant 16:557–564.

Tukey, H.B., Jr., H.B. Tukey, and S.H. Wittwer. 1958. Loss of nutrients by foliar leaching as determined by radioisotopes. Proc Am Soc Hortic Sci 71:496–506.

Turvey, N.D. 1974. Water in the nutrient cycle of a Papuan rain forest. Nature 251:414–415.

Tyree, S.Y. 1985. Gran's titrations: Inherent errors in measuring the acidity of precipitation. Atmos Environ 19:1389.

Ugolini, F.C., R. Minden, H. Dawson, and J. Zachara. 1977. An example of soil processes in the *Abies amabilis* zone of central Cascades, Washington. Soil Sci 124:291–302.

Ulrich, B. 1981. Eine okosystemare Hypothese über die Ursachen des Tannensterbens (*Abies alba* Mill.). Forstwiss Centralbl 100:228–236.

Ulrich, B. 1982. Effects of acid deposition. *In*: Acid Deposition, S. Beilke and A.J. Elshout (eds.), pp. 31–41. Boston: D. Reidel.

Ulrich, B. 1984. Effects of air pollution on forest ecosystems and waters—the principles demonstrated at a case study in Central Europe. Atmos Environ 18:621–628.

Ulrich, B., R. Mayer, P.K. Khanna, and J. Prenzel. 1978. Ausfilterung von Schwefelverbindungen aus der Luft durch einen Buchenbestand. Z Pflanzenernähr Bodenkd 141:329–335.

van Breeman, N., P.A. Burrough, E.J. Velthorst, H.F. van Dobben, T. de Wit, T.B. Ridder, and H.F.R. Reijnders. 1982. Soil acidification from atmospheric ammonium sulphate in forest canopy throughfall. Nature 299:548–550.

van Miegroet, H. and D.W. Cole. 1984. The impact of nitrification on soil acidification and cation leaching in a red alder ecosystem. J Environ Qual 13:586–590.

Vandenberg, J.J. and K.R. Knoerr. 1985. Comparison of surrogate surface techniques for estimation of sulfate dry deposition. Atmos Environ 19:627–635.

Vasudevan, C. 1982. Computer simulation of throughfall (COST): A deterministic model to simulate transport of acid deposition to a forested watershed. Ph.D. Dissertation, Rensselaer Polytechnic Institute, Troy, New York.

Verry, F.S. and D.R. Timmons. 1977. Precipitation nutrients in the open and under two forests in Minnesota. Can J For Res 7:112–119.
Voigt, G.K. 1960. Alteration of the composition of rainwater by trees. Am Midl Nat 63:321–326.
Waldman, J.M., J.W. Munger, D.J. Jacob, and M.R. Hoffmann. 1985. Chemical characterization of stratus cloud water and its role as a vector for pollutant deposition in a Los Angeles, California, USA pine forest. Tellus 37:91–108.
Waldman, J.M., J.W. Munger, D.J. Jacob, R.C. Flagan, J.J. Morgan, and M.R. Hoffman. 1982. Chemical composition of acid fog. Science 218:677–680.
Waller, H.D. and J.S. Olson. 1967. Prompt transfers of cesium-137 to the soils of a tagged *Liriodendron* forest. Ecology 48:15–25.
Waughman, G.J., J.R.J. French, and K. Jones. 1981. Nitrogen fixation in some terrestrial environments. *In*: Nitrogen Fixation, Vol. I, Ecology, W.J. Broughton (ed.), pp. 135–192. Oxford: Clarendon.
Weathers, K.C., G.E. Likens, F.H. Bormann, J.S. Eaton, W.B. Bowden, J.L. Andersen, D.L. Cass, J.N. Galloway, W.C. Keene, K.D. Kimball, P. Huth, and D. Smiley. 1986. A regional acid cloud/fog water event in the eastern United States. Nature 319:657–658.
Wedding, J.B., R.W. Carlson, J.J. Stuckel, and F.A. Bazzaz. 1976. Aerosol deposition onto plant leaves. Environ Sci Technol 9:151–153.
Wells, C.G. and J.R. Jorgenson. 1974. Nutrient cycling in loblolly pine plantations. *In*: Proceedings of 4th North American Forest Soils Conference, pp. 137–158, Laval University, Quebec, Canada.
Wells, C.G., A.K. Nicholas, and S.W. Buol. 1975. Some effects of fertilization on mineral cycling in loblolly pine. *In*: Mineral Cycling in Southeastern Ecosystems, F.G. Howell, J.B. Gentry, and M.H. Smith (eds.), pp. 754–764. Springfield, Virginia: National Technical Information Center.
Wells, C.G., D. Whigham, and H. Lieth. 1972. Investigations of mineral cycling in an upland Piedmont forest. J Elisha Mitchell Sci Soc 88:66–78.
Went, F.W. 1960. Blue hazes in the atmosphere. Nature 187:641–643.
Westman, W.E. 1978. Inputs and cycling of mineral nutrients in a coastal subtropical eucalypt forest. J Ecol 66:513–531.
White, E.J. and F. Turner. 1970. A method of estimating income of nutrients in catch of airborne particles by a woodland canopy. J Appl Ecol 7:441–461.
Will, G.M. 1955. Removal of mineral nutrients from tree canopies by rain. Nature 176:1180.
Will, G.M. 1959. Nutrient returns in litter and rainfall under some exotic conifers. New Zealand J Agric Res 2:719–734.
Will, G.M. 1968. The uptake, cycling and removal of mineral nutrients by crop of *Pinus radiata*. Proc New Zealand Ecol Soc 15:20–24.
Wind, E. 1979. Pufferkapazität in Koniferennadeln. Phyton (Hoth, Austria) 19:197–215.
Wisniewski, J. 1982. The potential acidity associated with dews, frosts, and fogs. Water Air Soil Pollut 17:361–377.
Witherspoon, J.P., Jr. 1964. Cycling of cesium–134 in white oak trees. Ecol Monogr 34:403–420.
Witkamp, M. 1970. Mineral retention by epiphyllic organisms. *In*: A Tropical Rain Forest, H.T. Odum and R.F. Pigeon (eds.), pp. H177–H180. Washington, D.C.: AEC.

Witkamp, M. and M.L. Frank. 1964. First year of movement, distribution, and availability of Cs–137 in the forest floor under tagged tulip poplars. Radiat Bot 4:485–495.

Wittwer, S.H. and M.J. Bukovac. 1969. The uptake of nutrients through leaf surfaces. *In*: Handbuch der Pflanzenernährung und Düngung, Band l, Pflanzenernährung, K. Scharrer and H. Linser (eds.), pp. 236–261. Vienna: Springer-Verlag.

Wittwer, S.H. and F.G. Teubner. 1959. Foliar absorption of mineral nutrients. Annu Rev Plant Physiol 10:13–32.

Wittwer, S.H., W.H. Jyung, Y. Yamada, M.J. Bukovac, R.D.S. Kannan, H.P. Rasmussen, and S.N. Haile Mariam. 1965. Pathways and mechanisms for foliar absorption of mineral nutrients as revealed by radioisotopes. *In*: Isotopes and Radiation in Soil-Plant Nutrition Studies, pp. 387–403. Vienna: International Atomic Energy Agency.

Wood, T. and F.H. Bormann. 1975. Increases in foliar leaching caused by acidification of an artificial mist. Ambio 4:160–162.

Woodwell, G.M. 1970. Effects of pollution on the structure and physiology of ecosystems. Science 168:429–433.

Yadav, A.K. and G.P. Mishra. 1980. Observations on the colors of stemflow and throughfall waters in forest trees. Geobios (Jodhpur) 7:148–150.

Yawney, H.W. and A.L. Leaf. 1971. Nutrient release from red pine crowns under artificial rain. Agron Abstr 1971:121.

Yawney, H.W., A.L. Leaf, and B.E. Leonard. 1978. Nutrient content of throughfall and stemflow in fertilized and irrigated *Pinus resinosa* Ait. stands. Plant Soil 50:433–445.

Zielinski, J. 1984. Decomposition in the pine forests of Niepolomice. *In*: Forest Ecosystems in Industrial Regions, W. Grodinski, J. Weiner, and P.F. Maycock (eds.), pp. 150–165. New York: Springer-Verlag.

Zinke, P.J. 1962. The pattern of influence of individual forest trees on soil properties. Ecology 43:130–133.

Zinke, P.J. 1966. Forest interception studies in the United States. *In*: Forest Hydrology, W.E. Sopper and H.W. Lull (eds.) pp. 137–161. New York: Pergamon Press.

3
Effects of Acidic Deposition on the Chemical Form and Bioavailability of Soil Aluminum and Manganese

JEFFREY D. WOLT

I. Introduction

Acidic deposition is hypothesized to cause changes in soil chemistry and to damage terrestrial ecosystems primarily through effects on soil acidification (Bloom and Grigal 1985, Environmental Resources Ltd. 1983, McFee et al. 1977, Norton 1979, Ruess 1983, Ulrich et al. 1980). The potential effects of acidic deposition on soils are perceived to be greater for forest ecosystems than for more intensively managed agricultural ecosystems, principally because of the limited nutrient status and strong aluminum-buffering of many forest soils (Comptroller General of the United States 1984, David and Driscoll 1984, Voigt 1980).

Conceptually, the indirect effect of acidic deposition on forest ecosystems is often equated with accelerated development of acid soil infertility, which is manifested in the sometimes concurrent factors of H, Al, and Mn toxicities and Ca, Mg, and Mo deficiencies. These factors are chemically interdependent. Individual measurements such as soil pH, base saturation, exchangeable Al, water-soluble Mn, and nutrient availability frequently do not provide an adequate explanation of the effects of acid soil infertility (Adams 1984).

The intent of this review is to consider the potential for acidic deposition-induced development of acid soil infertility through increased Al and Mn bioavailability in forest soils. The review proceeds on the premise that hypothesized effects of acidic deposition on soil Al and Mn will most likely be manifested in altered amounts and distribution of chemical forms of Al and Mn in soil solution. These effects on soil solution Al and Mn would have a direct bearing on bioavailability of Al and Mn as well as a subtler long-term influence on solid-phase pools of Al and Mn. Because acidic deposition effects are of greatest concern in forest ecosystems, this review emphasizes a forest soil perspective, but will necessarily review a great deal of information stemming from soil chemists' understanding of agricultural ecosystems.

Acidic deposition research has not yet provided a lucid explanation of the role of atmospheric deposition in soil chemical processes. Therefore, this review discusses the limited literature concerning acidic deposition within the broader context of literature concerning the soil chemistry and phytotoxicity of Al and Mn.

II. Aluminum and Manganese Biogeochemistry

A. Prevalence in the Soil Environment

Aluminum comprises from 1% to 30% (mean = 7.1%) of the soil (Bowen 1966), where it is found predominately as a component of a variety of aluminosilicate, oxyhydroxide, and nonsilicate minerals (Barnhisel and Bertsch 1982). Despite its prevalence in the soil environment, biochemically active Al is principally of concern only in acid environments where hexa-aqua Al [$Al(H_2O)_6^{3+}$], usually designated Al^{3+}, may reach toxic levels. Aluminum in representative soil solutions from temperate regions is typically <0.4 μmol l^{-1} (Bohn et al. 1979), but concentrations ranging from 23 to 410 μmol l^{-1} have been reported for acid, highly weathered surface soils (Kamprath 1984). Aluminum is reported to range up to 90 μmol l^{-1} in natural waters (Bowen 1966).

In comparison to Al, the Mn content of the soil is quite low, ranging from 0.1% to 4% (mean = 0.085%; Bowen 1966). Manganese is present in soils principally as oxyhydroxides of mixed valence (Lindsay 1979). These oxyhydroxides tend to accumulate in well-oxidized soils and their solubilities increase under reducing conditions (Bohn et al. 1979). Manganese may be present at the earth's surface in the 2+, 3+, and 4+ valence states, but only Mn^{2+} and Mn^{4+} are stable in soil solution (Gambrell and Patrick 1982), with Mn^{2+} predominating (Bohn et al. 1979, Lindsay 1979). The chemical activity of manganese is dependent on both pH and Eh. Manganese is reported to typically range from 2 to 20 μmol l^{-1} in most temperate region soil solutions (Bohn et al. 1979), but levels of 240 to 750 μmol l^{-1} have been found in soil solutions of Typic Hapludults on Coastal Plains sediments (Adams and Wear 1957, Morris 1949). In natural waters, Mn ranges from <0.01 to 2.4 μmol l^{-1} (Bowen 1966).

The principal reactions of Al and Mn in soils relate to their association with O^{2-} and OH^- ligands and their precipitation as oxyhydroxides. The solubility products of these oxyhydroxide precipitates have the general form $(M)(OH)^x$ (Bohn et al. 1979). The soil chemistry of Al has been studied extensively, whereas "the forms that Mn takes in soils can only be guessed" (Krauskopf 1972).

B. Aluminum in the Soil Solid Phase

Chemical reactivity of Al-bearing materials will be largely responsible for soil acidity (Thomas and Hargrove 1984), as well as for the control of Al concentration in soil solution. Therefore, the potential effects of acidic precipitation on Al bioavailability involve components of the soil solid phase that control soil solution pH and Al activity.

Aluminosilicate minerals such as feldspars, kaolins, smectites, and mica are the principal Al-bearing materials in most soils. Under most conditions in nature they control the concentration of Al in solution (Roberson and Hem 1969). These crystalline layer silicate minerals are composed of Si tetrahedral and Al octahedral sheets in varied arrangements and with differing degrees of isomorphous substitution. The aluminosilicate clays are chemically reactive because of their high surface-to-volume ratios and their permanent charge cation exchange capacities (CEC) resulting from isomorphous substitution.

The aluminosilicate minerals generally exhibit incongruent (incomplete) dissolution. When these minerals dissolve, silica is released to solution as silicic acid (H_4SiO_4) while Al is conserved in the solid phase as an oxyhydroxide. Therefore, silica is mobilized more rapidly than Al as soils are weathered and become more acid. If leaching conditions exist in a soil, Si is lost from the system and Al oxyhydroxides accumulate until they dominate the soil solid phase as gibbsite [$Al(OH)_3$]. Gibbsite has been frequently considered with regard to soil pH buffering and the control of Al^{3+} activity in soil solution (Lindsay 1979, Thomas 1974). The dissolution of gibbsite can be expressed as

$$Al(OH)_{3s} = Al^{3+} + 3OH^-$$

The solubility product is therefore

$$K_{sp} = (Al^{3+})(OH^-)^3$$

which can be written in terms of negative logarithms as

$$pK_{sp} = pAl + 3pOH$$

Substituting (14 − pH) for pOH, using the pK_{sp} for gibbsite (34.0; Kittrick 1966), and rearranging results in the relationship

$$pAl = 3pH - 8.0$$

which expresses Al^{3+} activity as a function of pH.

In practice this relationship does not hold for many temperate region soils. This may result if, in a given soil (i) gibbsite is not present as a stable solid-phase component; (ii) solutions have not achieved a steady state equilibrium with respect to the solid phase (May et al. 1979); (iii) the solubility of the Al oxyhydroxide mineral present is other than that of the reference mineral (Lindsay 1979); (iv) solution or solid-phase components have altered the form of Al oxyhydroxide present (Hsu 1979, Ng Kee Kwong and Huang 1981, Violante and Jackson 1981, Violante and Violante 1980); or (v) kinetic considerations preclude gibbsite control over

Despite these concerns, gibbsite solubility is still a useful point of departure for consideration of the solid-phase control of solution Al.

Other Al oxyhydroxides that have been postulated to control Al^{3+} activity in soil solution are nordstandite or pseudoboehmite, which may form in preference to gibbsite in the presence of SO_4^{2-}, Cl^-, or low molecular weight organic acids (Violante and Jackson 1981, Violante and Violante 1980). Solubility studies suggest boehmite (AlOOH) may control solution Al at pH > 6.7 (May et al. 1979).

Complex hydroxy Al polymers formed as interlayers between the lattices of 2:1 aluminosilicate minerals are another important solid-phase pool of Al (Jackson 1963, Rich 1968). Hydroxy-interlayered Al is strongly fixed and for the most part nonexchangeable. The formation of hydroxy Al polymers and their fixation into intralamellar spaces are strongly influenced by mono- and polyprotic acids (Goh and Huang 1984, Lind and Hem 1975, Violante and Jackson 1981), which may also hinder the normal precipitation reactions of Al oxyhydroxides (Goh and Huang 1985, Lind and Hem 1975, Ng Kee Kwong and Huang 1979a, Wang et al. 1983). Singh and Brydon (1967, 1970) found the stability of interlayered hydroxy Al to be dependent on the anion composition of the system. Sulfate destabilized Al polymers through formation of an Al-hydroxy-sulfate salt in the interlayer. The interlayer material disappeared with the subsequent formation of a crystalline Al-hydroxy-sulfate precipitate (Singh and Miles 1978).

Sparingly soluble Al-hydroxy-sulfate compounds, although unidentified as discrete minerals in soils, have been frequently associated with the control of solution pH and Al^{3+} activity in acid soils (Adams and Rawajfih 1977, Eriksson 1981, Khanna and Beese 1978, Rhodes and Lindsay 1978, van Breeman 1973, Wolt 1981, Wolt and Adams 1979). Nordstrom (1982) identified basaluminite $[Al_4SO_4(OH)_{10}{\cdot}5H_2O]$, alunite $[KAl_3(SO_4)_2(OH)_6]$, and jurbanite $[AlSO_4OH{\cdot}5H_2O]$, as Al-hydroxy-sulfate minerals capable of supporting elevated levels of Al in the solutions of acid soils. Control of solution Al by Al-hydroxy-sulfate minerals is possible where large quantities of H_2SO_4 are generated by (i) pyrite oxidation of mine spoils, (ii) sulfate oxidation following drainage of marine flood plains (van Breeman 1973), or (iii) anthropogenic acidic deposition (Eriksson 1981, Khanna and Beese 1978, Nordstrom 1982, Wolt 1981, Wolt and Lietzke 1982).

Allophane and imogolite are amorphous aluminosilicate gels that are intermediate products of weathering. They are responsible for large portions of pH-dependent CEC in many soils (Wada 1977). Allophane may be described as "a hydrous aluminosilicate mineral with short range order and with a predominance of Si-O-Al bonds" (Parfitt and Saigusa 1985). Imogolite consists of short-range, paracrystalline assemblies with the approximate composition $SiO_2{\cdot}Al_2O_3{\cdot}2.5H_2O$ (Wada 1977). Operationally, amorphous aluminosilicate gels are defined by the methods used in their extraction, which may vary considerably among investigators (Bohn et al. 1979, Wada 1977). Recent investigations indicate that these minerals are frequently involved in the eluviation of Al from the upper horizons of Spodosols and its subsequent accumulation in Bs and Bhs horizons (Farmer et al. 1980, Parfitt and Saigusa 1985). Amorphous aluminosilicates may be important for complexation of Al in humic horizons where Al release by primary mineral weathering exceeds organic matter accumulation (Parfitt and Saigusa 1985). Metastable cryptocrysalline aluminosilicates are postulated to control solution Al where dissolution kinetics of Al oxyhydroxide or aluminosilicate minerals are limiting (Paces 1978).

Complexation of Al with organic matter is an important factor controlling soil solution Al, the mobilization of Al in A horizons dominated by organic matter, and the subsequent movement of complexed Al into underlying mineral horizons (van Breeman and Brinkman 1978). In very acid soils, carboxyl sites of organic matter are largely complexed by Al^{3+} (Bloom et al. 1979b). At pH values <8.0, Al complexed with organic matter does not behave as an exchangeable cation. Therefore, Al-complexed organic matter contributes little to the pH-dependent CEC of soils (Thomas and Hargrove 1984).

Bloom et al. (1979a) considered Al binding by organic matter to be an important mechanism for control of Al^{3+} activity in the soil solution below pH 5. In their experiments, addition of 2% leaf humus to soil from the B horizon of an Inceptisol decreased soil solution Al^{3+} by 40%. The degree of Al binding with organic matter will vary among soils depending on the strengths of the carboxylic acid groups present (Bloom and McBride 1979, Hargrove and Thomas 1982).

Exchangeable acidity in mineral soil materials (i.e., that portion of soil acidity which is extractable with a neutral salt) is due predominately to Al^{3+} (Thomas and Hargrove 1984). Both the degree of Al saturation of exchange sites and the quantity of exchangeable Al present will be important to the Al-buffering capacity of a soil. Unfortunately, neutral salt-extractable Al^{3+} is an inconsistent measure of exchangeable Al^{3+}, especially in soil-surface horizons where organic matter may contribute significantly to CEC (Bloom et al. 1979b). The amount of Al^{3+} extracted by neutral salts will be a function of extraction time, and of cation composition and concentration of the extracting solution (Bloom et al. 1979b, Sivasubramaniam and Talibudeen 1972). Whereas exchangeable Al was once viewed as a unifying concept for the explanation of soil acidity, confounding factors such as polymer formation, hydroxy-Al interlayering of clay minerals, and reactions of Al with organic matter have obscured the relationship of exchangeable Al to soil acidity across diverse soil environments (Thomas 1988). Perhaps because of these factors, the degree of Al saturation has been an inadequate predictor of Al toxicity to plants (Adams 1984). Soil CEC with a high degree of Al saturation is suggested to be prerequisite to the control of soil solution Al^{3+} by gibbsite (Bohn et al. 1979).

C. Soil Solution Aluminum

The difficulty in assessing Al bioavailability arises partially from the distribution of total Al among a variety of hydrolytic species, ion pairs, and complexes in solution. Burrows (1977) states, "the form and concentration of aluminum in water depends on the pH and nature of substances dissolved in the receiving waters and, to a lesser extent, on the temperature and duration of exposure to the water." Fine-grained crystalline materials and colloidal organo-Al complexes present in soil solution further complicate the determination of biochemically active Al (Beck et al. 1974, Burrows 1977, Kennedy et al. 1974). The generally recognized toxic form of Al in acid systems is the Al^{3+} ion (Adams 1984, Foy 1974), while $Al(OH)_4^-$ is potentially toxic in strongly alkaline natural waters (Foy 1974, Freeman and Everhart 1971).

The hydrolytic reactions of hexa-aqua Al are important determinants of Al bioavailability (i.e., the free ionic activity of Al^{3+}) in soil solution, but their existence and distribution are by no means certain (Nair and Prenzel 1978). The important hydrolytic reactions of monomeric Al in an acid system can be represented as (Lindsay 1979):

	pK
$Al^{3+} + H_2O = AlOH^{2+} + H^+$	5.0
$Al^{3+} + 2H_2O = Al(OH)_2^+ + 2H^+$	9.3
$Al^{3+} + 3H_2O = Al(OH)_3^\circ + 3H^+$	15.0

where pK is the negative logarithm of the stability constant, and hexa-aqua Al $[Al(H_2O)_6^{3+}]$ and hydronium ion (H_3O^+) are represented as Al^{3+} and H^+, respectively. These relationships indicate that Al-hydroxy monomers will be the dominant Al species in slightly acid soil solution, while hexa-aqua Al will increasingly dominate at $<$pH 5.

In addition, a number of hydrolytic reactions involving the formation of Al-hydroxy polymers have been postulated (Bache and Sharp 1976, Nair and Prenzel 1978, Tsai and Hsu 1984). While there is no question that these polymeric Al species exist in dilute electrolytic solutions (Bache and Sharp 1976, Bertsch and Barnhisel 1985, Smith and Hem 1972), and that their occurrence in soil solution would prove useful in the explanation of certain soil chemical phenomena (Rich 1968, Richburg and Adams 1970), reliable thermodynamic data for stability constants as well as evidence for their presence in soil solution are lacking (Bache and Sharp 1976, Nair and Prenzel 1978). The distribution of Al among polymeric forms and the partition of Al between monomers and polymers in solution are kinetically controlled (Bersillon et al. 1980, Burrows 1977, Tsai and Hsu 1984). It is unlikely that Al-hydroxy species measured in

Hsu 1984). It is unlikely that Al-hydroxy species measured in soil extracts are representative of the hydrolysis species present in situ in soil solution. Ion speciation models (Adams 1971, Misra et al. 1974, Nair and Prenzel 1978) may be more effective for partitioning Al among hydrolytic species; but again, the lack of reliable thermodynamic data is a problem.

Bertsch et al. (1986a, 1986b) used ^{27}Al nuclear magnetic resonance (NMR) spectroscopy to identify the "Al_{13}" polymer [$A1O_4Al_{12}(OH)_{12}{}^{7+}$] as a consequential component of partially neutralized solutions containing >30 mmol l^{-1} Al. Bertsch (1987) subsequently concluded, however, "... that Al_{13} polymer formation is an artifact of synthesis procedure ... it would appear of little significance to secondary Al mineral formation in acidic weathering environments."

Formation of ion pairs between Al^{3+} and $SO_4{}^{2-}$ and F^- can reduce Al^{3+} activity in soil solutions substantially. Stability constants for Al-sulfate ion pairs are smaller than those for Al-fluoride ion pairs (Behr and Wendt 1962, Roberson and Hem 1969), but the greater abundance of $SO_4{}^{2-}$ in soil environments may increase the occurrence of Al-sulfate over Al-fluoride pairs (Burrows 1977). Fluoride addition to an acid Haplustalf from Western Australia was shown to increase total Al in displaced soil solution while leaving free Al^{3+} in solution relatively unchanged (Moore and Ritchie 1988). The major Al-fluoride ion pair present was a function of Al:F ratio with AlF^{2+} dominant at Al:F ratios >1 and $AlF_2{}^+$ dominant at Al:F ratios <1. Recent investigations of acid Haplorthod soil solutions and natural waters draining forested watersheds in the northeastern United States indicate F^- is responsible for complexing large fractions of total Al, despite the greater concentrations of $SO_4{}^{2-}$ relative to F^- in solution (David and Driscoll 1984, Driscoll and Newton 1985, Johnson et al. 1981). Notwithstanding kinetic constraints to Al-fluoride complexation, these ion pairs appear critical to the control of Al^{3+} activity (and therefore Al toxicity) in some watersheds identified as susceptible to acidic precipitation perturbation (Plankey et al. 1986).

Organometallic complexes are the least understood component of total Al in solution. When present, humic materials appear to complex considerable Al at pH 4.5 to 5.0, but as pH decreases, so does complexation (Krug and Frink 1983). Organically complexed Al is traditionally considered to be the dominant mechanism for Al mobilization from eluvial (E) to Bs or Bhs horizons in the podzolization process (Farmer et al. 1980, McFee and Cronan 1982, Nilsson and Bergkvist 1983), but current evidence suggests $AlSO_4{}^+$ (Evans and Zelazny 1987) and dissolved aluminosilicates (Childs et al. 1983, Farmer et al. 1980, Wada and Wada 1980) are involved as well. High molecular weight organic acids (fulvic and humic acids) are frequently identified with complexed Al in soil solution and natural waters, but in most instances the presence of these colloidal organic acids is inferred from measurements of dissolved organic matter (Beck et al. 1974, Nilsson and Bergkvist 1983), color absorbance (Budd et al. 1981), or other nonspecific measurements (Driscoll 1984, Krug and Isaacson 1984). Hay et al. (1985) reported fulvic acids to comprise an average of 46% of the total organic carbon in leachates from the upper 20 cm of a humoferric podzol. Evans (1986) partitioned dissolved organic carbon extracted from a Typic Paleudult Oe horizon into molecular weight classes by ultracentrifugation. Aluminum was complexed dominately by the 1000 to 2500 molecular weight fraction. Temporal variation in dissolved organic matter is a factor that greatly influences the Al chemistry of soil solutions (Nilsson and Bergkvist 1983).

Low molecular weight organic acids will also complex Al. The stability of complexes formed is in the approximate order:

$$\text{citric} > \text{oxalic} > \text{malic} > \text{tannic} > \text{aspartic} > \text{p-hydrobenzoic} > \text{acetic}$$

(Hue et al. 1985, Ng Kee Kwong and Huang 1979a, 1979b, Wang et al. 1983). These complexes will hinder Al oxyhydroxide precipitation in the same relative order (Ng Kee Kwong and Huang 1979a). The ability of low molecular weight organic acids to complex Al appears to be associated with the formation of 5- or 6-bond ring structures with Al, which is dependent on the relative position of OH/COOH groups along the main C chain (Hue et al. 1986). The kinds and amounts of these organic acids in soil solution are likely to be greatly influenced by microbial activity and will exhibit wide temporal fluctuation. Very little evidence documents the importance of low molecular weight organic acids in Al speciation; however, Hue et al. (1985) reported 76% to 93% of total soil solution Al in two acid subsoils was complexed with low molecular weight organic acids.

Total Al and free Al^{3+} concentrations in forest soil solutions are expected to vary extensively in space and time as a consequence not only of soil pH and degree of Al saturation but also because of variation in the kinds and concentrations of complexing ligands present. Aluminum-fluoride ion pairs will be more stable than Al-sulfate ion pairs (Behr and Wendt 1962; Roberson and Hem 1969), Al complexes with dissolved organic carbon will in general be stronger than with inorganic ligands (Ritchie et al. 1988), and Al complexes with humic or fulvic acid appear stronger than for simple carboxylic acids (Plankey and Patterson 1987; Ritchie et al. 1988; Sikora and McBride 1989).

Simple models of ligand-ligand competition for Al^{3+}, however, do not adequately describe the complex equilibria which govern Al speciation in natural waters (Plankey and Patterson 1988). Residence time of complexing ligands in the soil environment is an additional important determinant of which Al complexes will dominate in soil solution and will therefore have a controlling influence on the mobility and bioavailability of Al. Mechanistic studies of mixed ligand systems of fluoride and fulvic acid competing for Al have indicated fulvic acid to increase the initial rate of Al complexation with fluoride (Plankey and Patterson 1988). These studies have shown $AlOH^{2+}$ to be more reactive with all types of complexing ligands than is Al^{3+}. Introduction of complexing ligands to natural waters will shift the ratio $[AlOH^{2+}]$ / $[Al^{3+}]$ to favor $AlOH^{2+}$ and will increase the rate of Al complexation with all types of complexing ligands which are present.

As a consequence of Al speciation, only a relatively small fraction of total soil solution Al may be present as free Al^{3+} ion. Wolt (1981) found free Al^{3+} composed from 2% to 61% of total Al in soil solutions displaced from acid Hapludults where SO_4^{2-} was the dominant complexing ligand. Free Al^{3+} increased with decreasing solution pH and SO_4^{2-} activity. David and Driscoll (1984) found soil solutions from Haplorthod O and B horizons contained 6% to 7% and 26% to 28%, respectively, of total inorganic Al as Al^{3+}. Solution Al was associated primarily with organic and fluoride complexes. These examples indicate the inadequacy of total Al as a measure of Al bioavailability (i.e., Al^{3+} activity) in soil solution.

D. Measurement of Solution Aluminum

The measurement of Al in soil, nutrient, and dilute electrolytic solutions, in soil extracts, and in natural waters has proven quite problematic as evidenced by the number of analytical procedures employed. Burrows (1977) reviewed various methods that have been used for analysis of Al in water. Methods frequently employed for Al analysis of soil solutions and extracts have been summarized by Barnhisel and Bertsch (1982). Commonly used colorimetric methods utilizing triphenylmethane dyes (aluminon and Eriochrome cyanine R), which react with Al to form deep red colors (Frink and Peech 1962, McLean 1965), have been largely replaced by procedures that involve Al-hydroxyquinolate complexation, followed by solvent extraction and detection either spectrophotometrically or fluorometrically (Barnes 1975, Bloom et al. 1978, James et al. 1983, Motojima and Ishiwatari 1965). Although the latter procedures offer increased sensitivity over colorimetric procedures, the kinetics of the complexation reaction must be considered with respect to the Al species which are determined (Turner 1969, 1971, Turner and Sulaiman 1971). Kinetic reactions of Al with hydroxyquinolate (James et al. 1983), ferron (Jardine and Zelazny 1986, Batchelor et al. 1986), fluoride (Ares 1986), and pyrocatechol violet (Bartlett et al. 1987) have been used in efforts to distinguish Al species in solution.

It is clear that measurement of total Al in solution is entirely inadequate for interpretation of Al effects in aqueous systems. Aluminum can form any number of polymeric and monomeric species, pairs, and complexes in solution, and total solution Al is an inadequate indicator of Al bioavailability. Direct measurement of Al species in soil solution is difficult for several reasons. Solution aging (i.e., the time between solution sampling and analysis) will significantly change the nature and reactivities of Al species present (Smith and Hem 1972, Tsai and Hsu 1984). Shifts in Al speciation will likely depend on total Al concentration, rate of hydrolysis, anions present, temperature, and pH (Bache and Sharp 1976). Additionally, fine-grained Al-bearing solids can pass 0.45- and 0.22-μm-pore-size filter membranes leading to erroneous measurements of dissolved Al (Kennedy et al. 1974). Filter membranes can remove substantial quantities of Al from solution because of the presence of Al-complexing contaminants or the CEC of filter materials (Jardine et al. 1986).

Because the Al species present in soil solution or natural waters cannot be precisely ascertained, many researchers dealing with Al chemistry in natural systems determine and express Al speciation in relative terms (Driscoll 1984, Driscoll et al. 1980, James et al. 1983). Frequently Al is expressed as nonlabile (monomeric Al-organic complexes), labile (aquo-Al, Al-hydroxypolymers, ion pairs), and acid-soluble Al (polymers, colloids, and strong Al-organic complexes). Even with relative analyses, extraction of labile Al in the field has been necessary to avoid shifts between labile and nonlabile species (Driscoll and Newton 1985).

Uncertainties in direct measurement of Al species in solution make computation of Al speciation a more useful approach in many instances. James et al. (1983) found good agreement between labile Al measured by 8-hydroxyquinoline extraction and that predicted from ion speciation models. Computation of Al species entails measurement of total Al and other solution components, and calculation of Al species using an extrathermodynamic modeling routine (Adams 1971). Comprehensive soil solution analysis coupled with an extrathermodynamic modeling routine has been successfully employed to determine Al^{3+} activities toxic to plants (Adams and Lund 1966, Pavan et al. 1982).

Effective application of ion speciation models requires reliable thermodynamic data, which is frequently lacking in the case of Al-organic complexes and Al polymers. To circumvent this problem, reliable measurements of nonorganically complexed monomeric Al are desirable for input into ion speciation models. Hodges (1987) compared the effectiveness of hydroxyquinolate, ferron, chelating resins, ion exchange columns, and F complexation for the identification of Al species. All methods demonstrated limitations that influenced the distribution of Al between inorganic and organic species. Hydroxyquinolate, although effective, can degrade Al-organic complexes when low ratios of Al to dissolved organic carbon are present in solution (James et al. 1983, Adams and Hathcock 1984). Ferron is less sensitive than hydroxyquinolate because of a greater tendency to degrade Al-organic complexes. Chelating resins and ion exchange columns both appear to degrade large organic matter complexes, leading to poor estimates of Al-organic complexes. Methods involving F complexation require comprehensive knowledge of the Al-bearing solution being analyzed.

The occurrence of differing amounts of inorganic Al determined by hydroxyquinolate as compared to Erichrome cyanine R has been attributed to varied degrees of reactivity with dissolved organic carbon (Adams and Hathcock 1984). The presence of P may form Al-phosphate complexes of fine colloidal precipitates that will be very slowly reactive with hydroxyquinolate (DiPascale and Violante, 1986). Evans and Zelazny (1986) found crown ether to effectively determine monomeric Al in the presence of Al-organic complexes. The determination of reliable values for the complexation of Al with dissolved organic matter will result in ion speciation models that can effectively speciate Al from input values of total Al, thus avoiding the problem of analytical determination of nonorganically complexed monomeric Al. Plankey and Patterson (1987) have determined constants for the complexation of Al with fulvic acid, and this information has been built into an ion speciation model for more effective speciation of total solution Al (Wolt 1987).

E. Manganese in Soil Solid and Solution Phases and Its Measurement

Solid-phase Mn can take many forms in soils because of the complex chemistry imparted by its many valence states (Lindsay 1979). In natural waters, the solid phases MnO_2, Mn_2O_3, and Mn_3O_4 appear most important, while $Mn(OH)_2$ may be present at low electrode potentials (Eh) and $MnCO_3$ may be of consequence in high-salt systems (Hem 1972).

Pyrolusite (MnO_2) is the most stable Mn mineral under well-oxidized conditions, while Mn^{2+} is the most common solution species across a wide range of Eh and pH (Bohn et al. 1979, Lindsay 1979). The dissolution of MnO_2 leading to Mn^{2+} in solution is expressed as

$$MnO_{2s} + 4H^+ + 2e^- = Mn^{2+} + 2H_2O$$

which yields the relationship

$$Eh = 1.208 - 0.059[2pH - \tfrac{1}{2}pMn]$$

Bohn (1970) found that this relationship did not explain variation in solution Mn^{2+} in 1:10 soil–0.01 M $CaCl_2$ suspensions as a function of pH and Eh, when Eh was measured with a platinum (Pt) electrode or calculated from the O_2–H_2O redox couple. Lindberg and Runnels (1984) subsequently demonstrated that computed Eh values for multiple redox couples within complex systems, such as soil solutions, do not agree, nor do these values agree with the Eh for the entire system as measured by Pt electrode. Such internal disequilibrium among redox couples makes measured Eh an unsatisfactory parameter for describing redox equilibria in soils (Bohn 1968, 1969, Lindberg and Runnels 1984). By considering the equilibrium between MnO_2 and $MnCO_3$, Bohn (1970) was able to express Mn^{2+} activity independent of Eh as

$$pH - ½pMn = 3.4 - ½ \log pCO_2$$

Olomu et al. (1973) monitored changes in soil solution pH, Eh, and Mn in saturated soils over a 6-week period, during which Eh fluctuated from reducing to oxidizing conditions. Their observations were best explained by the dissolution of Mn_2O_3:

$$Mn_2O_{3s} + 6H^+ + 2e^- = 2Mn^{2+} + 3H_2O$$

In an oxygen-free environment, Fe^{2+} will begin to precipitate from soil solution at Eh-pH values lower than necessary for Mn^{2+} precipitation; Mn^{2+} in solution is decreased and tends to precipitate as a ferromanganous material that does not resolubilize with short-term shifts in Eh and pH (Collins and Buol 1970).

Sequential extraction techniques have been employed in an effort to better understand Mn distribution among soil fractions (Goldberg and Smith 1984, Jarvis 1984, Tokashiki et al. 1986). Estimation of the labile Mn pool using a radiotracer (^{54}Mn) in conjunction with sequential extraction, has indicated labile (i.e., potentially bioavailable) Mn to be present in water-soluble, exchangeable, organically bound, and easily reducible fractions (Goldberg and Smith 1984). The establishment of equilibrium between ^{54}Mn and soil Mn required from 30 min to 80 h for various soils, indicating the importance of kinetic considerations when assessing Mn bioavailability. Easily reduced Mn decreased as a result of soil drying, and other soil Mn fractions, including more resistant mineral phases, increased. Air-drying acid soil samples resulted in lowered redox potentials and increased soluble Mn concentrations in comparison to field-moist samples (Schwab 1989). Up to 38% and 54% of total Mn in surface and periodically flooded subsoils, respectively, has been associated with reducible oxyhydroxides, while exchangeable Mn^{2+} is of minor occurrence (Jarvis 1984). Acid, sandy lateritic soils under forest cover have been found to contain from 20% to 70% of total Mn in an easily reducible form (Sanningrahi et al. 1983). It appears, therefore, that transient changes in soil pH and Eh may result in acute Mn toxicity through shifts in the labile pool of soil Mn.

Manganese dioxide (MnO_2) may be important in environments rich in organic matter for the abiotic oxidation of humic acid precursors. These humic materials may have a large capacity for chelation and subsequent transport of polyvalent metals (Pohlman and McColl 1989). Polyhydroxyphenolic acids containing para- and ortho-OH groups were used as models of forest floor litter degradation products, and when reacted with MnO_2 rapidly oxidized to polymeric humic products with release of Mn^{2+} to solution (Pohlman and McColl 1989). Similarly, fulvic acid was found to act as a mild reducing agent leading to dissolution of MnO_2 (Waite et al. 1988). The reaction appeared to involve a rapid initial surface complexation forming an Mn^{4+}-fulvate complex on the surface of MnO_2 followed by the slower formation of Mn^{2+} and its subsequent release to solution.

Organic matter reportedly has a great capacity for complexation of Mn (Adams 1984), but organic Mn is probably important only in soils with appreciable organic matter content (Gambrell and Patrick 1982). In comparison with other trace metals, Mn is either weakly complexed or not complexed by organic matter (Bloom and McBride 1979, Gambrell and Patrick 1982, Olomu et al. 1973). The composition of organic matter has been strongly implicated to influence organic Mn contents of soils. Organic (O) horizons containing biologically active acid mull humic materials accumulate Mn, while O horizons containing moder and mor humus do not (Duchaufour and Rousseau 1960, Rousseau 1960).

As with Al, Mn forms a number of hydrolytic species and ion pairs in aqueous solutions (Lindsay 1979). In acidic soil systems, the dominant hydrolytic reactions of Mn are

	pK
$Mn^{2+} + H_2O = MnOH^+ + H^+$	10.9
$Mn^{2+} + 3H_2O = Mn(OH)_3^- + 3H^+$	34.0

and $MnCl^+$ and $MnSO_4^o$ are important ion pairs (Hem 1972, Lindsay 1979). The dominant solution species of Mn shift with changing redox of the system, depending on the solid phase controlling Mn solubility (Lindsay 1979). Sanders (1983) found that at pH <7.0, from 70% to 90% of soil solution Mn was present as Mn^{2+}. Small amounts of Mn were complexed as $MnSO_4^o$ at low pH. As pH decreased, free Mn^{2+} decreased, probably because it complexed with dissolved organic matter. This contrasts with the work of Geering et al. (1969), who found Mn^{2+} in displaced soil solutions of varied pH represented from 1% to 16% of total Mn in solution. Organic matter complexation with Mn^{2+} was suspected.

III. Aluminum and Manganese Phytotoxicity

This section reviews studies of aluminum and manganese bioavailability and phytotoxicity. It should be recognized, however, that increased bioavailability of Mn, an essential plant micronutrient, can be beneficial to plants, although the occurrence of Mn deficiency in forest trees is infrequently observed (Stone 1968). Similarly, Al may, in isolated instances, benefit plant growth through effects on availability of essential plant nutrients (Foy 1984).

A. Plant Adaptation to Aluminum and Manganese in the Soil Environment

Differential sensitivity to Al in the soil environment may influence forest tree distribution in the landscape (Loftus 1971, McCormick and Steiner 1978, Messenger 1975, Messenger et al. 1978, Steiner et al. 1984). Manganese toxicity may also influence species distribution on some landscapes (Winterhalder 1963). Since "there is good reason to suppose that the vegetation native to a soil area of some size must consist of species and genotypes fitted to the prevailing chemical environment" (Stone 1968), a principal concern relative to acidic deposition-induced changes in forest soils will be whether they are sufficient to shift species composition or reduce productivity in a landscape.

Plant tolerance to soil Al and Mn varies greatly across genera as well as within species. Toxicity of Al and Mn in agronomic crops and methods of screening crops for sensitivity to Al and Mn have been addressed in a number of reviews (see, for example, Wright 1976). Tolerance to Al in the plant environment appears to be controlled by a single gene in many crops, while Mn tolerance is more likely to be controlled by several genes (Reid 1976).

Populus hybrids and autumn-olive (*Elaeagnus umbellata* Thunb.) are many times more sensitive to Al in solution culture than are *Quercus*, *Betula*, or *Pinus* species (McCormick and Steiner 1978). Tree species dominating sites considered susceptible to acidic deposition exhibit Al tolerance in the approximate order

$$\text{beech} > \text{balsam fir} > \text{Norway spruce} \approx \text{red spruce},$$

when screened in solution culture (Scheir 1985, van Praag et al. 1985). Solution culture screening of *Picea* and *Pinus* species (Hutchinson et al. 1986) indicated tolerance of Al in the order

$$\text{white pine} > \text{jack pine} > \text{black spruce} > \text{white spruce} = \text{red spruce}$$

Variation in Al sensitivity among *Populus* hybrids appears to have an important genetic component. Greater tolerance to Al apparently allows superior growth in acid soils of *Populus* hybrid progeny from the section Tacamahaca, as compared with *Populus* hybrid progeny from the section Aegeiros (Steiner et al. 1984). Differential Mn and Al tolerance of cultivars within crop species can be the result of varied selection pressures associated with differing Al and Mn bioavailability among the environments in which particular cultivars were developed (Reid 1976). This may explain the wide variation in Al sensitivity expressed in *Populus* hybrids.

Paper birch (*Betula papyrifera* Marsh.) exhibits variation in Al tolerance among provenances but not among families within provenances (Steiner et al. 1980).

B. Gross Manifestation and Physiological Effects of Aluminum and Manganese Toxicity

Recent reviews by Foy (1974, 1976, 1983, 1984) have summarized the symptoms and physiology of Al and Mn toxicity. Recognizing symptoms of Al and Mn toxicity is complicated because several symptoms of acid soil infertility may be manifested concurrently in plants (Adams 1984,

Foy 1976). Manganese toxicity symptoms may occur in plants at stress levels that will have no detrimental effect on vegetative growth, whereas plant yields may be reduced substantially by Al toxicity without the occurrence of clearly identifiable symptoms in plant tops. As a consequence, plant growth reduction on acid soils may be frequently associated with Mn toxicity when in fact Al is the more important toxic agent (Foy 1976).

The gross symptoms of Al toxicity in plant tops are frequently similar to symptoms of P or Ca deficiency, or of Fe toxicity (Foy 1974, 1983), perhaps because of interactive effects of Al with these elements. Aluminum toxicity is more readily characterized by root morphology; Al-injured roots are frequently stunted, root tips are brown, lateral roots are thickened, fine roots are absent, and fungal infection is enhanced (Adams 1984, Foy 1974, Reid 1976). Root length, rate of root elongation, and weight of plant tops are reliable measures of Al toxicity, but root weight is not (Adams 1984, Adams and Lund 1966, McCormick and Steiner 1978, Pavan et al. 1982, Steiner et al. 1980, 1984). According to Foy (1976),

> For plants in general, excess Al has been reported to interfere with cell division in root tips and lateral roots, increase cell rigidity by cross linking pectins, reduce DNA replication by increasing the rigidity of the DNA double helix, fix P in less available forms in soils and on root surfaces, decrease root respiration, interfere with enzymes governing sugar phosphorylation and deposition of cell wall polysaccharides, and interfere with the uptake, transport, and use of several essential nutrient elements, including Ca, Mg, K, P, and Fe.

Small reductions in shoot growth were the only apparent symptoms of Al toxicity in the tops of red spruce (*Picea rubens* Sarg.) and balsam fir [*Abies balsamea* (L.) Mill.] seedlings in solution culture (Scheir 1985). Root weight was unaffected by Al at levels toxic to shoot growth, but Al toxicity was readily apparent in thickened, stunted, brown roots, and reduced root elongation. Root deterioration was associated with Al-induced destruction of epidermal and cortical cells. Toxic Al in solution culture reduced total seedling biomass of sugar maple (*Acer saccharum* Marsh.) and honey locust (*Gleditsia triacanthos* L.) (Thornton et al. 1986a, 1986b). Reduced root elongation was indicative of Al toxicity to honey locust, but secondary root production was a more sensitive indicator of phytotoxicity (Thornton et al. 1986b). Interestingly, for sugar maple, root elongation was increased with increasing Al in solution at levels that decreased seedling biomass (Thornton et al. 1986a). Van Praag et al. (1985) noted Al-induced reduction in the rate of secondary tissue formation in roots of Norway spruce [*Picea abies* (L.) Karst.] in nutrient solution culture, but found no effect of Al on the root morphology of beech (*Fagus sylvatica* L.). Red spruce, white spruce (*P. glauca*), and black spruce (*P. mariana*) exposed to phytotoxic levels of Al in solution culture exhibited brownish or blackened root systems, and yellowish or purplish needles characteristic of P deficiency. Aluminum was localized in the root cap and in the epidermal and outer cortical walls of older roots (Hutchinson et al. 1986). Northern red oak (*Quercus rubra* L.) grown in soils of varied Al content exhibited decreased fine root biomass and reduced fine root branching in response to elevated Al (Joslin and Wolfe 1989).

Manganese toxicity is clearly manifested in plant tops, although the symptoms frequently vary between plant species. Generally, leaf chlorosis and/or necrosis are observed, either in leaf margins or interveinally, and leaf shape may be distorted (Foy 1984). Internal bark necrosis of apple trees (*Malus sylvestris* Mill.) is associated with Mn toxicity (Miller and Schubert 1977), and Mn may be involved with white rot fungi in the delignification of infected wood in some forest tree species (Blanchette 1984). Plant yields are often not reduced at levels of Mn sufficient to produce visual symptoms of toxicity (Adams 1984, Foy 1974). Physiologically, Mn toxicity is most frequently associated with altered activities of a number of enzymes, particularly oxidases, that are Mn activated, and this may sometimes be related to altered plant hormonal activities (Foy 1984).

C. Aluminum and Manganese Uptake by Plants

Absolute bioavailability of an element to a plant is measured by the uptake of that element by the plant (i.e., if the element is present in plant tissue, it has therefore been available to the plant). Based on this criterion, tissue analysis is frequently used to gauge elemental availability to plants, particularly perennial species such as trees. Unfortunately, this approach to assessment of Al and Mn bioavailability has shown little promise because of (i) considerable variation in tree elemental uptake within and between species and across environments, (ii) variation in sampling methodology, and (iii) accumulation of Al and Mn in plant tops at high concentrations without evidence of toxicity.

Fernandez and Struchtemeyer (1984) sampled 1-year-old needles from the south-facing midcrown of even-aged stands of red spruce (*Picea rubens* Sarg.) growing on Haplorthods in eastern Maine. Foliar Mn averaged 0.17% (range, 0.05–0.45%; CV, 54%) and Al averaged 44 mg kg^{-1} (range, 20–90 mg kg^{-1}; CV, 35%). Their inability to correlate variation in foliar elemental contents with gross soil chemical characteristics led Fernandez and Struchtemeyer (1984) to conclude, "better approaches for diagnosing the availability of individual nutrients to trees growing on these soils must be developed before precise assessment of nutrient adequacy will be possible in these commercially important forested regions."

Two-year-old seedlings of Norway spruce [*Picea abies* (L.) Karst.] were sampled by van Praag and Weissen (1985) from mineral, hemiorganic, and organic (moder) horizons of acid brown and pseudo-gley soils in the Belgian Ardennes. Aluminum content of needles, stems, and roots ranged from 4 to 45, 3 to 94, and 41 to 210 mg kg^{-1}, respectively, under conditions where no Al toxicity was evident.

Ogner and Teigen (1980) examined Al and Mn uptake and tolerance of two Norway spruce clones differing in growth potential, biomass production, and elemental uptake. Rooted cuttings from each clone were grown for 3 years in pots containing mineral soil from an Umbric Dystrochrept. Availability and uptake of Al and Mn were increased by decreases in irrigation water pH from 5.4 to 2.5. Both Al and Mn were accumulated in higher concentrations in the needles and branches of the clone with greater growth potential. Aluminum ranged from 69 to 1855 and 20 to 177 mg kg^{-1} in needles and branches, respectively, while Mn ranged from 0.04 to 0.30% in needles and from 0.02 to 0.08% in branches. Growth was not reduced by increasingly acid irrigation water, and Al and Mn toxicity was not evident.

Investigators have had relatively greater success in relating Al toxicity to tissue Al concentrations in solution culture experiments where greater environmental control is possible. Aluminum concentrations of 40 and 140 mg kg^{-1} in newly expanded leaves of honey locust and sugar maple, respectively, were associated with 20% reductions in seedling biomass over no Al controls in solution culture experiments (Thornton et al. 1986a, 1986b). For honey locust, critical leaf Al identified after 35 days exposure (Thornton et al. 1986b) was not verified with exposures of $\leq$28 days when leaf Al ranged from 5 to 37 mg kg^{-1} and was not associated with biomass reductions (Thornton et al. 1986c). Hutchinson et al. (1986) reported 20% reduction in red spruce seedling biomass when shoot Al was $<$150 mg kg^{-1}. Thornton et al. (1987) were unable to relate red spruce seedling biomass to tissue Al in a study where leaf Al concentrations ranged from $<$20 to 160 mg kg^{-1}.

Wide variation in tissue Al and Mn concentrations in the absence of toxicity symptoms, as exemplified in the foregoing discussion, is the rule rather than the exception for forest tree species. Manganese concentrations in various parts of trees vary enormously within and between species, and the reported incidence of toxicity is quite low (Stone 1968). This is likely true for Al as well. Because forest trees, in general, appear insensitive to wide ranges in Al and Mn concentrations in foliage, tissue analysis is unlikely to prove diagnostically useful in assessment of potential Al and Mn toxicity in forest ecosystems. Correlation of any foliar analysis with elemental availability in forest soils will be complicated by elemental variation within individuals, populations, and species as influenced by season, tree age, plant part sampled, and location in the landscape (Powers 1984).

The ability to accumulate and cycle Al and Mn may be a factor influencing species dominance in some landscapes (Messenger 1975, Messenger et al. 1978, Stone 1968). Variation in soil solution Mn concentrations were ascribed to vegetative cycling of Mn by hardwoods growing on Spodosols in high-elevation catchments in the White Mountains leading to elevated Mn in Oa horizons (Driscoll et al. 1988). Eastern hemlock [*Tsuga canadensis* (L.) Carr.] and eastern white pine (*Pinus strobus* L.) have been identified as significant cyclers of foliar Al when compared to associated species on the same polypedon (Messenger 1975). Messenger et al. (1978) compared native hardwood stands with >25-year-old pine plantings on a Hapludalf and observed lowered soil pH and increased extractable Al in surface soils where pine was dominant. Mean Al concentrations in needles from the midcrown of jack pine [*Pinus banksiana* (Lamb.)], red pine (*P. resinosa* Alt.), and white pine were 604, 343, and 417 mg kg^{-1}, respectively, in comparison to mean concentrations of 115 and 91 mg kg^{-1}, respectively, for red oak (*Quercus rubra* L.) and white oak (*Q. alba* L.) leaves also sampled from midcrown.

D. Soil Indices of Aluminum and Manganese Bioavailability

Until recently, investigations of Al bioavailability in forest ecosystems and elsewhere have tended to concentrate on rather gross soil parameters such as soil pH (Hern et al. 1985, Lee et al. 1982), extractable Al (Johnson and Todd 1984, Lee et al. 1982, Stuanes 1983), degree of Al saturation (Lee et al. 1982), and total Al in leachates (Budd et al. 1981, Rutherford et al. 1985, Stuanes 1983). Not surprisingly, most of these investigations have been inconclusive and have not yielded information useful for analysis or management of forest ecosystems. Attempts to generalize agronomic research regarding Al toxicity have also been largely unsuccessful, which can be attributed to (i) genetic differences in test crops, (ii) varied procedures for extracting Al, (iii) variation in CEC of soils investigated, and (iv) failure to isolate Al as the sole factor adversely affecting plant growth (Adams 1984). In their effort to define "critical" levels of soil Al responsible for Al toxicity in cotton (*Gossypium hirsutum* L.), Adams and Lund (1966) found soil pH, exchangeable Al, degree of Al saturation, and total soil solution Al to be unsatisfactory indices of Al bioavailability; Al^{3+} activity, however, did satisfactorily describe the observed incidence of Al toxicity. Activity of Al^{3+} represents the fraction of soil solution Al^{3+} concentration that behaves as if in an ideal dilute solution. The computation of Al^{3+} activity accounts for the ionic strength of the solution after correction of total Al in solution for ion complexes and hydrolysis species.

Water-soluble Mn may be diagnostic of Mn availability, especially in acid soils or under reducing conditions. Care must be taken, however, to maintain soil conditions so that Mn status is not altered in the interval between sampling and analysis. Soils should not be dried or crushed before analysis, and determinations should be made as soon as possible after sampling (Gambrell and Patrick 1982). Measurement of soil solution

Mn has also been diagnostically useful in the assessment of Mn toxicity (Adams and Wear 1957, Morris 1949), but, as with water-soluble Mn, alteration of the Eh-pH environment can invalidate the values of solution Mn obtained. Analysis of various measures of soil solution Mn for prediction of Mn availability on periodically flooded acid sulfate soils indicated that foliar uptake of Mn was correlated to both soil solution Mn^{2+} activity and the ratio of Mn^{2+} to the sum of divalent cations (Moore and Patrick 1989). The best measure of Mn availability, however, was the ratio of Mn^{2+} activity to Fe^{2+} activity, which indicated competitive effects of these ions in soil solution and/or at the site of uptake by the plant.

E. Soil and Nutrient Solution Studies of Aluminum and Manganese Phytotoxicity

Agronomists have been attempting since the 1930s to relate plant growth in acid soils to total Al in soil solution (Adams 1984). These attempts failed because the correlation of total Al in soil solution with Al phytotoxicity was inconsistent among soils of varied pH and mineralogy.

Adams and Lund (1966) introduced the concept of Al^{3+} activity and used this to explain cotton root penetration into three acid Ultisols of differing mineralogy as well as into nutrient solutions. They found a common relationship between root penetration and Al^{3+} activity in soil solutions, and this relationship also held for root penetration and Al^{3+} activity in nutrient solutions. The threshold activity of Al^{3+} that was toxic to cotton roots was 1.5 μmol l^{-1}. (In its proper thermodynamic context, activity is a unitless quantity. However, activity is frequently expressed in units of concentration in soils literature, especially in literature dealing with soil fertility and plant nutrition, to connote "effective concentration." This convention is used herein.)

Pavan et al. (1982) studied effects of Al^{3+} activity on root and shoot growth of coffee (*Coffea arabica* L.) growing in acid Ultisols and Oxisols and in nutrient solutions. Both root and shoot growth were inhibited at a toxic threshold of 4.0 μmol l^{-1} Al^{3+} in soil or nutrient solution. Brenes and Pearson (1973) also defined root growth in three *Gramineae* species as a function of Al^{3+} activity in nutrient and soil solutions, where the toxic threshold appeared to be <9 μmol l^{-1}.

Free Al^{3+} activity alone may not be diagnostic of plant growth in acid soils; other components of acid soil infertility may be involved as well. The toxic effect of Al may sometimes be better expressed in terms of the ratio of soil solution Al^{3+} activity to the sum of the activities of other soil solution cations (e.g., Ca^{2+}) that may influence plant growth response (Adams 1984). Ulrich (1981b) suggested that Al toxicity to forest trees will be manifested with Ca:Al mole ratios <1 in soil solution. Aluminum antagonism of Ca uptake (Al-induced Ca deficiency) may be the most consequential manifestation of Al toxicity at $<$pH 5.5 (Foy 1984).

Attempts to better define thresholds for Al phytotoxicity have led to investigation of monomeric Al species, rather than Al^{3+} activity per se, as predictors of Al phytotoxicity. Adams and Hathcock (1984) added $Ca(OH)_2$, MgO, or $CaSO_4 \cdot 2H_2O$ to acid Paleudult subsoils sampled from cultivated fields and woodlands to isolate the differential effects of Ca and Al on cotton root growth. Aluminum toxicity was observed for some cultivated fields when total soil solution Al was 0.4 μmol l^{-1}, but not for woodlands where up to 11.5 umol l^{-1} total Al was detected by hydroxyquinolate complexation. Expression of soil solution Al as the sum of Al^{3+}, $AlOH^{2+}$, and $Al(OH)_2^+$ activities was not a diagnostically useful parameter for prediction of Al toxicity to cotton.

Ratios of P to Al were used by Alva et al. (1986) to alter total and monomeric Al in solutions of varied Ca concentration in which root elongation of soybean (*Glycine max* L.), subterranean clover (*Trifolium subterraneum* L.), sunflower (*Helianthus annus* L.), and alfalfa (*Medicago sativa* L.) were determined. Using the sum of the activities of monomeric Al species (Al^{3+}, $AlOH^{2+}$, $Al(OH)_2^+$, $Al(OH)_3^o$, and $AlSO_4^+$) as a predictor, they observed 50% reductions in root elongation as the sum of monomeric Al activities ranged from 12 to 17, <8 to 16, <7 to 15, and <5 to 10 μmol l^{-1} for soybean, sunflower, subterranean clover, and alfalfa, respectively. The critical threshold increased as Ca in solution increased, but a sum of monomeric Al activities $\geq$ 18 μmol l^{-1} was toxic to roots regardless of Ca level.

Wright and Wright (1987) used soil solution composition to evaluate subterranean clover growth on 13 acid surface soils from the southern Appalachians. Reduced root yields occurred when Al^{3+} activity in solution exceeded 3 μmol l^{-1}. The relationship between root growth and soil solution Al was not improved by using the sum of monomeric Al activities.

The phytotoxic threshold for barley (*Hordeum vulgare* L.) root elongation in nutrient solution culture was better predicted by Al^{3+} activity (1.5 μmol l^{-1}) than by the sum of total monomeric Al activities (Cameron et al. 1986). Addition of F or SO_4 to solution reduced Al phytotoxicity to the degree of Al complexation into F or SO_4 ion pairs.

In contrast to the preceding studies which support Al^{3+} as the species diagnostic of Al phytotoxicity to plant roots, nutrient culture studies with excised roots of a number of crop species identified Al-hydroxy polymers as the most highly toxic form of Al (Wagatsuma and Kaneko 1987). In these studies, nutrient solutions of varied composition and pH were used to develop treatments with varied Al species composition, and species distribution was determined by a modified ferron procedure. These results illustrate the difficulty in obtaining unambiguous results from studies of Al phytotoxicity, since the conclusions drawn are highly dependent on the ability of the analytical procedure utilized to differentiate Al species on the basis of the kinetics of ferron reaction with Al.

Recently, several investigators have considered tree growth response in relation to Al in either soil or nutrient solution. Van Praag and Weissen (1985) sampled >700 two-year-old seedlings of Norway spruce and associated surface soil horizons (predominately moder and dysmoder humus 0 horizons) in an attempt to relate seedling dry weight with soil solution Al^{3+} activity. Aluminum toxicity was not evident for soil solutions in which Al^{3+} activity ranged from 7.7 to 64.3 $\mu mol\ l^{-1}$. In related work, total Al of 3.3 mmol l^{-1} in nutrient solution was identified as the toxic threshold for Norway spruce seedlings, and a "somewhat higher" threshold was cited for beech (*Fagus sylvatica* L.) seedlings (van Praag et al. 1985). No effort was made to express nutrient culture results in terms of Al^{3+} activity for comparison against soil solution data. In nutrient solution culture, 0.19, 0.19, 0.37, 1.48, and 2.96 mmol l^{-1} total Al resulted in reduced biomass of red spruce, white spruce, black spruce, jack pine, and white pine, respectively (Hutchinson et al. 1986).

Scheir (1985) found >1.85 mmol l^{-1} total Al in nutrient solution to inhibit root elongation of red spruce and balsam fir seedlings. Aluminum was added as $Al_2(SO_4)_3 \cdot 18H_2O$, so it is important to recognize that complexed Al ranged from 17% to 63% of total Al present in solution. When total Al is corrected to Al^{3+} activity using an ion speciation model (Wolt 1987), the toxic threshold for Al^{3+} injury appears to be $\approx$ 0.3 mmol l^{-1} (Table 3.1). Thornton et al. (1987) concluded from solution culture experiments that the toxic threshold for root elongation of red spruce seedlings was 0.25 mmol l^{-1} total Al (or estimated Al^{3+} activity of 0.05 mmol l^{-1}). The six-fold difference in thresholds for Al^{3+} toxicity to red spruce seedlings as estimated by Scheir (1985) and Thornton et al. (1987) is reasonable when considering differences in experimental protocol and uncertainties in the estimation of Al^{3+} activity from their published re-

TABLE 3.1. Relative root elongation of red spruce and balsam fir in relation to Al^{3+} activity in nutrient solution.[a]

Nutrient solution aluminum[b]					Relative root elongation		
[Al] total					Red spruce		Balsam fir
mg l^{-1}	mmol l^{-1}	(Al^{3+}) mmol l^{-1}	[Al^{3+}] %	[$AlSO_4^+$] %	Exp 1 %	Exp 3 %	Exp 1 %
0	<10^{-7}	<10^{-7}	83	12	100	100	100
25	.93	.26	64	33	87	—	91
50	1.85	.40	55	43	87	77	76
100	3.70	.57	46	53	47	71	45
200	7.41	.82	37	61	39	61	50

[a]Calculated from the data of Scheir 1985.
[b]Total Al partitioned into Al^{3+}, $AlSO_4^-$, and Al-hydroxy species, 20% Clark's solution, pH 3.8.

sults. Red spruce seedlings grown in mixed B and C horizons of a Typic Fragiorthod were unaffected by total Al concentrations in soil solution ranging from 0.04 to 0.54 mmol l^{-1} (Ohno et al. 1988).

Honey locust root elongation in solution culture was reduced for $\geq$0.15 mmol l^{-1} total Al (estimated Al^{3+} activities of $\geq$0.04 mmol l^{-1}). Secondary root production appeared to be a more sensitive indicator of Al phytotoxcity, however, occurring at $\geq$0.05 mmol l^{-1} total Al in solution (Thornton et al. 1986b).

Total Al concentrations up to 3.0 mmol l^{-1} had no detrimental effect on root biomass production of red oak, American beech (*Fagus grandifolia* Ehrh.), or European beech (*F. sylvatica* L.), although shoot biomass of European beech was decreased as Al in solution increased (Thornton et al. 1989).

McCormick and Steiner (1978) used nutrient solution culture to measure the effect of total Al on root elongation of 6 genera and 11 species of trees. They found hybrid poplars (*Populus maximowiczii* × *trichocarpa* Schreiner and Stout) to be sensitive to >0.37 mmol l^{-1} total Al in solution. Autumn-olive (*Elaeagnus umbellata* Thumb.) was intermediate in sensitivity, while species of *Quercus*, *Betula*, and *Pinus* were relatively tolerant of solution Al. When the Al^{3+} activities in these solutions are estimated, the thresholds for Al^{3+} toxicity appear to be ≈ 0.1 and 0.4 mmol l^{-1} for hybrid poplar and autumn-olive, respectively. Because of the uncertain anion composition of nutrient solutions and the possibility for Al oxyhydroxide precipitation at higher levels of total Al in solution, the threshold tolerance for oak, birch, and pine is less certain (approximately 0.8 mmol l^{-1}).

The data available concerning forest tree response to Al in solution are restricted to these few experiments, whose objectives were not to define plant growth in terms of Al^{3+} activity per se. While more comprehensive studies incorporating soil and nutrient solution approaches are necessary, forest trees are apparently 10 to 100 times less sensitive to Al^{3+} in solution than are Al-sensitive agronomic crops, which generally exhibit Al toxicity at Al^{3+} thresholds of <37 μmol l^{-1} (Table 3.2). Organic matter in solution will substantially reduce the toxic effect of Al through Al complexation (Brogan 1964), so this aspect of Al bioavailability should be considered when defining soil solution Al effects on forest tree growth.

Nutrient and soil solution Mn appear to exhibit a toxic threshold for sensitive agronomic crops at concentrations of ≈ 0.2 mmol l^{-1} (Adams 1984, Adams and Wear 1957, Morris 1949). Nutrient solution experiments with birch (*Betula verrucosa* Ehrh.) indicated Mn toxicity occurred with >9 μmol l^{-1} in solution (Ingestad 1964). European fir seedlings in nutrient solution culture exhibited increased mortality when Mn exceeded 0.1 mmol l^{-1} (Rousseau 1960). Based on these few findings it is difficult to suggest levels of soil solution Mn that may be toxic to forest tree growth.

TABLE 3.2. Threshold of aluminum toxicity in experiments where root elongation was the measure of response.

Plant	(Al^{3+}) at phytotoxic threshold mmol l^{-1}	Rooting medium	Reference
Cotton	0.0015[a]	solution, soil	Adams and Lund 1966
Coffee	0.004[a]	solution, soil	Paven et al. 1982
Gramineae spp.	<0.009[a]	solution, soil	Brenes and Pearson 1973
Honeylocust	0.04[b]	solution	Thornton et al. 1986b
Hybrid poplar	0.1[b]	solution	McCormick and Steiner 1978
Red spruce	0.05[b]	solution	Thornton et al. 1987
Red spruce, balsam fir	0.3[b]	solution	Scheir 1985
Autumn-olive	0.4[b]	solution	McCormick and Steiner 1978
Pine, oak, birch	0.8[b]	solution	McCormick and Steiner 1978

[a](Al^{3+}) computed by authors.
[b](Al^{3+}) estimated from solution composition.

F. Toxicity of Aluminum and Manganese to Soil Microflora

Soil microflora mutualistically associated with plants may exhibit Al or Mn toxicity at lower levels of soil Al and Mn than would be toxic to the host plant itself. However, "to date there is little evidence indicating acid rain-induced Al and Mn toxicity inhibits soil microbial activity under field conditions" (Firestone et al. 1983).

Both Al and Mn can inhibit nodulation of legumes. *Rhizobium* species, in general, are more sensitive to Al than to Mn (Foy 1984). When 65 *Rhizobium* strains were screened, most demonstrated slower rates of growth in culture media with >50 μmol l^{-1} total Al; 40% of the strains screened exhibited no growth at this level of Al (Keyser and Munns 1979b). Aluminum toxicity was not ameliorated by increased Ca in the culture medium (Keyser and Munns 1979a). High Mn in the culture medium (200 μmol l^{-1}) was not as toxic as 25 to 50 μmol l^{-1} total Al (Keyser and Munns 1979a).

Hepper (1979) observed severe depression of spore germination and growth of *Glomus caledonium* (a vesicular-arbuscular mycorrhizal endophyte) with 2.6 μmol l^{-1} Mn in culture media; germination was nil with 25.5 μmol l^{-1} Mn present. Growth of *Aspergillus flavus* spores cultured in soil leachate was not inhibited by 3 to 470 μmol l^{-1} Mn. However, total Al concentrations in leachates greater than 600 μmol l^{-1} did have an inhibitory effect (Firestone et al. 1983). Ko and Hora (1972) used diluted 1:1 soil:water extracts from a Latisol and artificial media to culture

Neurospora tetrasperma. They observed inhibition of spore germination with greater than 24 μmol l^{-1} total Al in either soil extracts or artificial media containing various sources of Al. Thompson and Medve (1984) found mycelial growth of ectomycorrhizal fungi commonly associated with trees on acid sites to be more sensitive to Al than to Mn in culture media.

These data suggest toxicity to soil microbes from Al and Mn in soil solution at levels lower than would be toxic to forest trees. Data from these few investigations cannot, however, be considered diagnostic of Al and Mn toxicity in natural environments for a number of reasons: (i) growth response of microbes in culture media may not adequately reflect growth in the soil ecosystem, (ii) culture media do not realistically model soil solution chemical composition in most cases, and (iii) total Al, rather than biologically active Al (Al^{3+}), is used to describe growth responses.

Firestone et al. (1983) found that addition of fluoride to culture medium significantly reduced Al toxicity to *Aspergillus flavus*, as would be expected from formation of Al-fluoride ion pairs and consequent reduction in the fraction of total Al that would be biologically active. Ko and Hora (1972), however, reported no shift in the toxic threshold of Al for *Neurospora tetraspora* germination when Al was added as $AlCl_3 \cdot 6H_2O$ versus $Al_2(SO_4)_3 \cdot 18H_2O$, even though the formation of Al-sulfate ion pairs, in contrast to inconsequential Al-chloride ion pairs, would be expected to reduce Al bioavailability.

The need to study toxicity effects on soil microflora using better models of soil solution chemistry that consider Al^{3+} activity is demonstrated by the results of Thompson and Medve (1984). Their laboratory screening of fungi for Al toxicity was inconsistent with field observations of fungal occurrence and distribution on acid soils.

IV. Evaluation of Acidic Deposition Influences on Soil Aluminum and Manganese

The current literature addressing acidic deposition effects on soil falls predominately into four general categories:

1. Interpretive reviews discussing the postulated effects of acidic deposition on soils or watersheds
2. Models designed to predict the effects of acidic deposition on soils, watersheds, and natural waters
3. Field investigations on sites hypothesized to be adversely impacted by acidic deposition
4. Controlled field, laboratory, or greenhouse experiments in which simulated acidic deposition is applied to soils and plants.

The former two types of literature, of which this review is part, are hindered by the paucity of data available in the latter two categories. Available observational and experimental data, while insightful, must be treated cautiously because the subtle effects of acidic deposition are poorly understood in the context of the many concurrent processes in forest ecosystems.

Field investigations have proceeded with the a priori assumption that acidic deposition is a recent, anthropogenically induced phenomenon (Cronan 1980, Cronan and Schofield 1979, Eriksson 1981, Weaver et al. 1985). While this indeed may be the case, conclusions will be strongly biased by the implications of this assumption. Observational data collected to date are, for the most part, unamendable to statistical analysis and fail to consider the variability inherent to forest landscapes. Controlled experiments are generally based on unrealistic assumptions concerning the quantity and composition of acidic deposition impacting soils. Time scales important in acidic deposition processes have also been infrequently addressed.

Despite these individual limitations, the literature *in toto* provides useful generalizations concerning acidic depositioninduced effects on soils:

1. Acidic deposition probably most affects those soils with spodic or podzolic (less well defined, spodic-like) morphology (David and Driscoll 1984, McFee and Cronan 1982, Nilsson and Bergkvist 1983, Norton 1979, Rutherford et al. 1985).
2. Shallow, acidic, forest soils overlying hard rock in low-order, high-elevation watersheds are potentially most sensitive to acidic deposition (Cronan and Schofield 1979, Johnson et al. 1981, Nilsson and Bergkvist 1983).
3. Accelerated base cation leaching and mobilization of Al are the soil processes that may potentially be most affected by acidic deposition (McFee and Cronan 1982, Ruess 1983, van Breeman et al. 1984).
4. The long-term influence of increased inputs of acidic deposition into forest soils will likely be increased podzolization (David and Driscoll 1984, McFee and Cronan 1982, Norton 1979, Rutherford et al. 1985).

The degree to which acidic deposition-induced changes are indeed manifested depends on factors such as mineral acidity in deposition (van Breeman et al. 1984); time of exposure of soil to acidic deposition, both on an event (David and Driscoll 1984, Hooper and Shoemaker 1985) and cumulative basis (Cosby et al. 1985b); and the rate of acidic deposition in relation to rates of other acid-generating and acid-consuming processes in forested landscapes (Frink and Voigt 1977, Krug and Frink 1983).

A. Form and Composition of Atmospheric Deposition

Atmospheric deposition into ecosystems can be in the form of wet fallout, as rain, snow, hail, dew, fog, or frost, and in the form of dry deposition, as coarse particles, fine aerosols, or gases (Cowling and Linthrust 1981).

The chemical composition and acidity of atmospheric deposition varies extensively depending on the source, quantity, and rate of deposition, and the region considered.

Where acidic deposition occurs abundantly as snowfall, special problems of acidification and metal mobilization may be a consequence of snowmelt (David and Driscoll 1984, Hooper and Shoemaker 1985). Aspects of snowmelt that may affect soil Al and Mn mobility are the composition and properties of the snow, the weather conditions during the melting period, and the duration of contact between snowmelt and soil and vegetation (Seip 1978).

Mineral acids (H_2SO_4 and HNO_3) of anthropogenic origin are thought to acidify present-day precipitation in much of eastern North America and Western Europe (Galloway et al. 1976, van Breeman et al. 1984). Precipitation in remote sites away from industrial influence is predominately comprised of weak acids, most of which may be organic acids (Galloway et al. 1984).

The association of acidic deposition with strong mineral acids in the eastern United States appears to be an oversimplication, however. A minor organic acid component has been associated with acidic deposition in the northeastern United States (Cronan and Schofield 1979, Galloway et al. 1976). Frohlinger and Kane (1975) monitored rainfall at a site in Pennsylvania and found weak acids composed from 65% to >99% of total acidity measured. Others have reported weak acids to compose from 33% to 50% of rainfall acidity for some samplings in the eastern United States (Galloway et al. 1976, Lindberg et al. 1982).

Acidity of precipitation is modified considerably as it is intercepted by, and moves through, the forest canopy. Consequently, stemflow and throughfall reaching forest soils will differ considerably in acidity and chemical composition in comparison to incident deposition (see Parker, this volume).

Lindberg et al. (1982) found total acidity of incident precipitation and throughfall beneath an oak canopy to be the same. However, the contribution of weak acidity to total acidity was two times greater in throughfall than in incident precipitation. Similarly, Cronan and Schofield (1979) reported a nearly three-fold increase in organic acids in throughfall collected beneath fir in comparison to incident precipitation. The contribution of organic acids to total acidity in throughfall has been noted in Sweden as well (Nilsson and Bergkvist 1983). Throughfall is considered an important contributor of organic matter to forest soil that may influence Al complexation and mobilization in the soil profile (Malcom and McCracken 1968). Stemflow also alters the acidity of atmospheric deposition reaching the forest floor, with the degree of alteration depending on tree species, age, and size (Wolfe et al. 1987). Experiments intended to mimic acidic deposition effects on soils have generally used simulated rainfall containing strong mineral acids, but there is little reason to suspect that this would be the actual composition of forest canopy-modified deposition that encounters the forest floor.

In addition to the strong acid or proton-donating component of acidic deposition, which may indirectly influence soil Al and Mn, acidic deposition may have a more direct (although minor) effect through contribution of Al and Mn to the forest floor. Aluminum in incident precipitation is frequently <0.7 μmol l^{-1} (Christophersen and Seip 1982, Cronan 1980, David and Driscoll 1984), although values ranging from 1 to 4 μmol l^{-1} have been reported (Galloway et al. 1976). Throughfall may be enriched severalfold in Al content. Total Al in throughfall from forest canopies in eastern North America and Western Europe ranges from 1.5 to >4 μmol l^{-1} (Cronan 1980, Cronan and Schofield 1979, David and Driscoll 1984, Nilsson and Bergkvist 1983). The Al contained in incident deposition is most likely present as mineral particulates (such as clay particles), while Al in throughfall is almost entirely complexed with organic matter. Increased Al cycling through the forest canopy may represent the dominant route for acidic deposition-induced Al mobilization in forest ecosystems (David and Driscoll 1984).

Rainfall Mn concentrations are typically <0.1 μmol l^{-1} (Cronan 1980, Galloway et al. 1976). Manganese concentrations in throughfall are probably of the same relative magnitude as Al. Cronan and Schofield (1979) reported 2 μmol l^{-1} Mn in throughfall sampled under fir in the White Mountains of New Hampshire. Atmospheric deposition of Mn contributes <10% of the Mn flux to the forest floor under oak in East Tennessee during the growing season (Lindberg et al. 1982).

B. Acidic Deposition and Acid Buffering Capacity of Soils in Relation to Soil Aluminum and Manganese Bioavailability

It is not possible to make broad generalizations concerning the effects of acidic deposition on rates of soil acidification. This is a consequence of the many mechanisms by which acidic deposition may interact with the processes of soil acidification. Acidic deposition may contribute protons (H^+) to soils directly from strong mineral acids or indirectly from atmospheric components (SO_2, NH_4^+) that undergo acid-forming transformations in soils (van Breeman et al. 1984). The anion component of acidic deposition may accelerate acidification through increased rates of base cation leaching once base cations have been displaced from exchange sites by acid cations (Ruess 1983). In addition, atmospherically deposited salts may increase total salts in solution, thus lowering solution pH by the "salt effect" (Ruess 1983, Wiklander 1975). Nitric acid of atmospheric origin may have an inconsequential effect on soil acidification during periods of active plant growth, because plant uptake of NO_3^- will result in an equimolar release of basicity from the plant to soil solution that will neutralize the effect of H^+ contributed as HNO_3 (Nilsson et al. 1982). The degree to which any of these mechanisms is operative depends in large part on soil acid-buffering capacity.

The processes whereby soils are buffered against changes in acidity vary depending on the region of soil pH considered. At pH >5.0, cation exchange in conjunction with weathering of aluminosilicate minerals buffers pH; from pH 5.0 to 4.0, cation exchange is the predominant buffering process; and in the region of pH 4.0 to 2.8, Al buffers changes in soil pH (Bloom and Grigal 1985, Ulrich 1983). Acidic deposition with a large component of mineral acid will alter soil acidity if it is of sufficient magnitude to shift the acid-buffering range of a soil. Soils of low to intermediate acidity are probably the most sensitive to acidic deposition-induced changes to soil acid-buffering capacity (van Breeman et al. 1984).

From the standpoint of Al mobilization, soils of pH >5.0 that are buffered by cation exchange will not be susceptible to increased Al availability until control of soil pH is shifted downward into the region of Al buffering (Bloom and Grigal 1985, Cosby et al. 1985a, 1985b, Ruess 1983, Ulrich 1981a, van Breeman et al. 1984). In acid soils with low base saturation, rates of acidification are slow because of the influence of Al buffering. These soils may produce less acidity internally than is contributed as mineral acidity in acidic deposition; consequently, Al mobility may be enhanced by acidic deposition (van Breeman et al. 1984). Soils in disturbed ecosystems in various stages of vegetative succession may generate much greater quantities of internal acidity than would be contributed by deposition. These ecosystems would not be susceptible to deposition-induced acidification, and Al mobility would be a consequence of natural soil weathering processes (Krug and Frink 1983).

A large sulfate component in acidic deposition may influence soil Eh buffering in addition to soil acidity, because SO_4^{2-} once removed from atmospheric contact, may undergo reduction to H_2S (Norton 1979). The bioavailability of Mn^{2+}, which is dependent on the Eh-pH relationship, could thus be influenced by Eh flux as well as pH flux when acidic deposition percolates through the soil profile.

C. Transient or Event Effects of Acidic Deposition on Soil Aluminum and Manganese

During precipitation events, acidic deposition entering the soil is unlikely to reach equilibrium between liquid and solid phases as water rapidly infiltrates and percolates through the profile (Hooper and Shoemaker 1985). Residence time of water flowing through soil will be the crucial factor when assessing acute effects of acidic deposition on Al and Mn bioavailability (Johnson et al. 1981). Rapidly percolating soil water may not achieve a steady state with the soil mineral phase. In such cases, mechanistic rather than equilibrium considerations will determine the bioavailability of Al and Mn in leaching waters.

Lateral water flow may further reduce contact time of leaching waters with some soil mineral horizons. When significant lateral flow occurs in shallow soils overlying hard rock, organically complexed Al (and perhaps Mn as well) may be transferred to surface waters (Driscoll and Newton 1985) rather than undergoing precipitation in soil mineral horizons (McFee and Cronan 1982). Humic material may act as a solid-phase adsorbent controlling solution Al under conditions of lateral flow through organic surface soil horizons (Cronan et al. 1986).

If anthropogenic SO_4^{2-} is the dominant anion in soil water during transient flow (Cronan 1980), then Al and Mn release to leaching waters will likely involve rapid to moderately rapid events such as exchange reactions, mobilization of organometallic complexes, and possibly dissolution of amorphous Al and Mn oxyhydroxides (David and Driscoll 1984, McFee and Cronan 1982). Aluminum and Mn on exchange sites, in organometallic complexes, and in oxyhydroxide precipitates may represent a labile pool of soil Al and Mn released under conditions of transient flow. Because transient snowmelt and storm events may govern the majority of water flow in high-elevation watersheds sensitive to acidic deposition (Hooper and Shoemaker 1985, Johnson et al. 1981), the kinetics of Al and Mn release from this labile pool may be of more consequence than the thermodynamics of precipitation/ dissolution reactions.

Temporal fluctuations in total and free monomeric Al in soil solutions and surface waters under conditions of high flow will not be adequately explained by models that assume thermodynamic equilibrium between infiltration water and soil mineral phases (Hooper and Shoemaker 1985). Total and monomeric forms of Al in solution tend to increase during high flow conditions (David and Driscoll 1984, Hooper and Shoemaker 1985). This would be the consequence of (i) acidic deposition entering the soil with a corresponding depression in soil water pH, (ii) buffering of acidity through release of Al from labile pools, and (iii) decreased solubility of nonlabile organoaluminum complexes resulting in a proportionate reduction in the fraction of total solution Al present in organically complexed form (Hooper and Shoemaker 1985, James and Rhia 1984).

Column-leaching experiments designed to simulate snowmelt (high water flow) conditions have indicated no effect of anion composition (NO_3^- versus SO_4^{2-}) or acid concentration (pH 5 versus pH 3) on leachates moving through a sequence of forest floor litter and either alibic (E) or ochric (A) horizon (James and Rhia 1989). For a sequence of forest floor litter and spodic (Bs) horizon, increased leachate acidity did increase Al leachability in the presence of NO_3^- but not in the presence of SO_4^{2-}.

Under high flow conditions, Al^{3+} activity in soil solution has been reported to range from 7 to 30 μmol l^{-1} (Christophersen and Seip 1982,

Driscoll et al. 1980, Hooper and Shoemaker 1985). These levels of Al^{3+} activity may pose an environmental hazard to aquatic ecosystems (Cronan and Schofield 1979, Driscoll et al. 1980), but seem below the toxic Al thresholds for tree species (see Table 2). Acute instances of Al toxicity to trees would not appear to be a consequence of acidic deposition if these Al^{3+} activities are representative of waters percolating soils in low-order, acid-sensitive watersheds.

In well-drained soils in which percolating soil water makes prolonged contact with the soil mineral phase, soil solution Al and Mn activity should be controlled by dissolution/precipitation of discrete mineral phases. Models treating acidic deposition-induced acidification of soils and watersheds assume some crystalline form of gibbsite controls soil solution Al^{3+} (Christophersen and Seip 1982, Cosby et al. 1985a, 1985b). In some watersheds, gibbsite solubility adequately explains solution Al^{3+} (Budd et al. 1981, Johnson et al. 1981), but for others this does not appear to be the case (David and Driscoll 1984, Hooper and Shoemaker 1985, Nilsson and Bergkvist 1983). Column-leaching experiments with Spodosol Bs horizons sampled from the White Mountains indicate that solution residence times with mineral soil of 0.3 hours are sufficient to achieve equilibrium with respect to gibbsite (Dahlgren et al. 1989). The identification of gibbsite as a discrete mineral phase in soils and watersheds being modeled would better justify the association of solution Al^{3+} activity with gibbsite solubility.

D. Potential Effects of Anthropogenic Sulfate Deposition on Soil Aluminum

Potential chronic effects of acidic deposition on podzolization and on surface water chemistry relate to considerations of Al complexation, Al^{3+} activity, and the vertical and lateral transport of Al through soil horizons. Throughfall chemistry, soil solution pH, degree of S retention in soil horizons, and the dynamics of Al cycling through the forest canopy will all interact to determine total Al in soil solution, Al speciation, and Al mobility. The variable response of Al to acidification of O horizons indicates that generalizations concerning acidic deposition effects on O horizon Al chemistry will not be possible (James and Rhia 1984).

For many soils and watersheds considered to be sensitive to acidic deposition, the predominant chronic effect of acidic deposition appears to be a shift in the mineral phase that controls Al^{3+} activity in soil solution. For sulfur retentive soils, conditions of low pH and high sulfate input (analogous to the hypothesized effect of acidic deposition on soils) may favor the control of soil solution Al^{3+} activity by an Al-hydroxy-sulfate mineral (Nordstom 1982, van Breeman 1973, Wolt 1981). The reaction involves gibbsite weathering in an acid, high SO_4^{2-} environment and the formation of an Al-hydroxy-sulfate phase that would support increased levels of Al^{3+} in solution. Such a mechanism may explain elevated Al bioavailability in natural waters and soil solutions thought to be adversely affected by acidic deposition (David and Driscoll 1984, Evans and Zelazny 1987, Khanna and Beese 1978, Johnson et al. 1981, Nilsson and Bergkvist 1983, Urlich et al. 1980, Weaver et al. 1985).

A number of Al-hydroxy-sulfate minerals have been postulated to control soil solution Al. However, they represent minor components of the soil solid phase and thus have not been identified mineralogically. Nordstrom (1982) has summarized the evidence for the occurrence of Al-hydroxy-sulfates in soil systems as it relates to acidic deposition.

E. Adverse Effects of Aluminum and Manganese to Trees in Relation to Acidic Deposition

Research conducted for the past 20 years in the Solling Highlands of central Germany provides perhaps the best evidence for chronic effects of acidic deposition on soils and trees (Ulrich 1981a, Ulrich et al. 1980). Slight declines in soil solution pH in the upper soil profile of a Dystrocrept during the period 1966 to 1979 resulted in elevated total Al in soil solution (Ulrich et al. 1980), which is related to retention of SO_4^{2-} as an Al-hydroxy-sulfate mineral (Khanna and Beese 1978). A shift from SO_4^{2-} retention to SO_4^{2-} loss from this site, beginning in 1975, suggested to Khanna et al. (1987) that the Al-hydroxy-sulfate mineral that may have formed has begun to dissolve with continued atmospheric inputs of H^+ and SO_4^{2-}. Total soil solution Al, which was $<27\ \mu mol\ l^{-1}$ before 1973, remained in the range 37–74 $\mu mol\ l^{-1}$ during 1973 to 1979 (Ulrich et al. 1980). The elevated Al levels were considered sufficient to damage roots of beech (*Fagus silvatica*), although this has not been substantiated in field or greenhouse studies conducted elsewhere (Scheir 1985, van Praag and Weissen 1985).

Subsequent nutrient culture experiments with beech have served to clarify observations in the Solling Highlands (Hutterman and Ulrich 1984). Aluminum-induced Ca deficiency was observed as root necrosis and ultrastructural changes in beech seedlings when the Ca:Al molar ratio in solution culture was <1 (Hutterman and Ulrich 1984, Ulrich 1981b). It is therefore suggested that Ca:Al molar ratios <1 in the presence of free Al^{3+} will damage roots of spruce and beech (Matzner and Ulrich 1985).

Ulrich and co-workers view acidic deposition as a "predisposing stress" responsible for changes in ion balance in soil solution resulting in sometimes concurrent effects of P, Ca, and Mg deficiency and Al and Mn toxicity (Matzner and Ulrich 1985, Mayer and Ulrich 1977, Ulrich et al. 1980). Demonstrating the interrelationship of acidic deposition, ion balance in soil solution, and forest decline is complicated by variation in soil solution composition and fine root turnover both spatially and temporally (Matzner and Ulrich 1985). For example, beech seedling dieback is observed most often at the base of dominant beech within a stand where atmospheric deposition via stemflow is proportionately greater than for the tree stand in general (Hutterman and Ulrich 1984).

In contrast to the Solling Highlands, most other forests subject to acidic deposition have much lower levels of atmospheric S deposition and soil solution Al. Comparison of 10 forested catchments in western Europe and North America subject to S deposition ranging from 8 to 80 kg ha^{-1} yr^{-1}, indicated total monomeric Al [Al^{3+}, $AlOH^{2+}$, $Al(OH)_2^+$] to range from <1 $\mu mol\ l^{-1}$ in an Ultisol from the southeastern United States to >240 $\mu mol\ l^{-1}$ for an Inceptisol from Solling (Cronan et al. 1987). In Spodosols, the range in total monomeric Al was 15 to 80 $\mu mol\ l^{-1}$. Differences across locations were associated dominately with differing mechanisms of soil buffering and anion retention. Fine roots (<2 mm diameter) sampled from soil B horizons at sites in West Germany and New York had Ca:Al mole ratios near 0.3, which would suggest limited ability for root regeneration by the criteria of Ulrich et al. (1984).

In an oak-birch woodland in the Netherlands subject to high inputs of atmospheric $(NH_4)_2SO_4$, soil solutions from Inceptisol and Entisol root zones averaged 620 μmol Al l^{-1}, ranging from 100 to >1300 $\mu mol\ l^{-1}$ (Mulder et al. 1987). These levels of solution Al are considerably higher than reported elsewhere and are associated with increased soil solution NO_3^-, indicating nitrification as the dominant source of acid generation. Although soil solution Al was highest in the summer, Al transport was more consequential in the dormant season when water flux was the highest. Polymeric Al was always negligible, while monomeric Al was dominantly (>80%) Al^{3+}. Organically complexed Al composed from 30% to 50% of total monomeric Al in shallow depths (10 cm) and <20% of total monomeric Al at greater depths. Theoretically, the Al-hydroxy-sulfate, jurbanite, could control Al^{3+} activities in these soils, but this is not supported because of the lack of SO_4^{2-} retention by the soil.

Inconsistencies in observations of Al toxicity across experimental locations and over time are attributable to several interacting factors that may occur in acidic forest soils: (i) conditions favoring high N availability (Ulrich et al. 1980) or P deficiency (Johnson and Todd 1984) may increase the incidence of Al-induced damage to roots; (ii) wetting and drying cycles (Nilsson and Bergkvist 1983, Weaver et al. 1985) may favor release of Al from labile pools (Hooper and Shoemaker 1985); and (iii) Ca deficiency may lower the threshold for Al injury of roots (Ulrich 1981a, 1981b, van Praag and Weissen 1985). Other factors in acid soil infertility, such as Mn toxicity, have only been remotely addressed as having a potential role in acidic deposition-induced effects on forest soils.

V. Conclusions and Research Needs

Precipitation and other forms of atmospheric deposition are natural factors involved in the processes of soil weathering that lead to leaching of mobile elements and progressive acidification of soils. The chemical composition of deposition (particularly its strong mineral acid content) is a factor that may influence the rate of soil weathering. Anthropogenically generated mineral acidity may contribute to accelerated rates of mineral weathering, thereby affecting the chemical distribution and bioavailability of Al and Mn in forest soils. Acidic deposition, however, is only one of many concurrent stresses that have influenced forests in eastern North America and Western Europe in recent times. Consequently, it is difficult to isolate the subtle effects of acidic deposition from the natural processes of forest soil weathering and other concurrent anthropogenic stresses also influencing Al and Mn chemistry in soils.

Studies of soils and watersheds susceptible to acidic deposition-induced changes in Al and Mn chemistry are hindered by the inherent spatial variability in forest landscapes and the temporal variation in processes influencing Al and Mn bioavailability. The inadequacy of present schemes of soil sampling in acidic deposition studies is illustrated by the work of McFee and Stone (1965), who reported that $\approx$ 50 samples are needed to adequately define variability of soil horizon thickness within 405-m^2 plots. Quantification of temporal and spatial soil variability has received considerable recent attention (see, for example, Neilson and Bouma 1985). Determination of elemental bioavailability through sampling of soils and trees similarly requires intensive sampling in space and time (Powers 1984). Cosby et al. (1985a) stated, "... understanding the vertical distribution of metal cation concentrations in a soil column will most likely require a model based on physical and chemical processes that are highly variable in space." Observational data of soils in forest landscapes have not been collected to allow precise development of such models, so the "predictive capability of models is thus lost in the spectrum of spatial variability, not because the model failed, but because it does not accommodate reality" (Wilding and Drees 1983).

Coupled with the need for effective sampling strategies is the need for reliable statistical analysis in monitoring studies. Schweitzer and Black (1985) stated

> Statistics increasingly is recognized as an important component of soil and groundwater monitoring programs. In the design of these programs, reliance on subjective professional judgments unaccompanied by statistically based information has become less acceptable to enforcement agencies and the scientific community. The use of statistics is now considered necessary to determining the location of sampling sites, the frequency of sampling, and the representativeness of individual samples.

Most field studies of acidic deposition affects on soils have not met these criteria.

Although a very limited data base suggests that forest tree tolerance to Al and Mn toxicity appears to be orders of magnitude greater than current levels of Al and Mn bioavailability (Al^{3+} and Mn^{2+} activity) in forest soil solutions, research specifically addressing this concern is needed. Both field and controlled environment studies should recognize that Al^{3+} activity in soil solution (rather than total or labile Al) is the diagnostic measure of Al bioavailability. Controlled studies of seedling tree root responses to Al^{3+} activity in nutrient and soil solution should be conducted and should consider Ca $\times$ Al interactions. These experiments should be complemented with detailed investigations of root growth of both seedlings and mature trees in forest environments where soil solution composition is closely monitored. Once Al phytotoxicity is adequately defined with respect to root response to Al in solution, experiments designed to identify other measures of Al toxicity more readily determined in forest ecosystems will be in order. In studies of this type, it must be realized that Al and Mn toxicity per se may not adequately define tree growth on acid soils where several concurrent factors of acid soil infertility may operate.

Difficulties in defining Al species in soil solutions are compounded by the variety of analytical procedures used to determine labile and nonlabile forms of solution Al. Driscoll (1984) has described procedures that have been widely used in acidic deposition studies, but these procedures have been frequently modified by other investigators. Such procedural modifications hamper strict cross comparisons among many of the published research findings. If useful conclusions are to be drawn from soil solution research, uniform sampling and analytical procedures must be used, particularly with respect to determination of labile Al. In any event, it is unlikely that Al species in soil extracts are representative of the species present in situ in soil solution. Comprehensive soil solution analysis and the use of ion speciation models may be a more effective method for partitioning total solution Al among the various species present.

Conceptual vagueness further hampers the ability to effectively interpret literature concerned with the relationship of solution Al to soil geochemical processes and phytotoxicity. The literature is replete with examples in which authors misconstrue total concentrations, free ion concentration, and ion activities when presenting their own data or interpreting the data of others. Such misrepresentations and misinterpretations of data (for example when total or free ion concentrations are presented as ion activities) will limit refinement in scientists' abilities to relate Al chemistry to forest tree response.

Experiments modeling acidic deposition effects on forest soils have frequently used simulated acidic precipitation rather than simulated throughfall. These research results are therefore overstated because they fail to recognize that a significant portion of throughfall acidity may be in the form of organic acids which will influence complexation and mobilization of Al and Mn in the forest floor. The modification of incident precipitation by tree canopies needs to be investigated more extensively, and experiments simulating acidic deposition to soils should consider throughfall as well as incident precipitation.

Despite the long-standing opinion of many foresters that Mn toxicity is an operative feature in many forest landscapes (Stone 1968), very little information currently addresses Mn bioavailability to forest trees and the potential influence of acidic deposition on soil Mn. Even though Al mobilization remains the dominant concern with regard to acidic deposition-induced changes in forest soil chemistry, Mn also deserves research attention because of the limited knowledge of its potential role.

References

Adams, F. 1971. Ionic concentrations and activities in soil solution. Soil Sci Soc Am Proc 35:420–426.

Adams, F. 1984. Crop response to lime in the southern United States. *In*: Soil Acidity and Liming, 2d Ed., F. Adams (ed.), pp. 211–265. Agronomy Monogr. 12, Madison, Wisconsin: American Society of Agronomy.

Adams, F. and P.J. Hathcock. 1984. Aluminum toxicity and calcium deficiency in acid soil subhorizons of two Coastal Plains soil series. Soil Sci Soc Am J 48:1305–1309.

Adams, F. and Z.F. Lund. 1966. Effect of chemical activity of soil solution aluminum on cotton root penetration of acid subsoils. Soil Sci 101:193–198.

Adams, F. and Z. Rawajfih. 1977. Basaluminite and alunite: A possible cause of sulfate retention by acid soils. Soil Sci Soc Am J 41:686–692.

Adams, F. and J.I. Wear. 1957. Manganese toxicity and soil acidity in relation to crinkle leaf of cotton. Soil Sci Soc Am Proc 21:305–308.

Alva, A.K., D.G. Edwards, C.J. Asher, and F.P.C. Blamey. 1986. Effects of phosphorus/aluminum molar ratio and calcium concentration on plant response to aluminum toxicity. Soil Sci Soc Am J 50:133–137.

Ares, J. 1986. Identification of aluminum species in acid forest soil solutions on the basis of Al:F reaction kinetics: 1. Reaction paths in pure solutions. Soil Sci 141:399–407.

Ares, J. and W. Ziechman. 1988. Interactions of organic matter and aluminum ions in acid forest solutions: Metal complexation, flocculation, and precipitation. Soil Sci 145:437–447.

Bache, B.W. and G.S. Sharp. 1976. Soluble polymeric hydroxy-aluminum ions in acid soils. J Soil Sci 27:167–174.

Barnhisel, R. and P.M. Bertsch. 1982. Aluminum. *In*: Methods of Soil Analysis, Part 2, Chemical and Microbiological Properties, 2d Ed., A.L. Page et al. (eds.), pp. 275–300. Agronomy Monograph 9, Madison, Wisconsin: American Society of Agronomy.

Barnes, R.B. 1975. The determination of specific forms of aluminum in natural water. Chem Geol 15:177–191.

Bartlett, R.J., D.S. Ross, and F.R. Magdoff. 1987. Simple kinetic fractionation of reactive aluminum in soil "solutions". Soil Sci Soc Am J 51:1479–1482.

Batchelor, B., J.B. McEwen, and R. Perry. 1986. Kinetics of aluminum hydrolysis: Measurement and characterization of reaction products. Environ Sci Technol 20:891–894.

Beck, K.C., R.H. Reuter, and E.M. Perdue. 1974. Organic and inorganic geochemistry of some Coastal Plain rivers of the southeastern United States. Geochim Cosmochim Acta 38:341–364.

Behr, B. and H. Wendt. 1962. Schnelle Ionenreaktionen in Lösugen. I. Die Bildung des aluminiumsulfatokomplexes. Z Elektrochem 66:223–228.

Bersillon, J.L., P.H. Hsu, and F. Fiessinger. 1980. Characterization of hydroxy-aluminum solutions. Soil Sci Soc Am J 44:630–634.

Bertsch, P.M. 1987. Conditions for Al_{13} polymer formation in partially neutralized aluminum solutions. Soil Sci Soc Am J 51:825–828.

Bertsch, P.M. and R.I. Barnhisel. 1985. Speciation of hydroxy-Al solutions by chemical and Al NMR method. Agron Abstr 77:145.

Bertsch, P.M., W.J. Layton, and R.I. Barnhisel. 1986b. Speciation of hydroxy-aluminum solutions by wet chemical and aluminum -27 NMR methods. Soil Sci Soc Am J 50:1449–1454.

Bertsch, P.M., G.W. Thomas, and R.I. Barnhisel. 1986a. Characterization of hydroxy-Al solutions by ^{27}Al NMR spectroscopy. Soil Sci Soc Am J 50:825–830.

Blanchette, R.A. 1984. Manganese accumulation in wood decayed by white rot fungi. Phytopathology 74:725–730.

Bloom, P.R. 1983. The kinetics of gibbsite dissolution in nitric acid. Soil Sci Soc Am J 47:164–168.

Bloom, P.R. and D.F. Grigal. 1985. Modeling soil response to acidic deposition in nonsulfate adsorbing soils. J Environ Qual 14:489–495.

Bloom, P.R. and M.B. McBride. 1979. Metal ion binding and exchange with hydrogen ions in acid-washed peat. Soil Sci Soc Am J 43:687–692.

Bloom, P.R., M.B. McBride, and R.M. Weaver. 1979a. Aluminum organic matter in acid soils: buffering and solution aluminum activity. Soil Sci Soc Am J 43:488–493.

Bloom, P.R., M.B. McBride, and R.M. Weaver. 1979b. Aluminum organic matter in acid soils: Salt-extractable aluminum. Soil Sci Soc Am J 43:813–815.

Bloom, P.R., R.M. Weaver, and M.B. McBride. 1978. The spectrophotometric and fluorometric determination of aluminum with 8-hydroxyquinoline and butyl acetate extraction. Soil Sci Soc Am J 42:713–716.

Bohn, H.L. 1968. Emf of inert electrodes in soil suspension. Soil Sci Soc Am Proc 32:211–215.

Bohn, H.L. 1969. The EMF of platinum electrodes in dilute solutions and its relation to pH. Soil Sci Soc Am Proc 33:639–640.

Bohn, H.L. 1970. Comparisons of measured and theoretical Mn^{+2} concentrations in soil suspensions. Soil Sci Soc Am Proc 34:195–197.

Bohn, H.L., B.L. McNeal, and G.A. O'Connor. 1979. Soil Chemistry. New York: Wiley.

Bowen, N.J.M. 1966. Trace Elements in Biochemistry. London: Academic Press.

Brenes, E. and R.W. Pearson. 1973. Root responses of three *Gramineae* species to soil acidity in an Oxisol and Ultisol. Soil Sci 116:295–302.

Brogan, J.C. 1964. The effect of humic acid on aluminum toxicity. *In*: Proceedings of 8th International Congress on Soil Science, pp. 227–233, Bucharest, Romania.

Budd, W.W., A.H. Johnson, J.B. Huss, and R.S. Turner. 1981. Aluminum in precipitation, streams, and shallow groundwater in the New Jersey Pine Barrens. Water Resour Res 17:1179–1183.

Burrows, W.D. 1977. Aquatic aluminum: Chemistry, toxicology, and environmental prevalence. CRC Crit Rev Environ Control 7:167–216.

Cameron, R.C., G.S.P. Ritchie, and A.D. Robson. 1986. Relative toxicities of inorganic aluminum complexes to barley. Soil Sci Soc Am J 50:1231–1236.

Childs, C.R., R.L. Profitt, and R. Lee. 1983. Movement of aluminum as an inorganic complex in some podzolized soils, New Zealand. Geoderma 29:139–155.

Christophersen, N. and H.M. Seip. 1982. A model for streamwater chemistry at Birkens, Norway. Water Resour Res 18:977–996.

Collins, J.F. and S.W. Buol. 1970. Effects of fluctuations in the Eh-pH environment on iron and/or manganese equilibria. Soil Sci 110:111–117.

Comptroller General of the United States. 1984. Report to Congress: An Analysis of Issues Concerning Acid Rain. GAO/RCED–85–13. Washington, D.C.: U.S. General Accounting Office.

Cosby, C.J., G.M. Hornberger, J.N. Galloway, and R.F. Wright. 1985a. Modeling the effects of acid deposition: Assessment of a lumped parameter model of soil water and steamwater chemistry. Water Resour Res 21:51–63.

Cosby, B.J., G.M. Hornberger, N.J. Galloway, and R.F. Wright. 1985b. Time scales of catchment acidification. Environ Sci Technol 19:1144–1149.

Cowling, E.B. and R.A. Linthurst. 1981. The acid precipitation phenomenon and its ecological consequences. BioScience 31:649–654.

Cronan, C.S. 1980. Solution chemistry of a New Hampshire subalpine ecosystem: A biogeochemical analysis. Oikos 34:272–281.

Cronan, C.S. and C.L. Schofield. 1979. Aluminum leaching response to acid precipitation effects on high-elevation watersheds in the Northeast. Science 204:304–305.

Cronan, C.S., W.J. Walker, and P.R. Bloom. 1986. Predicting aqueous aluminum concentrations in natural waters. Nature 324:140–143.

Cronan, C.S., J.M. Kelly, C.I. Schofield, and R.A. Goldstein. 1987. Aluminum geochemistry and tree toxicity in forests exposed to acidic deposition. *In*: Acid Rain: Scientific and Technical Advances, R. Perry et al. (eds.) pp. 649–656. London: Selper Ltd.

Dahlgren, R.A., C.T. Driscoll, and D.C. McAvoy. 1989. Aluminum precipitation and dissolution rates in Spodosol Bs horizons in the Northeastern USA. Soil Sci Soc Am J 53:1045–1052.

David, M.B. and C.T. Driscoll. 1984. Aluminum speciation and equilibria in soil solutions of a Haplorthod in the Adirondack Mountains (New York, U.S.A.). Geoderma 33:297–318.

Di Pascale, G. and A. Violante. 1986. Influence of phosphate ions on the extraction of aluminum by 8-hydroxyquinoline from OH-Al suspensions. Can J Soil Sci 66:573–579.

Driscoll, C.T. 1984. A procedure for the fractionation of aqueous aluminum in dilute acidic waters. Int J Environ Anal Chem 16:267–283.

Driscoll, C.T. and R.M. Newton. 1985. Chemical characteristics of Adirondack lakes. Environ Sci Technol 19:1018–1024.

Driscoll, C.T., R.D. Fuller, and D.M. Simone. 1988. Longitudinal variation in trace metal concentrations in a northern forested ecosystem. J Environ Qual 17:101–107.

Driscoll, C.T., Jr., J.P. Baker, J.J. Bisogni, Jr., and C.L. Schofield. 1980. Effect of aluminum speciation on fish in dilute acidified waters. Nature 284:161–164.

Duchaufour, P. and L.Z. Rousseau. 1960. Les phénomènes d'intoxication des plantules de résineux par le manganèse dans les humus forestiers. Rev For France 11:835–847.

Environmental Resources Limited. 1983. Acid Rain: A Review of the Phenomenon in the EEC and Europe. New York: Unipub.

Eriksson, E. 1981. Aluminum in groundwater possible solution equilibria. Nord Hydrol 12:43–50.

Evans, A., Jr. 1986. Effects of dissolved organic carbon and sulfate on aluminum mobilization in forest soil columns. Soil Sci Soc Am J 50:1576–1578.

Evans, A., Jr. and L. W. Zelazny. 1986. Determination of inorganic mononuclear aluminum by selected chelation using crown ethers. Soil Sci Soc Am J 50:910–913.

Evans, A., Jr. and L.W. Zelazny. 1987. Effects of sulfate additions on the status of exchangeable aluminum in a Cecil soil. Soil Sci 143:410–417.

Farmer, V.C., J.D. Russell, and M.L. Berrow. 1980. Imogolite and proto-imogolite allophane in spodic horizons: Evidence for a mobile aluminum silicate complex in podzol formation. J Soil Sci 31:673–684.

Fernandez, I.J. and R.A. Struchtemeyer. 1984. Correlations between element concentrations in spruce foliage and forest soils. Commun Soil Sci Plant Anal 15:1243–1255.

Firestone, M.K., K. Killham, and J.G. McColl. 1983. Fungal toxicity of mobilized soil aluminum and manganese. Applied Environ Microbiol 48:556–560.

Foy, C.D. 1974. Effects of aluminum on plant growth. *In*: The Plant Root and Its Environment, E.W. Carson (ed.), pp. 601–642. University Press of Virginia, Charlottesville.

Foy, C.D. 1976. General principles involved in screening plants for aluminum and manganese tolerance. *In*: Proceedings of Workshop on Plant Adaptation to Mineral Stress in Problem Soils, pp. 255–267. Beltsville, Maryland, 22–23 November 1976. Cornell University Agricultural Experimental Station, Ithaca, New York.

Foy, C.D. 1983. The physiology of plant adaptation to mineral stress. Iowa State J Res 57:355–391.

Foy, C.D. 1984. Physiological effects of hydrogen, aluminum, and manganese toxicities in acid soil. *In*: Soil Acidity and Liming, 2d Ed., F. Adams (ed.), pp. 57–97. Agronomy Monograph 12, Madison, Wisconsin: American Society Agronomy.

Freeman, R.A. and W.H. Everhart. 1971. Toxicity of aluminum hydroxide complexes in neutral and basic media to rainbow trout. Trans Am Fish Soc 100:644–658.

Frink, C.R. and M. Peech. 1962. Determination of aluminum in soil extracts. Soil Sci 93:317–324.

Frink, C.R. and G.K. Voigt. 1977. Potential effects of acid precipitation on soils in the humid temperate zone. Water Air Soil Pollut 7:371–388.

Frohlinger, J.O. and R. Kane. 1975. Precipitation: Its acidic nature. Science 189:455–457.

Galloway, J.N., G.E. Likens, and E. Edgarton. 1976. Acid precipitation in the northeastern U.S.—pH and acidity. Science 194:722–733.

Galloway, J.N., G.E. Likens, and M.E. Hawley. 1984. Acid precipitation: Natural versus anthropogenic components. Science 226:829–831.

Gambrell, R.P. and W.H. Patrick, Jr. 1982. Manganese. *In*: Methods of Soil Analysis, Part 2, Chemical and Microbiological Properties, 2d Ed., A.L. Page et al. (eds.), pp. 313–322. Agronomy Monograph 9, Madison, Wisconsin: American Society of Agronomy.

Geering, H.R., J.F. Hodgson, and C. Sdano. 1969. Micronutrient cation complexes in soil solution. IV. The chemical state of manganese in soil solution. Soil Sci Soc Am Proc 33:81–85.

Goh, T.B. and P.M. Huang, 1984. Formation of hydroxy-Al-montmorillonite complexes as influenced by citric acid. Can J Soil Sci 64:411–421.

Goh, T.B. and P.M. Haung. 1985. Changes in the thermal stability and acidic characteristics of hydroxy-Al-montmorillonite complexes formed in the presence of citric acid. Can J Soil Sci 65:519–522.

Goldberg, S.P. and K.A. Smith. 1984. Soil manganese: E values, distribution of manganese-54 among soil fractions, and effects of drying. Soil Sci Soc Am J 48:559–564.

Hargrove, W.L. and G.W. Thomas. 1982. Titration properties of Al-organic matter. Soil Sci 134:216–225.

Hay, G.W., J.H. James, and G.W. vanLoon. 1985. Solubilization effects of simulated acid rain on the organic matter of forest soil: preliminary results. Soil Sci 139:422–430.

Hem, J.D. 1972. Chemical factors that influence the availability of iron and manganese in aqueous systems. Geol Soc Am Bull 83:443–450.

Hepper, C.M. 1979. Germination and growth of *Glomus caledonius* spores: The effects of inhibitors and nutrients. Soil Biol Biochem 11:269–277.

Hern, J.A., G.K. Rutherford, and G.W. vanLoon. 1985. Chemical and pedogenetic effects of simulated acid precipitation on two eastern Canadian forest soils. I. Nonmetals. Can J For Res 15:839–847.

Hodges, S.C. 1987. Aluminum speciation: A comparison of five methods. Soil Sci Soc Am J 51:57–64.

Hooper, R.P. and C.A. Shoemaker. 1985. Aluminum mobilization in an acidic headwater stream: Temporal variation and mineral dissolution disequilibria. Science 229:463–465.

Hsu, P.H. 1979. Effect of phosphate and silicate on the crystallization of gibbsite from OH-Al solutions. Soil Sci 127:219–226.

Hue, N.V., G.R. Craddock, and F. Adams. 1985. Effect of organic acids on aluminum toxicity in subsoils. Agron, Abstr 77:148.

Hue, N. V., G. R. Craddock, and F. Adams. 1986. Effect of organic acids on aluminum toxicity in subsoils. Soil Sci Soc Am J 50:28–24.

Hutchinson, T. C., L. Bozic, and G. Munoz-Vega. 1986. Responses of five species of conifer seedlings to aluminum stress. Water Air Soil Pollut 31:283–294.

Hutterman, A. and B. Ulrich. 1984. Solid phase-solution-root interactions in soils subjected to acid deposition. Philos Trans R Soc London 305:353–368.

Ingestad, T. 1964. Growth and boron and manganese status of birch seedlings grown in nutrient solution. *In*: Plant Analysis and Fertilizer Problems IV, C.

Bould et al. (eds.), pp. 169–173. East Lansing, Michigan: American Society of Horticulture Science.

Jackson, M.L. 1963. Aluminum bonding in soils: A unifying principle in soil science. Soil Sci Soc Am Proc 27:1–10.

James, B.R., C.J. Clark, and S.J. Rhia. 1983. An 8-hydroxyquinoline method for labile and total aluminum in soil extracts. Soil Sci Soc Am J 47:893–897.

James, B.R. and S.J. Rhia. 1984. Soluble aluminum in acidified organic horizons of forest soils. Can J Soil Sci 64:637–646.

James, B.R. and S.J. Rhia. 1989. Aluminum leaching by mineral acids in forest soils: I. Nitric-sulfuric acid differences. Soil Sci Soc Am J 53:259–264.

Jardine, P. M. and L. W. Zelazny. 1986. Mononuclear and polynuclear aluminum speciation through differential kinetic reactions with ferron. Soil Sci Soc Am J 50:895–900.

Jardine, P. M., L. W. Zelazny, and A. Evans, Jr. 1986. Solution aluminum anomalies from various filtering materials. Soil Sci Soc Am J 50:891–894.

Jarvis, S.C. 1984. The forms of occurrence of manganese in some acidic soils. J Soil Sci 35:421–429.

Johnson, D.W. and D.E. Todd. 1984. Effects of acid irrigation on carbon dioxide evolution, extractable nitrogen, phosphorus, and aluminum in a deciduous forest soil. Soil Sci Soc Am J 48:664–666.

Johnson, N.M., C.T. Driscoll, J.S. Eaton, G.E. Likens, and W.H. McDowell. 1981. "Acid rain," dissolved aluminum, and chemical weathering at the Hubbard Brook Experimental Forest, New Hampshire. Geochim Cosmochim Acta 45:1421–1437.

Joslin, J.D. and M.H. Wolfe. 1989. Aluminum effects on northern red oak seedling growth in six forest soil horizons. Soil Sci Soc Am J 53:274–281.

Kamprath, E.J. 1984. Crop response to lime on soils in the tropics. *In*: Soil Acidity and Liming, 2d Ed., F. Adams (ed.), pp. 349–368. Agronomy Monograph 12, Madison, Wisconsin: American Society of Agronomy.

Kennedy, V.C., G.W. Zellwerger, and B.F. Jones. 1974. Filter pore-size effects on the analysis of Al, Fe, Mn, and Ti in water. Water Resour Res 10:785–790.

Keyser, H.H. and D.N. Munns. 1979a. Effects of calcium, manganese, and aluminum on growth of rhizobia in acid media. Soil Sci Soc Am J 43:500–503.

Keyser, H.H. and D.N. Munns. 1979b. Tolerance of rhizobia to acidity, aluminum, and phosphate. Soil Sci Soc Am J 43:519–523.

Khanna, P.K. and F. Beese. 1978. The behavior of sulfate on salt input in podzolic brown earth. Soil Sci 125:16–22.

Khanna, P.K., J. Prenzel, K.J. Meiwes, B. Ulrich, and E. Matzner. 1987. Dynamics of sulfate retention by acid forest soils in an acidic deposition environment. Soil Sci Soc Am J 51:446–452.

Kittrick, J.A. 1966. The free energy of formation of gibbsite and $Al(OH)_4^-$ from solubility measurements. Soil Sci Soc Am Proc 30:595–597.

Ko, W.H. and F.K. Hora. 1972. Identification of the Al ion as a soil fungitoxin. Soil Sci 113:42–45.

Krauskopf, K.B. 1972. Geochemistry of micronutrients. *In*: Micronutrients in Agriculture, J.J. Mortvedt et al. (eds.), pp. 7–40. Madison, Wisconsin: Soil Science Society of America.

Krug, E.C. and C.R. Frink. 1983. Acid rain on acid soil: A new perspective. Science 221:520–525.

Krug, E.C. and P.J. Isaacson. 1984. Comparison of water and dilute acid treatment on organic and inorganic chemistry of leachate from organic-rich horizons of an acid forest soil. Soil Sci 137:370–378.

Lee, E.H., H.E. Heggestad and J.E. Bennett. 1982. Effects of sulfur dioxide fumigation in open-top field chambers on soil acidification and exchangeable aluminum. J Environ Qual 11:99–102.

Lind, C.J. and J.D. Hem. 1975. Effects of organic solutes on chemical reactions of aluminum. U.S. Geological Survey Water Supply Paper 1827-G.

Lindberg, R.D. and D.D. Runnels. 1984. Groundwater redox reactions: An analysis of equilibrium state applied to Eh measurements and geochemical modeling. Science 225:925–927.

Lindberg, S.E., R.C. Harris and R.R. Turner. 1982. Atmospheric deposition of metals to forest vegetation. Science 215:1609–1611.

Lindsay, W.L. 1979. Chemical Equilibria in Soils. New York: Wiley.

Loftus, N.S. 1971. Yellow-poplar root development on Hartsells subsoil. U.S.D.A. Forest Service Research Note SO–131.

Malcom, R.L. and R.J. McCracken. 1968. Canopy drip: A source of mobile soil organic matter for mobilization of iron and aluminum. Soil Sci Soc Am Proc 38:834–838.

Matzner, E. and B. Ulrich. 1985. 'Waldsterben': our dying forests. II. Implications of the chemical soil conditions for forest decline. Experientia 41:578–584.

May, H.M., P.A. Helmke, and M.L. Jackson. 1979. Gibbsite solubility and thermodynamic properties of hydroxy-aluminum ions in aqueous solution at 25° C. Geochim Cosmochim Acta 43:861–868.

Mayer, R. and B. Ulrich. 1977. Acidity of precipitation as influenced by the filtering of atmospheric sulphur and nitrogen compounds—its role in the element balance and effect on soil. Water Air Soil Pollut 7:409–416.

McCormick, L.H. and K.C. Steiner. 1978. Variation in aluminum tolerance among six genera of trees. For Sci 24:565–568.

McFee, W.W. and C.S. Cronan. 1982. The action of wet and dry deposition components of acid precipitation on litter and soil. *In*: Acid Precipitation Effects on Ecological Systems, F.M. D'Itri (ed.), pp. 435–451. Ann Arbor, Michigan: Ann Arbor Science.

McFee, W.W., J.M. Kelly and R.H. Beck. 1977. Acid precipitation effects on soil pH and base saturation of exchange sites. Water Air Soil Pollut 7:401–408.

McFee, W.W. and E.L. Stone. 1965. Quantity, distribution, and variability of organic matter and nutrients in a forest podzol in New York. Soil Sci Soc Am Proc 29:432–436.

McLean, E.O. 1965. Aluminum. *In*: Methods of Soil Analysis, Part 2, Chemical and Microbiological Properties, C.A. Black et al. (eds.), pp. 978–998. Agronomy Monograph 9, Madison, Wisconsin: American Society of Agronomy.

Messenger, A.S. 1975. Climate, time, and organisms in relation to podzol development in Michigan sands. II. Relationships between chemical element concentrations in mature three foliage and upper humic horizons. Soil Sci Soc Am Proc 39:698–702.

Messenger, A.S., J.R. Kline, and D. Wilderotter. 1978. Aluminum biocycling as a factor in soil change. Plant Soil 49:703–709.

Miller, S.S. and O.E. Schubert. 1977. Plant manganese and soil pH associated with internal bark necrosis in apple. Proc W Va Acad Sci 49:97–102.

Misra, U.K., R.W. Blanchar and W.J. Upchurch. 1974. Aluminum content of soil extracts as a function of pH and ionic strength. Soil Sci Soc Am Proc 38:897–902.

Moore, C.S. and G.S.P. Ritchie. 1988. Aluminum speciation and pH of an acid soil in the presence of fluoride. J Soil Sci 39:1–8.

Moore, P.A., Jr. and W.H. Patrick, Jr. 1989. Manganese availability and uptake by rice in acid sulfate soils. Soil Sci Soc Am J 53:104–109.

Morris, H.D. 1949. The soluble manganese content of acid soils and its relation to the growth and manganese content of sweet clover and lespedeza. Soil Sci Soc Am Proc 13:362–371.

Motojima, K. and N. Ishiwatari. 1965. Determination of microamount of aluminum, chromium, copper, iron, manganese, molybdenum, and nickel in pure water by extraction photometry. J Nucl Sci Technol 2:13–17.

Mulder, J., J.J.M. van Grinsven, and N. van Breemen. 1987. Impacts of acid atmospheric deposition on woodland soils in the Netherlands: III. Aluminum chemistry. Soil Sci Soc Am J 51:1640–1646.

Nair, V.D. and J. Prenzel. 1978. Calculations of equilibrium concentration of mono- and polynuclear hydroxyaluminum species at different pH and total aluminum concentrations. Z Pflanzenernaehr Bodenkd 141:741–751.

Neilson, D.R. and J. Bouma (eds.). 1985. Soil spatial variability. *In*: Proceedings of a Workshop of the International Soil Science Society and the Soil Science Society of America, 30 Nov–1 Dec 1984, Las Vegas, Nevada. Netherlands: Pudoc Wageningen.

Ng Kee Kwong, N.F. and P.M. Huang. 1979a. Surface reactivity of aluminum hydroxides precipitated in the presence of low molecular weight organic acids. Soil Sci Soc Am J 43:1107–1113.

Ng Kee Kwong, N.F. and P.M. Huang. 1979b. The relative influence of low-molecular weight, complexing organic acids on the hydrolysis of aluminum. Soil Sci 128:337–342.

Ng Kee Kwong, N.F. and P.M. Huang. 1981. Comparison of the influence of tannic acid and selected low-molecular-weight organic acids on precipitation products of aluminum. Geoderma 26:179–193.

Nilsson, S.I. and B. Bergkvist. 1983. Aluminum chemistry and acidification processes in a shallow podzol on the Swedish west coast. Water Air Soil Pollut 20:311–329.

Nilsson, S.I., H.G. Miller and J.D. Miller. 1982. Forest growth as a possible cause of soil and water acidification: An examination of concepts. Oikos 30:40–49.

Nordstrom, D.K. 1982. The effect of sulfate on aluminum concentrations in natural waters: Some stability relations in the system Al_2O_3-SO_3-H_2O at 298 K. Geochim Cosmochim Acta 46:681–692.

Norton, S.A. 1979. Changes in chemical processes in soils caused by acid precipitation. Water Air Soil Pollut 7:389–400.

Ogner, G. and O. Teigen. 1980. Effects of acid irrigation and liming on two clones of Norway spruce. Plant Soil 57:305–321.

Ohno, T., E.I. Sucoff, M.S. Ehrich, P. Bloom, C.A. Buschena, and R.K. Dixon. 1988. Growth and nutrient content of red spruce seedlings in soil amended with aluminum. J Environ Qual 17:666–672.

Olomu, M.D., G.J. Racz, and C.M. Cho. 1973. Effect of flooding on the Eh, pH, and concentrations of Fe and Mn in several Manitoba soils. Soil Sci Soc Am Proc 37:220–224.

Paces, T. 1978. Reversible control of aqueous aluminum and silica during irreversible evolution of natural waters. Geochim Cosmochim Acta 42:1487–1493.

Parfitt, R.L. and M. Saigusa. 1985. Allophane and humus-aluminum in Spodosols and Andepts formed from the same volcanic ash beds in New Zealand. Soil Sci 139:149–155.

Pavan, M.A., F.T. Bingham, and P.F. Pratt. 1982. Toxicity of aluminum to coffee in Ultisols and Oxisols amended with $CaCO_3$, $MgCO_3$, and $CaSO_4 \cdot 2H_2O$. Soil Sci Soc Am J 46:1201–1207.

Plankey, B.J. and H.H. Patterson. 1987. Kinetics of aluminum-fulvic acid complexation in acidic waters. Environ Sci Technol 21:595–601.

Plankey, B.J. and H.H. Patterson. 1988. Effect of fulvic acid on the kinetics of aluminum fluoride complexation in acidic waters. Environ Sci Technol 22:1454–1459.

Plankey, B.J., H.H. Patterson and C.S. Cronan. 1986. Kinetics of aluminum fluoride complexation in acidic waters. Environ Sci Technol 20:160–165.

Pohlman, A.A. and J.G. McColl. 1989. Organic oxidation and manganese and aluminum mobilization in forest soils. Soil Sci Soc Am J 53:686–690.

Powers, R.F. 1984. Estimating soil nitrogen availability through soil and foliar analysis. *In*: Forest Soils and Treatment Impacts, E.L. Stone (ed.), pp. 353–379. Proceedings 6th North America Forest Soils Conference, June 1983. Knoxville: University of Tennessee.

Reid, D.A. 1976. Genetic potentials for solving problems of soil mineral stress: Aluminum and manganese toxicities in the cereal grains. *In*: Proceedings of Workshop on Plant Adaptation to Mineral Stress in Problem Soils, pp. 55–64, Beltsville, Maryland, 22–23 November 1976. Ithaca, New York: Cornell University Agricultural Experimental Station.

Rhodes, E.R. and W.L. Lindsay. 1978. Solubility of aluminum in soils of the humid tropics. J Soil Sci 29:324–330.

Rich, C.I. 1968. Hydroxy interlayers in expansible layer silicates. Clays Clay Miner 16:15–30.

Richburg, J.S. and F. Adams. 1970. Solubility and hydrolysis of aluminum in soil solutions and saturated-paste extracts. Soil Sci Soc Am Proc 34:728–734.

Ritchie, G.S.P., M.P. Nelson, and M.G. Whitten. 1988. The estimation of free aluminum and the competition between fluoride and humate anions for aluminum. Commun Soil Sci Plant Anal 19:857–871.

Roberson, C.E. and J.D. Hem. 1969. Solubility of aluminum in the presence of hydroxide, fluoride, and sulfate. U.S. Geological Surv Water Supply Paper 1827-C.

Rousseau, L.Z. 1960. De l'influence du type d'humus sur le développement des plantules sapins dans les Vosges. Ann Ec Nat Eaux For Stn Rech Expér. 17:13–118.

Ruess, J.O. 1983. Implications of the calcium-aluminum exchange system for the effect of acid precipitation on soils. J Environ Qual 12:591–595.

Rutherford, G.K., G.W. vanLoon, and J.A. Hern. 1985. Chemical and pedogenetic effects of simulated acid precipitation on two eastern Canadian forest soils. II. Metals. Can J For Res 15:848–854.

Sanders, J.R. 1983. The effect of pH on the total and free ionic concentration of manganese, zinc, and cobalt in soil solutions. J Soil Sci 34:315–323.

Sannigrahi, A.K., P.C. Bishagee and S.K. Gupta. 1983. Distribution of different forms of iron and manganese in some lateritic soils of West Bengal under different forest vegetations. Ind Agric 27:85–91.

Scheir, G.A. 1985. Response of red spruce and balsam fir seedlings to aluminum toxicity in nutrient solutions. Can J For Res 15:29–33.

Schwab, A.P. 1989. Manganese-phosphate solubility relationships in an acid soil. Soil Sci Soc Am J 53:1654–1660.

Schweitzer, G.E. and S.C. Black. 1985. Monitoring statistics: an important tool for groundwater and soil studies. Environ Sci Technol 19:1026–1030.

Seip, H. 1978. Acid snow—snowpack chemistry and snowmelt. *In*: Effects of Acid Precipitation on Terrestrial Ecosystems, T.C. Hutchinson and M. Havas (eds.), pp. 77–94. New York: Plenum Press.

Sikora, F.J. and M.B. McBride. 1989. Aluminum complexation by catechol as determined by ultraviolet spectrophotometry. Environ Sci Technol 23:349–346.

Singh, S.S. and J.E. Brydon. 1967. Precipitation of aluminum by calcium hydroxide in the presence of Wyoming bentonite and sulfate ions. Soil Sci 103:162–168.

Singh, S.S. and J.E. Brydon. 1970. Activity of aluminum hydroxy sulfate and the stability of hydroxy aluminum interlayers in montmorrillonite. Clays Clay Miner 7:114–124.

Singh, S.S. and N.M. Miles. 1978. Effect of sulfate ions on the stability of an aluminum-interlayered Wyoming bentonite. Soil Sci 126:323–329.

Sivasubramaniam, S. and O. Talibudeen. 1972. Potassium-aluminum exchange in acid soils. I. Kinetics. J Soil Sci 23:163–176.

Smith, R.W. and J.D. Hem. 1972. Effect of aging on aluminum hydroxide complexes in dilute aqueous solutions. U.S. Geological Survey Water Supply Paper 827-D.

Steiner, K.C., J.R. Barbour, and L.H. McCormick. 1984. Response of *Populus* hybrids to aluminum toxicity. Forest Sci 30:404–410.

Steiner, K.C., L.H. McCormick, and D.S. Canavera. 1980. Differential response of paper birch provenances to aluminum in solution culture. Can J For Res 10:25–29.

Stone, E.L. 1968. Microelement nutrition of forest trees. *In*: Forest Fertilization Theory and Practice, pp. 132–175. Tennessee Valley Authority, National Fertilizer Development Center, Muscle Shoals, Alabama.

Stuanes, A.O. 1983. Possible indirect long-term effects of acid precipitation on forest growth. Aquilo Ser Bot 19:50–63.

Thomas, G.W. 1974. Chemical reactions controlling soil solution electrolyte concentration. *In*: The Plant Root and Its Environment, E.W. Carson (ed.), pp. 483–506. University Press of Virginia, Charlottesville.

Thomas, G.W. 1988. Beyond exchangeable aluminum: Another ride on the merry-go-round. Commun Soil Sci Plant Anal 19:833–856.

Thomas, G.W. and W.L. Hargrove. 1984. The chemistry of soil acidity. *In*: Soil acidity and liming, 2d Ed., F. Adams (ed.), pp. 3–56. Agronomy Monograph 12, Madison, Wisconsin: American Society of Agronomy.

Thompson, G.W. and R.J. Medve. 1984. Effects of aluminum and manganese on the growth of ectomycorrhizal fungi. Applied Environ Microbiol 48:556–560.

Thornton, F.C., M. Schaedle, and D.J. Raynal. 1986a. Effect of aluminum on growth of sugar maple in solution culture. Can J For Res 16:892–896.

Thornton, F.C., M. Schaedle, and D.J. Raynal, and C. Zipperer. 1986b. Effect of aluminum on honeylocust (*Gleditsia triacanthos* L.) seedlings in solution culture. J Exp Bot 37:775–785.
Thornton, F.C., M. Schaedle, and D.J. Raynal. 1986c. Effects of aluminum on growth, development, and nutrient composition of honeylocust (*Gleditsia triacanthos* L.) seedlings. Tree Physiol 2:307–316.
Thornton, F.C., M. Schaedle, and D.J. Raynal. 1987. Effects of aluminum on red spruce seedlings in solution culture. Environ Exp Bot 27:489–498.
Thornton, F.C., M. Schaedle, and D.J. Raynal. 1989. Tolerance of red oak and American and European beech seedlings to aluminum. J Environ Qual 18:541–545.
Tokashiki, Y., J. B. Dixon, and D. C. Golden. 1986. Manganese oxide analysis in soils by combined x-ray diffraction and selective dissolution methods. Soil Sci Soc Am J 50:1079–1084.
Tsai, P.P. and P.H. Hsu. 1984. Studies of aged OH-Al solutions using kinetics of Al-ferron reactions and sulfate precipitation. Soil Sci Soc Am J 48:59–65.
Turner, R.C. 1969. Three forms of aluminum in aqueous systems determined by 8-quinolinolate extraction methods. Can J Chem 47:2521–2527.
Turner, R.C. 1971. Kinetics of reactions of 8-quinolinol and acetate with hydroxyaluminum species in aqueous solutions. 2. Initial solid phases. Can J Chem 49:1688–1690.
Turner, R.C. and W. Sulaiman. 1971. Kinetics of reactions of 8-quinolinol and acetate with hydroxyaluminum species in aqueous solutions. 1. Polynuclear hydroxy-aluminum cations. Can J Chem 49:1683–1687.
Ulrich, B. 1981a. ökologische Gruppierung von Böden nach ihrem chemischen Bodenzustand. Z Pflanzenernaehr Bodenkd 144:289–305.
Ulrich, B. 1981b. Destabilisierung vom Waldökosystem durch Akkumulation von Luftverunreinigungen. Forst Holzwirt 21:525–532.
Ulrich, B. 1983. Soil acidity and its relation to acid deposition. *In*: Effects of Accumulation of Air Pollutants in Forest Ecosystems, B. Ulrich and J. Pankrath (eds.), pp. 127–146. Dordrecht, Netherlands: D. Reidel.
Ulrich, B., R. Mayer, and P.K. Khanna. 1980. Chemical changes due to acid precipitation in a loess-derived soil in Central Europe. Soil Sci 130:193–199.
Ulrich, B., K.J. Meiwes, N. Konig, and P.K. Khanna. 1984. Untersuchungsverfahren and Kriterien zur Bewertung der Versauerung und ihrer Folgen in Waldböden. Forst Holzwirt 39.
van Breeman, N. 1973. Dissolved aluminum in acid sulfate soils and acid mine waters. Soil Sci Soc Am J 37:694–697.
van Breeman, N. and R. Brinkman. 1978. Chemical equilibria and soil formation. *In*: Soil Chemistry. A. Basic Elements, G.H. Bolt and M.G.M. Bruggenwert (eds.), pp. 141–170. Amsterdam: Elsevier.
van Breeman, N., C.T. Driscoll, and J. Mulder. 1984. Acidic deposition and internal proton sources in acidification of soils and water. Nature 307:599–604.
van Praag, H.J. and F. Weissen. 1985. Aluminum effects on spruce and beech seedlings. I. Preliminary observations on plant and soil. Plant Soil 83:331–338.
van Praag, H.J., F. Weissen, S. Sougnez-Remy, and G. Carletti. 1985. Aluminum effects on spruce and beech seedlings. II. Statistical analysis of sand culture experiments. Plant Soil 83:339–356.

Violante, A. and M.L. Jackson. 1981. Clay influence on the crystallization of aluminum hydroxide polymorphs in the presence of citrate, sulfate, and chloride. Geoderma 25:199–214.

Violante, A. and P. Violante. 1980. Influence of pH, concentration, and chelating power of organic anions on the synthesis of aluminum hydroxides and orthohydroxides in nitric acid. Soil Sci Soc Am J 47:164–168.

Voigt, G.K. 1980. Acid precipitation and soil buffering capacity. *In*: Proceedings of an International Conference on Ecological Impacts of Acid Precipitation, Sandefjord, Norway. 11–14 March 1980. SNSF Project, Oslo, Norway.

Wada, K. 1977. Allophane and imogolite. *In*: Minerals in Soil Environments, J.B. Dixon and S.B. Weed (eds.), pp. 603–638. Madison, Wisconsin: Soil Science Society of America.

Wada, S. and K. Wada. 1980. Formation, composition and structure of hydroxyaluminosilicate ions. J Soil Sci 31:457–467.

Wagatsuma, T. and M. Kaneko. 1987. High toxicity of hydroxy-aluminum polymer ion to plant roots. Soil Sci Plant Nutr 33:57–67.

Waite, T.D., I.C. Wrigley, and R. Szymczak. 1988. Photoassisted dissolution of colloidal manganese oxide in the presence of fulvic acid. Environ Sci Technol 22:778–785.

Wang, M.K., M.L. White and S.L. Hem. 1983. Influence of acetate, oxalate, and citrate anions on precipitation of aluminum hydroxide. Clays Clay Miner 31:65–68.

Weaver, G.T., P.K. Khanna, and F. Beese. 1985. Retention and transport of sulfate in a slightly acid forest soil. Soil Sci Soc Am J 49:746–750.

Wiklander, L. 1975. The role of neutral salts in the ion exchange between acid precipitation and soil. Geoderma 14:93–105.

Wilding, L.P. and L.R. Drees. 1983. Spatial variability and pedology. *In*: Pedogenesis and Soil Taxonomy. I. Concepts and Interactions, L.P. Wilding et al. (eds.), pp. 83–116. New York: Elsevier.

Winterhalder, E.K. 1963. Differential resistance of two species of Eucalyptus to soil manganese. Aust J Sci 25:363–364.

Wolfe, M.H., J.M. Kelly, and J.D. Wolt. 1987. Soil pH and extractable sulfate-sulfur distribution as influenced by tree species and distance from the stem. Soil Sci Soc Am J 51:1042–1046.

Wolt, J.D. 1981. Sulfate retention by acid-sulfate polluted soils in the Copper Basin area of Tennessee. Soil Sci Soc Am J 45:283–287.

Wolt, J. 1987. Soil Solution: Documentation, source code, and program key. Tennessee Agricultural Experimental Station Research Report No. 87–19.

Wolt, J.D. and F. Adams. 1979. The release of sulfate from soil-applied basaluminite and alunite. Soil Sci Soc Am J 43:118–121.

Wolt, J.D. and D.A. Lietzke. 1982. The influence of anthropogenic sulfur inputs upon soil properties in the Copper Basin region of Tennessee. Soil Sci Soc Am J 46:651–656.

Wright, M.J. (ed). 1976. Proceedings of a Workshop on Plant Adaptation to Mineral Stress in Problem Soils, 22–23 November 1976, Beltsville, Maryland. Ithaca, New York: Cornell University Agricultural Experimental Station.

Wright, R.J. and S.F. Wright. 1987. Effects of aluminum and calcium on the growth of subterranean clover in Appalachian soils. Soil Sci 143:341–348.

4
Rates of Nutrient Release by Mineral Weathering

G. NORMAN WHITE, STEVEN B. FELDMAN, and
LUCIAN W. ZELAZNY

I. Introduction

Acidic deposition is capable of altering the *geochemistry* of several elements important to forest productivity (Reuss and Johnson 1986). *Long-term effects* of acidic deposition may include *accelerated depletion* of *base cations* and *increased concentrations* of *aluminum in soil solutions.* Forest soils, which generally do not receive fertilizers and limestone, are therefore more susceptible to such effects than cultivated soils.

Fortunately, most forest soils have substantial capacity to resist effects of acidic deposition. *Mechanisms of resistance* and their effectiveness vary among soils, thus precluding the possibility of a consistent relationship between *deposition rates* and *soil impacts.* Assessing the likelihood and timescale of impacts requires site-specific quantitative information about deposition rates, *impact mechanisms,* and resistance mechanisms.

Mineral weathering is an important mechanism of soil resistance to effects of acidic deposition. *Weathering reactions* are sinks for protons and sources of plant nutrients. Moreover, weathering reaction rates may increase in *response to acid inputs,* thus perhaps compensating for deposition-induced accelerations in acidification processes. Incomplete knowledge of weathering rates is a major source of uncertainty in *assessments of potential effects* of acidic deposition on forest soil fertility and tree nutrition.

In this review, rates of nutrient release via mineral weathering are examined in detail. A general review of processes and rate-controlling factors is provided to elucidate approaches to the study of mineral weathering. A critical review of the various laboratory and field methods of estimating nutrient release rates follows to supply insight into existing experimental techniques. The applicability of such studies to the study of forest processes is then examined.

II. Processes of Mineral Weathering

Weathering has been defined in diverse ways by authors with contrasting backgrounds and interests. Reiche (1950) defined weathering as "the response of materials which were in equilibrium within the lithosphere to conditions at or near its contact with the atmosphere, the hydrosphere, and perhaps still more importantly, the biosphere." Keller (1955) preferred to delete the words "which were in equilibrium" because the period of equilibrium in which rocks formed may have been only momentary. Ollier (1984) believed Reiche's definition put too much stress on the biosphere because biological weathering operates through physical and chemical processes fundamentally similar to nonbiological weathering. He defined weathering as "the breakdown and alteration of materials near the earth's surface to products that are more in equilibrium with newly imposed physico-chemical conditions." This definition suggests that physical weathering would result in materials more in equilibrium with surface conditions, but large unstable mineral particles are rendered less rather than more stable by the decrease in size accompanying physical weathering. With these and other definitions in mind, we will consider weathering to be the response of materials to physical and chemical conditions at or near the earth's surface.

In general, weathering reactions can be ascribed to purely physical or chemical processes. Physical weathering is the breakdown of materials by mechanical processes (Ollier 1984). These mechanical processes may involve changes in pressure or temperature, mechanical collapse, abrasion, or other disturbances. No matter what processes cause physical weathering, the end result is production of smaller particles without change in particle composition. This is not true of chemical weathering processes, which involve chemical reactions of minerals with air and aqueous solutions. The final product of chemical weathering depends on a variety of factors including time. Chemical processes that may result in weathering include oxidation-reduction, hydrolysis, and chelation.

A. Physical Weathering

Physical weathering by itself does not release nutrients for plant growth. Physical weathering does, however, increase potential chemical alteration and nutrient release by physically decreasing the size of weathering materials, thereby increasing the surface area available for attack by weathering agents. Ollier (1984) divided physical weathering into 10 processes: loading-unloading, crystal growth, insolation weathering, moisture swelling, irreversible water absorption (slaking), abrasion, mechanical collapse, colloid plucking, cavitation, and soil ripening.

Physical weathering by loading-unloading is probably the most macroscopic form of weathering, with effects covering large areas. Loading-unloading phenomena can be expressed in many ways such as sheeting, spalling, rock blisters (A-tents), and load-induced fracturing. *Sheeting* occurs in response to pressure release associated with the erosion of overburden materials. Most rocks are formed deep within the earth with considerable geostatic loading. When erosion unloads this pressure, the underlying rock expands. As it is most often impossible to expand laterally, the general response is the formation of unloading domes in which the upper layers of rock break off as sheets. Sheeting may also occur when high horizontal compressive stresses are applied by tectonics. *Spalling* is the process whereby rocks break into relatively thin, commonly curved fragments by exfoliation. Spalling is similar to sheeting but on a smaller scale. *Rock blisters* are formed when lateral expansion causes a rock to crack and be pushed up. The resulting formation often resembles the A-tents used in camping. *Load-induced fracturing* is a type of mechanical weathering most often associated with isolated boulders in a landscape; the boulder splits at the air-ground contact because of pressure differences among parts of a boulder in contact with the ground and adjacent parts not in ground contact. This process results in pestle-shaped fragments. In general, loading-unloading phenomena do not in themselves enhance chemical weathering because the scale of these processes does not result in significant changes in exposed surface area.

Volume changes from crystal growth can cause tremendous stress resulting in physical weathering of rock. Ollier (1984) differentiated three main divisions in weathering by crystal growth: frost weathering, salt weathering, and chemical alteration. *Frost weathering* involves the formation of ice in pores and preexisting cracks in a rock, resulting in very great pressures from the 9% expansion of water at freezing. These pressures are sufficient to crack rocks. This type of weathering is more significant in temperate climates where repeated cycling above and below freezing can accelerate weathering rates. *Salt weathering* is also climate related, being important in semiarid and arid areas and virtually nonexistent in more humid-temperate areas. Salt weathering occurs when soluble salts such as gypsum and halite precipitate in cracks, resulting in sufficient pressures to cause disruption of a rock. Physical weathering by salts can also occur because of the higher thermal expansion of salts compared to the rocks in which they are contained. This difference is especially significant in areas of high temperatures including rock walls heated by the sun. In addition, hydration of these salts can result in the production of very high pressures, which can easily disrupt rock structure. Volume changes resulting in physical alteration of rocks can also occur simultaneously with *chemical alteration* processes. The significant volume changes accompanying chemical alterations can result in a peeling-off of weathering rinds and self-generation of new surfaces on which further chemical weathering processes can act.

Repeated temperature changes can also cause a rock to break apart. If these temperature changes are caused by solar radiation, the process is called insolation weathering. *Insolation weathering* is a result of two characteristics of rocks: their expansion on heating and their poor thermal conductivity. This combination causes stresses to develop along the thermal gradient from a rock interior to the solar-heated exterior. Heterogeneity of minerals in rocks can also cause heat-induced stress, which results from differences in specific heats and coefficients of thermal expansion along with color variations. Darker minerals, for example, absorb heat faster than lighter minerals, thus producing temperature differences and thermal stress. The inability of a partially buried boulder to expand when heated because of confinement can cause it to break at the soil surface. Drastic temperature changes caused by forest fires can also produce large thermal gradients in rock structures that result in rapid physical weathering.

Simple reversible water adsorption with accompanying volume changes (i.e., *moisture swelling*) also may be sufficient to cause physical weathering. These volume changes are of the same magnitude as those resulting from thermal expansion. The exact mechanism of moisture swelling is unknown, but the effects are more important than formerly realized.

Slaking, as a weathering process, is similar to moisture swelling except that the effects of water adsorption are irreversible. This type of swelling is the result of water adsorption by expansible minerals. Slaking is very important in the weathering of mudstones, shales, and slates, which contain high quantities of expansible phyllosilicates.

Minerals and rocks also may be worn away by simple *mechanical abrasion.* Abrasion is most important when material is transported by glaciers, wind, or water. In environments where materials are being actively transported, this type of weathering may dominate; in environments where transport is not occurring, the process may be negligible.

Mechanical collapse and *colloid plucking* are both processes that can be locally important but are generally not significant modes of physical weathering. Rock weathering may occur as the result of mechanical collapse in cases where rock is undercut by erosion. Colloid plucking has been suggested as the name for the process whereby a film of clay drying on a rock pulls grains out of the rock surface.

Cavitation is a weathering process resulting in sudden changes in pressure caused by bubble collapse in fast-moving water. These sudden and violent bubble collapses produce a shock like a hammer blow on a very small area. Cavitation has been calculated to occur at very high water velocities such as at the base of high waterfalls (Ollier 1984).

The last of Ollier's (1984) modes of weathering is soil ripening. This process is the first stage of weathering in the evolution of soil from newly drained alluvium. Physical ripening involves the loss of water by compaction, evaporation, and transpiration. The process is mostly irreversible and results in the inability of the soil to absorb water to the extent of the unripe soil.

B. Chemical Weathering

Chemical weathering occurs because rocks and minerals are rarely in equilibrium with the water compositions, temperatures, and pressures found in near-surface environments (Birkeland 1974). Products of chemical weathering are more stable in the environment in which they occur than their precursors. Changes can be very slight and involve little more than oxidation of Fe^{2+}, or be so intense as to leave little clue as to the precursor.

The first step in studying chemical weathering is to determine which minerals will persist in a given environment. A stable mineral is one that is thermodynamically inert in the surrounding solution. The solution stability of a mineral can be determined by comparing its solubility product with the actually solution ion-activity product (Berner 1971). For example, kaolinite dissolution can be described by the following reaction:

$$Al_2Si_2O_5(OH)_4 + 6H^+ \rightleftharpoons 2Al^{3+} + 2H_4SiO_4 + H_2O$$

The mineral solubility product, K_{sp}, for this reaction is

$$K_{sp} = \frac{(Al^{3+})^2\,(H_4SiO_4)^2}{(H^+)^6}$$

where the parentheses denote solution activities. The actual solution ion-activity product, IAP, with respect to kaolinite is similarly defined as

$$IAP = \frac{(Al^{3+})^2\,(H_4SiO_4)^2}{(H^+)^6}$$

and is determined by substituting the measured ion activities into this equation. The state of saturation of a solution, determined by comparing the IAP to the literature value for K_{sp}, is defined as

$IAP > K_{sp}$ for solution supersaturation

$IAP = K_{sp}$ for solution saturation

$IAP < K_{sp}$ for solution undersaturation.

If the solution is supersaturated, the solid (kaolinite in the example) is likely to precipitate. If the solution is undersaturated, the solid should dissolve. If the solution is saturated with respect to the solid, either rapid precipitation or dissolution of the solid is controlling the elemental activities in solution. A high degree of undersaturation can result from a small initial amount of mineral present, a slow rate-controlling step in dissolution, or a fast rate of solute removal by precipitation of another mineral, ion exchange, fast flushing (leaching), or plant uptake of ions in solution.

In humid environments, a mineral is more likely to be stable with increasing depth below the soil surface, because the original source for the weathering solution is rainfall, which is undersaturated with respect to most minerals. This dilute solution acquires the solutes released by weathering as it percolates downward, picking up alkali cations, alkaline earths, and Si while losing protons (Velbel 1985). A slow flushing rate of this downwardly percolating water results in a solution that will be closer to thermodynamic equilibrium at increased depth.

Three simultaneous processes can be considered to be taking place in the weathering environment (Loughnan 1969). These processes are (1) breakdown of parent minerals with release of cations and Si, (2) removal in solution of some of the released constituents, and (3) reconstitution of the constituents to form new minerals which are stable or metastable in the environment.

The breakdown of minerals to yield their ionic constituents can proceed by several reaction pathways including hydrolysis, hydration, oxidation, carbonation, chelation, complexation, ion exchange, and simple dissolution. Some minerals are only affected by one of these processes, but many can be affected simultaneously by several.

Hydrolysis is clearly the most important process in the chemical weathering of silicates and aluminosilicates (Birkeland 1974, Loughnan 1969, Ollier 1984). Hydrolysis is the chemical reaction between a mineral and water that takes place whenever a mineral is in contact with water. All that is required is a source of H^+ and exposed atoms or ions on mineral surfaces that hydrate in solution resulting in a surface charge. For every H^+ consumed by mineral surfaces in the process, one cation equivalent must be released to maintain electroneutrality. The result is an initial increase in pH until acidic conditions develop from the removal of basic cations in leaching waters and the release of H^+ to solution.

The classic example for surface hydrolysis is the reaction of fresh orthoclase with water, which can be written as

$$KAlSi_3O_8 + H_3O^+ \rightleftharpoons HAlSi_3O_8 + K^+_{(aq)} + H_2O$$

where $KAlSi_3O_8$ represents unaltered orthoclase faces and $HAlSi_3O_8$ represents the altered product. This altered product is unstable on the surface of the mineral. The importance of this layer on further weathering is highly controversial (Holdren and Berner 1979, Nickel 1973).

Hydrolysis reactions can be chemically modeled by equilibrium chemistry with the residue layer acting only as a rate-limiting factor in dissolution. The fact that hydrolysis reactions consume H^+ makes solution pH highly significant (Loughnan 1969). The source of H^+ in solution can be (1) redox reactions such as the oxidation of pyrite, (2) the reaction of CO_2 with water to form H_2CO_3, (3) plant roots, (4) the pH-dependent charge of mineral surfaces and humic substances, (5) cation hydrolysis, (6) organic decomposition products, or (7) inputs from atmospheric deposition. Soil pH, however, cannot be considered an independent variable, but must be viewed as a function of the composition and structure of parent minerals, the removal of bases by plant uptake and leaching, the rate of biomass degradation, and the nature and cation exchange properties of residual mineral products (Loughnan 1969).

Hydration-dehydration processes involve the addition or removal of water from a mineral yielding a new product. Such reactions can be exothermic and therefore occur spontaneously. An example of hydration-dehydration is the formation of gypsum ($CaSO_4 \cdot 2H_2O$) and anhydride ($CaSO_4$) by adding or removing water:

$$CaSO_4 \cdot 2H_2O \rightleftharpoons CaSO_4 + 2H_2O$$

Other important examples of hydration-dehydration reactions in humid environments are the responses of goethite and hematite (Chesworth 1972), halloysite and metahalloysite (Grim 1968), and boehmite and gibbsite (Tardy 1982) to changes in water activity in small pores and relative humidity. These reactions do not directly result in the release of nutrients to solution and are probably not significant in overall chemical weathering (Birkeland 1974). However, as previously indicated, the large increase in volume accompanying hydration reactions is an important process in physical weathering as the reaction produces pressure that may aid in rock disintegration (Birkeland 1974, Ollier 1984).

Oxidation-reduction reactions involve the reaction of certain ions and compounds with O_2 and other electron acceptors resulting in a change in valence. In the environment, these reactions are most important for Fe, Mn, S, some other less common transition metals and anions, and organic matter. These reactions are highly dependent on the redox potential and pH of the soil solution. In soils, redox potential is primarily determined by the amount of O_2 and CO_2 dissolved in soil water and by soil organic matter (Ollier 1984). Under conditions of increasing acidity, many species that would normally exist in the oxidized state can remain reduced and potentially mobile.

Carbonation reactions directly involve CO_2 dissolved in solution. The presence of CO_2 and H_2CO_3 in soil water is believed to greatly facilitate the cation exchange process and is critical to the dissolution of carbonates (Ollier 1984). The mechanism of the reaction of H_2CO_3 with noncarbonates is not well understood, but numerous observations suggest that H_2CO_3 is a much stronger solvent than its acidity alone would indicate (Ollier 1984).

Chelation and complexation processes involve the reaction of ions with organic matter or other ions in such a way as to allow an ion to be stable at concentrations and pH conditions well above levels where it would normally precipitate. Chelation involves the holding of an ion, for example, Fe or Al, as a metallic coordination center within an organic ring structure (Schnitzer 1969). This process usually involves the formation of multiple bonds between the metal and the organic ligand (Birkeland 1974). Free ions in mineral structures can also be attacked directly by chelating agents (Tan 1980). Low molecular weight organic acids, especially humic acid and fulvic acid, are organic matter constituents involved in the chelation process. Fulvic acid, however, is considered to be the most significant organic chelating agent in soils (Schnitzer and Kodama 1976). Birkeland (1974) suggested that chelation may be responsible for a considerable amount of weathering, possibly more than hydrolysis.

Complexation reactions are inorganic reactions involving the strong pairing of an anion with a cation. Complexation is regarded to be of only minor significance in the overall weathering scheme, but it has been suggested that many chemical changes in soils caused by acid precipitation may be attributable to complexation reactions (Cronan 1980a,b).

Mineral transformations may occur by a simple ion exchange process (Birkeland 1974). This type of weathering is most important in phyllosilicates where the replacement of one cation by another in the interlayer region may result in a different mineral. This process occurs by a mass-action phenomenon and is controlled to a large extent by the ionic character of the soil solution.

Twidale (1976) suggested simple dissolution to be essential to all chemical weathering processes. All minerals are soluble to some degree, although more often some other process controls dissolution. Some salts such as halite and gypsum dissolve readily while other minerals such as quartz dissolve relatively slowly. Quartz is considered rather soluble by many authors (Birkeland 1974, Chesworth 1973), yet tends to persist to late stages of weathering relative to other resistant minerals through a combination of slow reaction kinetics and, more importantly, a higher initial content of coarse particles in the parent material.

All other factors being equal, many authors propose a difference in ion mobilities as controlling reaction rates. Ion mobilities represent the relative differences in leaching rates that would be observed for different elements when they are released to soil solution. Ion mobilities are affected by the rate of fixation by adsorption, ion exchange, and precipitation reactions occurring in soil solution. Ions that are not strongly adsorbed, or which exist in soil solutions undersaturated with respect to some easily precipitated constituent, would be highly mobile. Chesworth (1973) believed that the most mobile ions in soil solutions are Na, K, and Mg. Calcium may or may not be mobile depending upon the CO_2 content. He considered Si, Fe, and Al to be relatively immobile but not static. Birkeland (1974) agreed in principle to this classification, except that he considered Si to be more intermediate in mobility.

Reconstruction of the ionic constituents of weathered minerals into secondary products can have a significant effect on the availability of nutrients to trees and the rate of nutrient loss from the rooting zone via leaching. Most of the factors that affect dissolution also affect the precipitation of new minerals. In addition to these factors, the initial presence or absence of the product mineral is also important. In the absence of this mineral, a certain degree of supersaturation is required for precipitation to occur (Berner 1971). Once crystallization commences and crystals grow larger, the level of supersaturation required for further precipitation diminishes. The initial presence of mineral seeds allows precipitation to occur at all levels above saturation.

III. Rate-Controlling Factors in Nutrient Release and Weathering

The wide variety of authigenic (secondary) minerals found at the earth's surface provides an insight into the diversity of weathering environments that can result from the interaction of many factors through time. Differences in these factors influence the physical and chemical processes of weathering, which in turn cause mineral transformations and nutrient release to proceed at variable rates. Even within narrowly confined areas, differences in the weathering environment with depth often result in the differentiation of isotropic parent materials into sequences of soil horizons.

Jackson and Sherman (1953) identified mineral weathering rates as being controlled by the following factors: (a) temperature, (b) leaching and internal soil drainage, (c) soil reaction, (d) redox potential, (e) biotic effects, (f) particle size and specific surface effects, and (g) the specific weatherability of minerals. In this section, discussion includes these components as well as the influence of chemical kinetics, topography, and anthropogenic effects. Finally, the relationship of weathering studies to Jenny's (1941) soil-forming factors is considered.

A. Temperature

The temperature of a system influences whether or not a given chemical or biochemical reaction will occur, the nature of reaction end products, and the rate at which the reaction will proceed (Stumm and Morgan 1981). Temperature affects evapotranspiration rates (and therefore moisture availability), microorganism activity, and the rate of biomass production and decomposition. Although laboratory studies can isolate the various components of climate, the effects of temperature on mineral weathering are not as obvious under actual field conditions.

High average temperatures generally increase chemical weathering rates, assuming that moisture is not a limiting factor. van't Hoff's rule states that for every 10°C rise in temperature, the velocity of a chemical reaction increases by a factor of two to three. This greatly influences the rate of initial alteration of primary minerals by affecting the kinetic favorability of hydrolysis reactions (Chesworth 1972, Mattigod and Kittrick 1980). Low average temperatures create a severely restricted weathering environment in which chemical reaction rates are decreased and minerals that are otherwise highly susceptible to weathering become more stable (Kodama and Foscolos 1981, Loughnan 1969). Physical weathering processes dominate in areas with cold, dry climates. Chemical weathering in these regions is minimal (Carroll 1970).

Lanyon and Hall (1979) studied the effects of temperature variation on the chemical weathering of four contrasting materials: a coarse crystalline forsteritic olivine, a fine-grained micaceous and poorly sorted sandstone, a slightly acid brown shale, and a microcline specimen. They subjected each of these materials to dissolution in 0.01 N $AlCl_3$ at pH 3 with constant temperatures of 3°, 25°, and 47°C, and a diurnal cycle of alternating temperatures averaging 25°C. Concentrations of ionic components recorded over a 90-day period were found to consistently increase with increasing temperature, although not at the same rate for all four materials. Differences in the magnitude of dissolution for each material were attributed to differences in structural and crystallographic characteristics of the minerals undergoing alteration. The constant temperature of 25°C resulted in significantly less dissolution than the alternating temperature regime for each of the materials studied. These findings indicate that mineral weathering and nutrient release rates in forest ecosystems are greatly affected by microclimate (i.e., slope aspect and position) and vegetative cover, especially in steeply sloping areas. In addition, higher rates of mineral dissolution would be expected to occur following intensive harvesting because soil temperature fluctuations are greater under these conditions than under forest cover.

Field studies have supported conclusions drawn from the laboratory studies described above. Research has shown that soils on southerly slopes (in the northern hemisphere) have higher mean annual temperatures and more highly weathered profiles than soils on northern slopes. Losche et al. (1970) studied the effects of slope aspect on soil development in southwestern North Carolina, and concluded that kaolinite and hydroxy-interlayered vermiculite were dominant on north-facing slopes while gibbsite was dominant on the warmer south-facing slopes. Franzmeier et al. (1969) observed a similar effect on steeply sloping soils on the Cumberland Plateau of eastern Kentucky and Tennessee. Argillic horizons were found on south-facing slopes and weakly developed cambic horizons were found on opposing northerly slopes.

Although chemical reactions are slower in cold environments, there is evidence to support the fact that a considerable amount of chemical weathering can indeed take place with temperatures near freezing, especially in moist areas. Reynolds (1971) reported on the formation of pedogenic vermiculite and gibbsite from the weathering of metadiorite and biotite-quartz diorites in the South Cascade Glacier region of northwestern Washington. Although gibbsite is generally considered to be restricted to more tropical environments, intense leaching during winter snowmelt can favor Si removal and the formation of Fe- and Al-oxides and hydroxides. Minerals in rock surfaces above 0°C are subject to dissolution, especially when slope, aspect, and exposure are favorable. Runoff waters coming into contact with mechanically fractured rocks having high surface areas can further enhance chemical weathering in these environments. Seasonal soil freezing or permafrost can, however, restrict water movement through the weathering profile and limit chemical weathering reactions to areas near the surface (Carroll 1970).

The various weathering indices discussed here and elsewhere in the literature (Birkeland 1984, Brewer 1976) express general trends in weathering and nutrient release that can be related to temperature. The ratio of partly altered parent minerals to highly weathered residues in soils, for example, generally decreases along a transect from periglacial areas through temperate to tropical areas. This generalization, however, is complicated by factors such as nonuniform lithology, differences in the stability of the mineral species involved, grain-size differences, and the effects of the specific weathering environment. In a compilation of data obtained from several residual soils derived from igneous and metamorphic rocks on the unglaciated southeastern Piedmont Plateau, Jenny (1941) noted a significant increase in average clay content associated with mean annual temperature. After the adjustment of all clay values to an equivalent moisture basis, the clay content of comparable soils was found to increase exponentially with increasing temperature and linearly with increasing moisture.

B. Leaching and Internal Soil Drainage

Moisture is involved in most physicobiogeochemical processes in soils. Water removes the soluble products of weathering, thereby driving chemical reactions toward completion. The availability of water also governs the depth of the wetting front and the duration of wet-dry and shrink-swell cycles. The rate of water percolation has been cited as the single most important factor in controlling the chemistry of the weathering environment and therefore the rate of parent mineral breakdown and the genesis of secondary products (Kellogg 1941, Loughnan 1969). Velbel (1985) agreed, stating that flushing rates relative to reaction rates determine streamwater chemistry with reaction equilibrium of little importance.

Weathering intensity is greatest in high rainfall and temperature areas where water movement through the soil is at a maximum (Allen and Fanning 1983). However, to predict the influence of moisture regime on weathering rates and nutrient release, it is important to consider not only the annual distribution of precipitation, but also losses from potential evapotranspiration throughout the year. This approach provides a more comprehensive picture of the soil moisture regime and the seasonal amounts of water available for leaching (Yaalon 1983). It also takes into account any effect of annual precipitation maximums coinciding with annual temperature maximums to reduce total moisture flux.

Infiltration of water into the soil is facilitated by the presence of vegetation, which slows surface runoff, improves soil structure, and increases available water-holding capacity. After entering the soil, the depth that leaching water will penetrate depends on the amount of available water that the soil horizons can hold, which in turn depends on soil texture and the amount of humified organic matter present. Other factors such as subsoil and substrate permeability, landscape position, and groundwater flowpaths influence leaching rates in weathering environments.

High leaching rates in a temperate climate generally result in a shift in clay mineralogy from cation-rich phyllosilicates to more siliceous minerals. Barshad (1966) assessed the frequency distributions of clay minerals as a function of precipitation by sampling soils along the Sierra Nevada range from Death Valley to the northern border of California. Mean annual precipitation values in the study areas ranged from 13 to 230 cm. To minimize the effects of other climatic influences, samples were collected at various altitudes to maintain temperatures between 10°C and 15°C. Smectite minerals and other 2:1 phyllosilicates high in basic cations were found to be confined to areas receiving limited rainfall where the products of dissolution could accumulate in the soil. Kaolinite was dominant in higher rainfall areas where leaching was more pronounced.

Intense leaching ultimately results in the solubilization and removal of Si, leaving only a residual accumulation of Fe- and Al-oxides weathered from parent minerals (Birkeland 1984, Yaalon 1983). These resistant materials, which are characteristic of soils of the humid tropics, persist near the site of weathering and become progressively concentrated in the soil as noncrystalline or crystalline oxides, hydroxides, and oxyhydroxides. As weathering and leaching progress, nutrient depletion is accompanied by an increase in H^+ concentration (pH decreases), especially in soils with low buffering capacities.

C. Soil Reaction

Low pH values in the weathering environment contribute to the rapid dissolution of minerals by increasing the rate of hydrolysis reactions and by governing species solubility (Kittrick 1969). As discussed in previous sections, the persistence of ions in solution is also highly dependent on soil moisture regime and leaching rates. Free drainage conditions thus contribute to acid weathering processes by removing basic cations. Impeded drainage, on the other hand, can result in an increase in pH by limiting cation removal and by the reduction of Fe^{3+} compounds which consumes H^+.

The pH of the weathering environment also has a pronounced influence on mineral stability by controlling the solubility equilibria of aluminosilicate minerals. In general, Al is more soluble than Si in an acid medium. In a neutral medium, Al is insoluble whereas Si retains its solubility. In an alkaline medium, the two solubilities are similar and increase at the same rate (Millot 1970). Within the normal pH range of many soils (pH 5 to 7), most nutrient cations exist in a soluble state or are located on the soil exchange complex whereas Fe and Al exist as noncrystalline or crystalline oxides, hydroxides, and oxyhydroxides that persist close to the site of weathering. Silica sources in soils dissolve slowly in most pH ranges because of slow reaction kinetics, but they become increasingly soluble with decreasing crystallite size (Wilding et al. 1977). While the stability of many soil minerals is affected by pH, others such as rutile and zircon are little affected because of insolubility of their cationic constituents (Ti and Zr, respectively).

Many laboratory studies have examined the dissolution rates of soil minerals over a wide range of pH values. Nickel (1973) determined the weathering rates of 12 common soil minerals at pH values of 0.2, 3.6, 5.6, and 10.6. The solubility of the minerals was determined by the solubility of their individual components which increased as pH decreased. Similar results were reported by Grandstaff (1986), who studied the dissolution rate of forsteritic olivine in 0.1 M KCl solutions over the pH range of 3 to 5 and at temperatures between 1° and 49°C. Most of his experiments could be explained by a linear (zero-order) rate law in which mineral dissolution was virtually congruent (see Section III.H., this chapter, *Chemical Kinetics*). Over all temperature ranges, the rate of olivine dissolution increased with increasing H^+ activity. Because a decrease in 1 pH unit multiplies the concentration of H^+ in the weathering zone by 10, it should be evident that mineral dissolution rates are heavily dependent on the pH of the surrounding medium.

D. Redox Potential

Redox potential also affects the mobility of ions in soils. Although many nutrient elements can undergo oxidation-reduction reactions, the oxidation of Fe^{2+} to Fe^{3+} is the most significant redox reaction involved in weathering. This reaction occurs in oxidizing environments at pH levels above 2.5. Under these conditions, Fe^{2+} loses an electron, which creates a charge and size imbalance in the mineral lattice structure resulting in the departure of other ions to maintain electroneutrality. High levels of microbial activity and large amounts of readily decomposable organic matter can also affect redox potential by influencing O_2 availability. Under reducing conditions, insoluble species that would otherwise precipitate can remain mobile in solution.

Although the oxidation of Fe^{2+} is generally regarded as an important mechanism in the accelerated weathering of most ferromagnesian minerals (Courbe et al. 1981), some controversy exists as to its overall significance to the weathering rates of other minerals, especially phyllosilicates. Oxidation reactions in the phyllosilicates are affected by leaching of soil solution components because low ionic strength solutions enhance lattice expansion and K release, thereby facilitating structural Fe^{2+} oxidation. To illustrate this point, Amonette et al. (1985) used various oxidants (O_2, NaOCl, H_2O_2, and Br_2) in aqueous salt solutions to examine the oxidation of structural Fe^{2+} in biotite and the corresponding expansion of the mineral lattice at 25°C. Under laboratory conditions, O_2 had minor oxidizing ability relative to H_2O_2 and Br_2. Salt treatments that minimized lattice expansion and K exchange did not result in detectable amounts of oxidized Fe^{2+}. The authors suggested that the importance of structural Fe^{2+} oxidation must always be evaluated in terms of the environmental conditions.

E. Biotic Factors

Biotic factors affect many weathering processes through their influence on the physical and chemical properties of the weathering environment. Biological processes can alter and be altered by each of the rate-determining factors already mentioned. The production and decomposition of organic matter is a key feature of biological processes that greatly affects mineral weathering.

Vegetation exerts many important effects on mineral weathering. For example, the growth and movement of plant roots contributes to the mechanical weathering of rocks and minerals and affects the overall composition of soil solution by the absorption and release of soluble inorganic and organic ions. Plant roots also can participate in direct ion exchange reactions with mineral surfaces, resulting in an increased rate of hydrolytic alteration.

Vegetation improves water relationships and soil aeration through effects of root penetration and organic matter inputs on soil structure. Strongly developed soil structure has a beneficial effect on water drainage and leaching rates as well as providing an improved environment for the oxidation of reduced species. The presence of a canopy cover also minimizes fluctuations in air and soil temperatures, reduces runoff and erosion, and influences the chemical nature of water that reaches the soil surface via throughfall and stemflow.

The presence of organic acids produced during the decomposition of leaf litter and plant roots has been shown to accelerate mineral weathering rates (Huang and Keller 1970). Organic acids can be separated into two groups on the basis of chemical reactivity: (1) those compounds such as formic and oxalic acid in which the acidic characteristics are attributed only to carboxylic functional groups, and (2) those compounds such as humic and fulvic acid in which both carboxylic and phenolic functional groups are predominant. Acids of the first category display dissociation constants comparable to strong inorganic acids. However, acids of the second group can exert both an electrostatic interaction with ions as well as a complex/chelation interaction (Tan 1986). Weathering in the presence of organic acids may proceed at a faster rate than would be expected in the presence of inorganic acids alone (Huang and Keller 1970).

In forest ecosystems, base cations accumulate near the soil surface where they are deposited in throughfall and litterfall after uptake from the soil by vegetation (Ovington 1965). As a result, acid conditions may develop in lower portions of the rooting zone of forest soils. Weathering thus tends to occur faster under forest than under grassland vegetation where nutrient cycling occurs deeper in the profile.

Spatial variation of soil-surface properties near individual trees is great near tree stems and decreases toward the edge of the canopy (Gersper and Holowaychuk 1970a,b, Zinke 1962). Zinke (1962) studied soil variation along radial transects from the stem to the edge of the canopy of several tree species. Similar patterns were found for all species studied. Exchangeable bases, cation exchange capacity (CEC), total N content, and pH all increased from the bole outward with maximum values occurring toward the edge of the area covered by the crown. These patterns may be because water inputs per unit area (and thus leaching rates) may be much greater for soils receiving stemflow than for those not receiving stemflow (Gersper and Holowaychuk 1971). The observed patterns might also be attributable to differences in the chemistry of bark litter and leaf litter (Zinke 1962). Bark litter is more acid and lower in base cations and total N content than leaf litter, thus affecting the distribution of nutrients around trees. Variations in soil properties around trees undoubtedly influence rates of mineral weathering in forest soils.

The activity of microorganisms, including bacteria, fungi, and algae, greatly contributes to the rate of nutrient release from minerals. In addition to being largely responsible for the decomposition of organic matter and the production of organic acids, soil microbes contribute to the degradation and formation of minerals though the following processes: (1) enzymatic reactions, (2) acidolysis by microbially formed acids, (3) oxidation-reduction reactions, (4) complexation and chelation reactions, and (5) nutrient uptake (Eckhardt 1985). Biochemical weathering by microorganisms is especially important in the initial stages of weathering, particularly in harsh environments such as deserts and arctic areas.

Microorganism populations and activity vary with temperature and moisture conditions. The optimum temperature for microbial activity is between 25° and 37°C, but many species can exist between 10° and 45°C. The optimum moisture content for most aerobic microorganisms is between 50% and 70% of the soil saturation value (Carroll 1970). Soil microorganisms are generally found in most abundance in the rhizosphere and above the depth to which humic material has penetrated.

Chemical and physical weathering induced by lichens has also been recognized and may be important in severe environments (Birkeland 1984, Carroll 1970, Loughnan 1969). Lichens are often the initial colonizers of rocks at high altitudes and latitudes. They excrete chelating agents that have been reported to result in higher rates of mineral weathering than the lichen-free areas of the same parent material (Jackson and Keller 1970).

F. Particle Size and Specific Surface Effects

The rate of mineral weathering is commonly assumed to be inversely proportional to particle size (Jackson and Sherman 1953, Loughnan 1969) because specific surface area increases with decreasing particle size, resulting in a greater overall area for chemical reactions to occur. For example, quartz is generally persistent in weathering environments, but it becomes highly soluble in the clay (<2-μm) fraction (Wilding et al. 1977). Colloidal quartz is seldom found in great abundance in soils except in those soils derived from volcanic parent materials or those containing large amounts of quartz in larger size fractions.

The influence of particle size on mineral weathering rates has been extensively studied on the micas. These minerals commonly alter to vermiculite by the exchange of interlayer K and related lattice expansion (Mortland 1958, Reed and Scott 1962, von Reichenbach and Rich 1969). The exchange of interlayer cations proceeds from lateral edges to the center of the mica particles along a moving boundary between the K-depleted, expanded outer regions and the unaltered mica core (von Reichenbach 1972). This mechanism is largely controlled by the diffusion of ions to and from the exchange front and the particle periphery. Nutrient release rates are therefore affected by the distance K ions must travel to get to the edge of the particle. With decreasing particle size, rates of K exchange are enhanced by shorter diffusion paths and larger specific surface areas.

Although nutrient release by mineral weathering generally increases with decreasing particle size, the opposite situation can sometimes exist, especially in the phyllosilicates. von Reichenbach and Rich (1969) reported an increase in K selectivity (inhibition of release) with decreasing particle size in muscovite samples leached with a 0.1 N $BaCl_2$ solution. They found that in addition to particle thickness, K selectivity depended on the uniformity of the exchange front. With decreasing particle size (<0.2-μm), the width of this exchange front approached the dimension of the particle radius, resulting in the complete expansion of the individual interlayers in a stable configuration. In this manner, the wedge-like bending of the muscovite layers associated with the moving weathering front interface was reduced and the exchange of K inhibited.

G. Specific Weatherability of Minerals

Chemical weathering rates depend on the specific (or relative) weatherabilities of the minerals undergoing alteration. Elements of crystal structure (i.e., the nature of cleavage planes and fractures, the presence of twin boundaries, lattice defects, and the number and nature of interatomic bonds) greatly influence the nature of mineral surface reactions. In general, ferromagnesian minerals weather more rapidly than feldspars, which in turn weather more rapidly than quartz and other siliceous minerals (Loughnan 1969). Weathering agents initially gain entry into crystals at sites of cleavage planes, fractures, and twin boundaries where surface ions have unsatisfied valences or where bonds are strained and weakened by lattice dislocations and substitutions (Eggleton 1986). The rate of cation depletion and structural rearrangement after the initiation of weathering depends on a complex array of biological and chemical factors which interact to either block these diffusion pathways or keep them open to weathering agents (Eggleton 1986).

All silicate minerals have, as their basic structural unit, the Si tetrahedron in which Si^{4+} is in fourfold coordination with O. The classification of silicate minerals and their relative resistance to weathering is related to the degree of Si tetrahedron linkage and the amount of basic cations per unit cell of the crystal structure (Loughnan 1969). In general, ferromagnesian mineral stability is of the order: olivine < pyroxene < amphibole < biotite. For nonferromagnesian minerals, the order of stability is: Ca-plagioclase < Na-plagioclase < K-feldspar < muscovite < quartz (Goldich 1938) (Figure 4.1).

A close structural relationship between a primary mineral and its alteration product has been reported to result in the rapid growth of the secondary phase, which favors the blockage of diffusion avenues thus inhibiting further weathering (Eggleton 1986). In cases where no crystallographic continuity exists between a primary mineral and its secondary product, the new phase must reprecipitate from solution, leaving open the pathways for entry of weathering agents into the crystal. Feldspars would be expected to weather rapidly under these circumstances despite the strong interlinking network of Si tetrahedrons because they lack octahedrally coordinated cations that can rapidly rearrange to form layer silicate sheets. As a result, secondary alteration products must be recrystallized from noncrystalline aluminosilicate components released

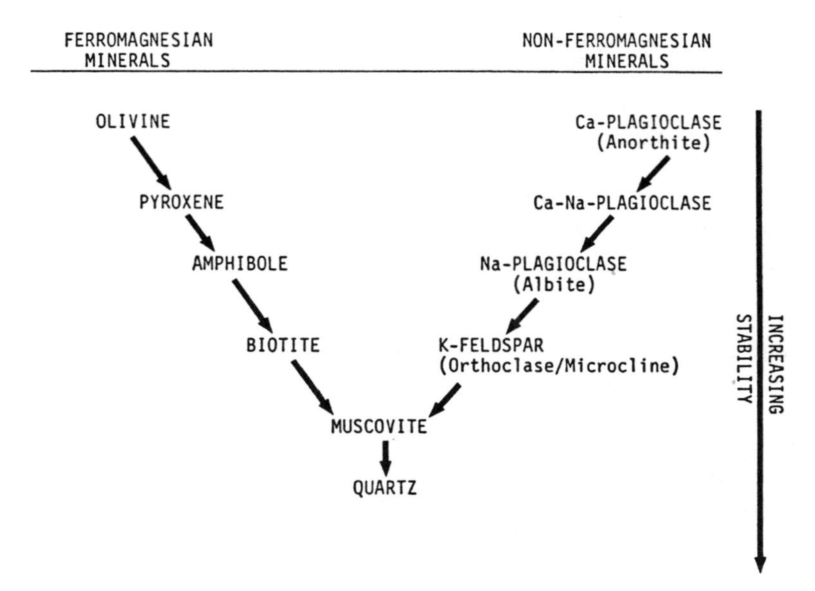

FIGURE 4.1. A weathering stability series (after Goldich 1938).

during weathering. The slow formation of secondary minerals leaves diffusion channels open and subject to increased attack.

H. Chemical Kinetics

Considerable controversy is recorded in the literature as to whether mineral dissolution is a congruent or incongruent process (Berner 1981, Sparks 1989, Velbel 1986, Wollast 1967, Wollast and Chou 1985). Congruent dissolution occurs when the ratios of elements released by weathering are equivalent to the ratios present within the mineral undergoing dissolution. Mineral weathering processes ultimately tend toward congruent dissolution, but are often incongruent in the initial stages (Nickel 1973, Schott and Berner 1985, Wollast and Chou 1985). The controversy stems, in part, from early studies that were not continued long enough to observe congruent weathering (Helgeson 1971, Luce et al. 1972, Wollast 1967). Researchers concluded from these early studies that a resistant layer was formed by the selective leaching of cations from mineral surfaces during weathering. Further dissolution was believed to be hindered by the presence of this resistant layer.

Most studies of mineral weathering have used equilibrium or quasi-equilibrium models (Helgeson et al. 1969), but more recently chemical kinetics studies have been used to characterize mechanisms of dissolution. Diffusion control of reaction rates is thought to result when the incongruent removal of cations relative to Si causes a protective residual layer to develop on mineral surfaces. Reaction rates are believed to be controlled by diffusion of soluble components through this layer. Diffusion controlled weathering is characterized by rapid nonspecific attack such that uniform etching occurs over the entire mineral surface with the consequent rounding of grains (Berner 1981). In contrast, surface-reaction control of mineral dissolution is thought to be largely congruent, occurring slowly at crystallographic defects and points of excess energy, and resulting in characteristic etch pits and "fresh mineral" compositions on weathered mineral surfaces. The actual rate-limiting mechanism controlling mineral dissolution in a given environment depends on the specific nature of the weathering zone. Solution pH, structural properties of the minerals undergoing alteration, oxidation-reduction conditions, and flushing rates all play a major role in the kinetics of weathering, especially under field conditions.

In a comprehensive acid-base dissolution study of a variety of common soil minerals, Nickel (1973) reported that a cation-depleted layer formed on mineral surfaces because of a high rate of solid-state diffusion of cations. Given enough time, a steady-state thickness of this depleted region was obtained with the rate of diffusion of cations from the unaltered mineral surface equaling the rate of dissolution and retreat of the outer, cation-depleted layer. Thus, congruent dissolution ultimately was achieved. Other laboratory studies of weathering kinetics indicated that when silicates were leached with buffered aqueous solutions, the release of Si to solution decreased parabolically with time (Helgeson 1971, Luce et al. 1972, Wollast 1967). It was suggested in these studies that dissolution proceeded incongruently with the consequent formation of a cation-depleted aluminous protective layer that inhibited the diffusion of ions to and from mineral surfaces.

More recent experimental evidence has suggested that surface-reaction control, rather than diffusion control, is the rate-limiting step in silicate mineral weathering resulting in a constant rate of mineral dissolution over time (Berner 1978 and 1981, Holdren and Berner 1979, Velbel 1984). Petrovic and others (1976) suggested that the rapid initial increase in Si determined in the earlier research (Helgeson 1971, Nickel 1973, Wollast 1967) probably was an artifact resulting from the preferential dissolution of ultrafine and strained particles produced by sample grinding. The use of HF to remove ultrafine particles before leaching resulted in linear, rather than parabolic, release of Si over time. Because no cation-depleted layer was found on mineral surfaces, these researchers concluded that solid-state diffusion control of reaction rates in feldspar weathering was overstressed.

The presence or absence of a cation-depleted layer on mineral surfaces has also been a source of controversy in chemical kinetics studies of weathering rates. Nickel (1973) calculated a depleted layer thickness of several hundred angstroms, which was believed to be sufficiently thick to cause incongruent dissolution in the initial stages of weathering. Conversely, the measurements of Berner et al. (1980), Holdren and Berner (1979), and Schott and Berner (1985) using x-ray photoelectric spectroscopy (XPS) suggested a much thinner depleted layer. It was the contention of these later researchers that the cation-depleted layer was insufficiently thick to affect the rate of cation diffusion. For that reason, they suggested that weathering is a congruent process controlled by surface-reaction kinetics.

Similar conclusions were also reached by Berner et al. (1980), who found that linear kinetics explained pyroxene and amphibole dissolution when ultrareactive fine particles were removed with HF. The pyroxenes showed some cation depletion on their surfaces, but this altered surface layer extended to a depth of only a few angstroms and was not considered to be diffusion limiting. The dissolution rate of these minerals was reported to be controlled by surface chemical reactions.

Schott and Berner (1985) found similar dissolution kinetics for enstatite, diopside, and bronzite (Mg, Ca-Mg, and Mg-Fe pyroxenes, respectively) leached with a pH 6 solution. They reported incongruent release of Ca and Fe, relative to Si, from diopside and bronzite, respectively, followed by congruent dissolution after approximately 100 h. The preferential release of Ca and Fe was partly explained by weaker bonding energies of these cations relative to other cations in structurally dissimilar sites. The change to congruent dissolution over time was attributed to the formation of a cation-depleted surface layer of constant thickness. Dissolution of the outer surface of this layer was thought to occur at the same rate as diffusion of cations to the surface such that a steady-state thickness was maintained. Thus, the rate of diffusion from the unaltered mineral surface became slower, eventually equaling retreat of the outer layer. Although they observed a cation-depleted residual layer, the authors suggested that the concept of a classical diffusion gradient breaks down in this case because the surface layer was only one or two unit cells (10–20 angstroms) in thickness.

The more recent tendency in the literature to discard the hypothesis of diffusion control results from failure to identify a significant residual layer with modern spectroscopic methods. However, some researchers believe that this tendency may not be entirely justified. Wollast and Chou (1985) studied the dissolution kinetics of albite (Na-plagioclase) using a continuous flow-cell reaction chamber in contrast to the batch-type reaction chambers used in some previous studies (Wollast 1967). Because ions in solution are maintained at low levels in flow-cell reaction vessels, it is possible to rule out the effects of secondary precipitates as an influence on further dissolution. In this study, release rates of Si, Na, and Al over pH values ranging from 2.1 to 5.6 were initially very high, but decreased rapidly with time. Wollast and Chou concluded that dissolution resulted in the formation of a cation-depleted layer which increased in thickness over decreasing pH ranges. The release of cations continued nonstoichiometrically until a steady-state, congruent stage developed in which diffusion, hindered by buildup of the surface layer, became equal to retreat of the layer. The authors emphasized that this steady-state dissolution is the main process controlling feldspar weathering in the natural environment. As long as flushing rates are greater than reaction rates, secondary precipitates would not be expected to cause surface-reaction control of further dissolution.

In short, the proponents of surface-reaction control suggested that the rapid increase in Si reported in earlier research was an artifact resulting from the preferential dissolution of ultrafine and strained particles produced by sample grinding (Berner 1981, Holdren and Berner 1979, Petrovic et al. 1976). Those still contending that weathering rate is controlled by diffusion through a cation-depleted layer note that many of the studies conducted by proponents of surface-reaction control have initial incongruent stages (e.g., Berner et al. 1980, Schott and Berner 1985) and attribute the thinness of the measured depleted layer in those studies to crystallographic disturbances created by the HF pretreatment used to remove ultrafine particles (Wollast and Chou 1985).

It is probable that both surface-reaction and diffusion-controlled processes are governing mineral dissolution in natural weathering environments. Velbel (1984) collected samples of garnet from fresh rock, weathered saprolite, soil, and stream sediments from the North Carolina Piedmont. The samples from fresh rock and saprolite were taken from below the zone of biological activity and were found to have Fe and Al coatings on mineral grains. These coated samples were ultrasonically treated to clean their surfaces and examined with a scanning electron microscope. The treated garnets showed smooth unetched surfaces at the contact with the coatings, suggesting diffusion control with uniform attack by weathering agents over all parts of the garnet surface. Garnets from the soil environment lacked surface coatings and were "densely pocked with numerous well-formed etch pits" suggesting surface-reaction control. The absence of coatings was attributed to the presence of biologically produced organic chelating compounds in the soil. It should be noted that this process is probably most effective in surface soil horizons and may not be significant lower in the profile or in soils with low amounts of organic chelating agents. Nevertheless, this experiment illustrated that soil solution chemistry is determined largely by biological processes that must be accounted for if the results of laboratory weathering studies are to be used to estimate nutrient release rates in the field.

The controversy surrounding incongruent vs. congruent dissolution and surface-reaction vs. diffusion control of mineral weathering will probably be debated for some time to come. However, we believe that its overall significance to weathering studies is minimal because most mineral grains in natural weathering environments should be in the congruent weathering stage. Rate-controlling mechanisms governing mineral weathering are not easily identified in the laboratory, much less in heterogenous systems such as soils. Weathering rates and concentrations of solubilized elements are influenced by both surface-reaction rates and transport between microscopic regions (Skopp 1986). The effects of surface layers on the kinetics of weathering reactions has been a controversial topic that remains unresolved.

I. Topography

Topography influences mineral weathering through drainage, microclimate, and the transport of dissolved elements to lower areas of accumulation on the landscape (Birkeland 1984). Soil moisture storage capacity and leaching conditions are affected by landscape position as rates of surface runoff, infiltration rates, and proximity to a high water table vary considerably on hillslopes. Slope orientation also affects microclimate by influencing temperature and moisture availability, which in turn affect chemical reaction rates, amount and composition of the vegetative canopy, and the associated type and amount of organic decomposition products. Soils on southern exposures tend to have deeper profiles and higher clay contents than their counterparts on northern aspects because of higher ambient temperatures and accelerated weathering rates (Franzmeier et al. 1969, Losche et al. 1970). This effect of slope aspect is particularly pronounced at higher elevations and in steeply sloping areas (Daniels et al. 1987).

Soils formed in downslope depositional areas where nutrient recharge is significant often have a higher proportion of weatherable minerals when compared to higher landscape positions. This results, in part, from the influx of dissolved elements available for recrystallization, impeded drainage conditions, and higher ionic strength soil solutions.

J. Anthropogenic Effects

The effects of man as an influence on the chemistry of the weathering environment cannot be overlooked. Wet and dry deposition of anthropogenic sulfur and nitrogen oxides can cause substantial changes in soil solution chemistry, including reduced pH and increased concentrations of sulfate, nitrate, aluminum ions, and base cations. These changes can result in increased rates of base cation depletion from the soil exchange complex, alterations in soil organic matter pools, and increased potential for Al toxicity to plant roots and aquatic ecosystems (Reuss and Johnson 1986). The rate of base cation depletion depends on the magnitude of the soil anion exchange complex, the reservoir of exchangeable bases present in the soil, the rate of replenishing by weathering processes and biological recycling, and the level of deposition loading (Reuss and Johnson 1986). The rate of weathering, as we have seen previously, depends on numerous interacting physical and chemical factors, several of which may be altered directly or indirectly by atmospheric deposition. It is conceivable that weathering reaction rates would increase in response to acid inputs, thus perhaps compensating to some extent for deposition-induced accelerations in acidification processes.

Acidic deposition is, of course, only one of several anthropogenic influences on soil nutrient reserves. Mineral weathering processes are also important to the continued productivity of areas that have been subjected to intensive harvesting. Intensive harvesting of vegetation causes the export of base cations and disruption of nutrient cycling chains. Rapid biomass accumulation and low returns through litterfall during regrowth after harvest may cause further soil acidification. Intensive harvesting of forested tracts also may impact nutrient release via mineral weathering in woodlands by causing local climatic modification, including changes in soil temperature and moisture regimes. Erosion losses of both mineral and organic matter from soils during intensive harvesting can result in a significant lowering of soil CEC and thus the degree to which these soils are buffered against changes in mineral solubility and further cation losses by leaching. Soils inherently low in weatherable minerals are especially sensitive to the impact of intensive harvesting practices.

Neary et al. (1984) examined the effects of various harvesting practices on the nutrient status of pine sites in the southeastern United States. On a flatwoods site in Florida, harvesting of slash pine (*Pinus elliottii* Engelm. var. elliottii) removed substantial percentages of macronutrient supplies contained in aboveground biomass and the upper 20 cm of soil. Complete aboveground biomass harvesting removed 7% of total N, 36% of biomass + soil extractable (B + EX) P, 45% of B + EX K, 39% of B + EX Ca, and 26% of B + EX Mg. Stem removal alone depleted 4% of total N and 21%, 28%, 30%, and 19% of B + EX P, K, Ca, and Mg, respectively. Leaching and runoff losses of nutrients from clearcut pine sites were found to be short-term and relatively minor effects, but erosion was found to be equal to levels normally found only in cultivated fields. The largest effects on nutrient capital were found to result from windrowing and piling of residual biomass, where displacements of N, P, K, Ca, and Mg were four- to fivefold greater than removals associated with harvesting.

K. Relationship To Soil-Forming Factors

Processes of physical and chemical weathering that affect nutrient release rates in soils cannot be specifically equated with conceptual models of soil formation and therefore must be considered independently. The best-known of these models is the state-factor analysis of Jenny (1941), who attempted to develop quantitative relationships between soil properties and the processes responsible for their development. Jenny (1941) defined the functional relationship:

$$S = f(cl,o,r,p,t,...)$$

where

S = the state of the soil system (i.e., whole soil properties)
cl = climate
o = organisms
r = topography/relief
p = parent material
t = time.

The dots indicate additional unspecified factors. According to Jenny (1941), changes in a given soil property could be attributed to one state factor if all other state factors could be held constant. The soil-forming factors were thus assumed to vary independently. However, it is difficult to isolate the effects of any one state factor because they are all at least partially interdependent (Smeck et al. 1983). Observed soil-forming rates also may not be applicable in describing the current state of all soils because of constantly changing climatic conditions through time and because of the flux of materials in landscapes. Additionally, the level of detail at which biogeochemical processes are considered in Jenny's soil-forming factors is not adequate for the study of nutrient release rates by mineral weathering. Jenny's climate factor, for example, considered regional climatic variation (above the canopy), but did not take into account the potentially important microclimatic influences of vegetation, slope, and aspect on temperature and moisture regimes. Soil morphological features respond slowly to changing environmental conditions, but changes in the composition of the soil solution, which affect the adsorption and desorption of ions, mineral weathering, and nutrient release, occur much more rapidly. Future models designed to estimate nutrient release rates for particular forest stands will almost certainly require detailed site-specific information on a variety of vegetation, soil, hydrologic, and topographic variables.

IV. Methods of Estimating Release Rates

There are no known models for predicting absolute rates of nutrient release through weathering. A plethora of factors affecting weathering rate prevent easy calculation of absolute rates. Other factors being equal, relative mineral weathering rates can be proposed using solubilities as a model. In general, the methods that have been applied to the determination of mineral weatherability and weathering rates vary widely in approach. In the most general approach, mineral stability is assumed to be mirrored by the presence or absence of the mineral in rocks of increasing age. This conceptual method has been used to rank minerals according to relative weatherability but it cannot provide absolute weathering rates. Dissolution studies can provide weathering rates under specific laboratory conditions. Laboratory solution studies measure the effects of different chemical weathering agents under controlled conditions, but they are simplified and not easily applied to field conditions. Column leaching studies result in a closer approximation of weathering under field conditions. These studies more closely mimic soil leaching processes and allow for simulation of longer term weathering by increasing leaching rates.

The effects of plants on weathering can be determined in greenhouse studies where plants are grown in pots with one or more minerals as the sole source of nutrients. While these studies can generate important information on the weathering effects of plants, it is not convenient or even possible to simulate actual field conditions in the greenhouse. For example, it is impossible to imitate ecosystem diversity in a greenhouse study. In addition, greenhouse studies commonly use young plants rather than the older, more mature plants often found in the field. To properly determine field weathering rates, measurement in the field is essential. Each type of study has unique advantages, but no single study has overcome all the problems inherent with its application and design.

A. Conceptual Methods

Many methods have been proposed for estimating the degree of weathering or the potential for weathering. The earliest approach involved observation of the persistence of minerals in rocks of different ages. Pettijohn (1941) published the most widely accepted table of mineral persistence (Table 4.1). His results generally agree with later models based on more sophisticated methods (Brewer 1976) except for a few minerals. The reason for this misplacement of some minerals by Pettijohn (1941) is that the effects of diagenesis (i.e., mineral transformations caused by low temperature and pressure processes) on the mineralogy of sedimentary rocks was not recognized at the time of his study. Since the publishing of Pettijohn's table in 1941, many other authors have proposed their own sequences of relative stabilities based on mineral provenance data resulting in only minor changes to the original table (see Brewer 1976 for a good review).

Reiche (1943) took a chemical approach and devised a "weathering potential index" based on the chemical formulas of minerals. His empirical equation is:

$$\mathrm{WPI} = \frac{100 \times \text{moles}\ (\mathrm{Na_2O} + \mathrm{K_2O} + \mathrm{CaO} + \mathrm{MgO} - \mathrm{H_2O})}{\text{moles}\ (\mathrm{Na_2O} + \mathrm{K_2O} + \mathrm{CaO} + \mathrm{MgO} + \mathrm{SiO_2} + \mathrm{Al_2O_3} + \mathrm{Fe_2O_3})}$$

TABLE 4.1. Relative rates of mineral weathering as reported by Nickel (1973) compared to Pettijohn's (1941) persistence and Reiche's (1943) weathering potential.

Mineral[a] persistence	Weathering[b] potential	Sequences of decreasing dissolution stability			
		pH 0.2	pH 3.6	pH 5.6	pH 10.6
Muscovite	Muscovite	Quartz	Zircon, rutile	Zircon, rutile	Zircon, rutile
Rutile	Tourmaline				Almandine, staurolite
Zircon	(Rutile)	Rutile	Muscovite	Muscovite	Quartz
Tourmaline	Zircon	Disthene, muscovite	Quartz	Disthene	Hornblende
Almandine	Disthene	Tourmaline	Hornblende	Tourmaline, staurolite	Apatite
Apatite	Quartz	Zircon	Disthene	Quartz	Staurolite
Staurolite	Alamandine	Hornblende	Albite	Epidote, albite	Disthene
Disthene	Staurolite	Staurolite	Staurolite	Hornblende	Epidote
Epidote	Albite	Epidote	Tourmaline	Almandine	Hornblende
Hornblende	Epidote	Albite	Epidote	Apatite	
	hornblende	Almandine	Almandine		
	(apatite)	Apatite	Apatite		

[a]Selected from Pettijohn (1941) to be equivalent mineralogically to Nickel (1973)
[b]Calculated order using equation from Reiche (1943). For rutile, Ti assumed equivalent to but more stable than Si. For apatite, P assumed equal to Si.

where WPI is the weathering potential index of silicate minerals. This approach was based on the premise that high cation levels increase weatherability. The index is successful to some degree but fails for some mineral groups such as zeolites, possibly due to the different bonding mechanism for zeolitic water in this mineral group.

B. Laboratory Studies

Laboratory studies of mineral weathering can be divided into two main classes, that is, those using inorganic ions and those using organic ions. The inorganic ion studies can be subdivided into those involving (1) acid or base dissolution, (2) catalysis by exchangeable cations, and (3) the hydrolysis of minerals by water leaching.

Nickel (1973) performed the most complete acid-base study of the weathering rates of common minerals of sand and silt size. He determined the weathering rates of albite, muscovite, quartz, apatite, almandine, disthene, epidote, hornblende, rutile, staurolite, tourmaline, and zircon at pH 0.2, 3.6, 5.6 and 10.6. All minerals initially dissolved in an incongruent manner (nonstoichiometrically), but with time all reactions except those of phyllosilicates approached stoichiometric proportions. Reaction rates were found to be much faster during the first incongruent reaction stage (usually lasting about 5 days) than in the subsequent congruent reaction stage. His results indicated that ferromagnesian and aluminosilicate minerals released Si faster and at higher concentrations than quartz. He proposed that this resulted from dissolution of other components from the lattice.

Nickel also reported that relative mineral dissolution rates were not constant over the wide range in pH used in the study, indicating that dissolution rates of minerals were determined by the solubility of their individual components as influenced by pH. These findings indicate that laboratory dissolution studies using acid concentrations higher than those found in natural environments may lead to an overestimation of weathering rates. There is generally good agreement between Nickel's (1973) relative solubility study, Pettijohn's (1941) persistence index, and Reiche's (1943) weathering potential index (see Table 4.1). However, large differences in relative weatherability are observed when effects of pH are taken into account. Many minerals such as zircon, tourmaline, albite, and apatite show large changes in stability as pH changes while others, such as rutile, are little affected by pH.

In other acid dissolution studies, Gilkes et al. (1973a,b) examined the weathering rate of oxidized biotite in 0.1, 0.01, and 0.001 M HCl as affected by the degree of oxidation and K content of the solution. Oxidation of structural Fe^{2+} resulted in increased resistance of biotite to alteration. Dissolution was faster as the acid concentration increased and was severely depressed by the presence of high solution K, suggesting vermiculite as an intermediate phase in the dissolution of the oxidized biotite.

The effects of high salt contents on dissolution rates of phyllosilicates have been investigated. Phyllosilicate weathering is sensitive to salt content because the interlayer cation is exchangeable with salts in solution. Studies using micas have shown that trioctahedral members exchange interlayer cations more easily than dioctahedral members. The release of K from oxidized biotite by NaCl has been studied by Gilkes et al. (1973a). They found that K release decreased as the fraction of Fe^{2+} oxidized increased. The more highly oxidized biotites were also found to retain x-ray diffraction characteristics of biotite at levels of K loss where less oxidized biotites altered to vermiculites.

Robert (1973) performed a similar study on biotite, glauconite, illite, sericite, and muscovite. He found that expansion of biotite by NaCl resulted in oxidation of some structural Fe^{2+}. The mineral to which a mica transformed depended on the charge location. Octahedrally charged micas altered to high-charged smectites or low-charged vermiculites depending upon the magnitude of layer charge. Vermiculite was found to have resulted from the alteration of tetrahedrally charged micas. Abudelgawad et al. (1975) and El-Amamy et al. (1982) also studied the alteration of micas to vermiculite by K release to NaCl solutions. They found that most of the K was removed from glauconite in the initial 40 to 80 h with solution K concentration becoming constant after 80–160 h. Smaller particle-size fractions released K faster, in contrast to the findings of von Reichenbach and Rich (1969), wherein small particle size resulted in greater total release with other factors remaining constant.

Changes of soil water chemistry accompanying irrigation of saline and sodic soils have been extensively studied to determine the consequences of ion exchange on clay mineral weathering. Data from such studies are applicable to changes in the soil exchange complex accompanying acid deposition since both result in cation depletion. These studies have focused on the effects of cation hydrolysis resulting from leaching of salts from saline soils. The general result of leaching homoionic clays has been the decomposition of the clays. When the solution concentration of exchangeable ions becomes low enough, H^+ present in solution becomes the preferred exchangeable ion. Exchangeable H^+ penetrates into the structure and displaces octahedral Al and Mg, which become adsorbed on exchange sites (Frenkel and Suarez 1977). In CO_2-free systems, a linear relationship between electrical conductivity and the square root of time suggests diffusion control for the reaction. The rate-limiting step for Ca-montmorillonite appears to be Ca-H exchange (Frenkel and Suarez 1977).

A similar study by Oster and Shainberg (1979) resulted in two linear segments in a plot of electrical conductivity vs. the square root of time. They suggested that the two linear segments resulted from the presence of $CaCO_3$. They concluded that the rate of Ca, Mg, and K release from silicates, and the hydrolysis of exchangeable Na and Ca occurred at sufficiently rapid rates to prevent the preparation of homoionic Na soils. Suarez and Frenkel (1981) subjected Na- and Ca-saturated samples from two soils and three standard clays to leaching in an attempt to determine whether response to cation hydrolysis was related to the cation originally on the exchange complex. They found that the saturating ion affected reaction rate with Na-saturated samples reacting fastest, but differences among samples were greater than those from cation saturation.

Frenkel et al. (1983) determined the effects of exchangeable cations on hydrolysis rates by mixing Na-, Ca-, and Mg-saturated montmorillonitic and kaolinitic soil samples to different exchangeable Na percentages (ESP). They found that reaction rates were approximately equivalent for Ca- and Mg-saturated samples. Intermediate saturations reacted slower than Na-saturated samples but faster than the Ca- or Mg-saturated samples. Frenkel et al. considered the intercept of a plot of the electrical conductivity vs. the square root of time to be a measure of the relative contribution of the first reaction to the overall weathering process. Suspension percentage effects and the straight-line portion of the curves implied diffusion-controlled reaction rates. They proposed a two-step mechanism to describe mineral weathering by hydrolysis. The first step involved a rapid cation exchange reaction that depended on the nature of the exchangeable cations, CEC, dominant type of weatherable minerals, and solution CO_2 levels. A slow, diffusion-controlled reaction, independent of exchangeable cations that followed was surface area and CO_2 dependent. This reaction was probably the interaction of exchangeable H^+ with the octahedral layer, resulting in release of octahedral cations.

The effects of organic chemicals and soil organic matter extracts on mineral dissolution rate have been extensively studied because of the marked effect of plants on weathering rates. Huang and Keller (1970) studied the weathering of olivine, augite, muscovite, labradorite, and microcline by deionized water, CO_2- charged water, and four organic acids (acetic, aspartic, salicylic, and tartaric). All dissolution reactions proceeded rapidly for the first 24 h and then slowed to near constant rates after 5 to 21 days. The order of concentration of cations dissolved from the minerals was independent of the acid solution used, and was affected only by the cation concentration in the original mineral. The organic acids acted as solution pH buffers that prevented the rapid pH changes observed in unbuffered systems. Chelation of Fe and Al by organic acids caused higher equilibrium concentrations of other cations and Si in solution. Huang and Keller concluded that weathering with organic acids may result in a different order of mineral stability than the traditional Goldich (1938) sequence, which otherwise agrees with observed stability with respect to inorganic dissolution. The dissolution reactions were either congruent or incongruent depending upon the mineral and solution used. They made no weatherability comparisons between minerals.

Iskandar and Syers (1972) studied the bulk dissolution characteristics of biotite and basalt using solutions containing synthetic lichen complexes and other organic chelates. Lichen complexes invariably released more Ca than Mg, Fe, or Al from the silicates. The reactions were largely the result of metal complexation (chelation) rather than acid reactions. Citric, salicylic, P-hydroxybenzoic acids, and EDTA were more efficient at cation removal than lichen compounds.

Berthelin and Belgy (1979) studied microbial degradation of two granites and a granitic sand. When they compared the results of 22-week perfusion studies for sterile and unsterilized systems, they found that microbes dissolved large amounts of mineral elements, solubilized much of the ferromagnesian minerals (mainly biotite), and destroyed primary chlorite and vermiculite. Biotite weathered through an intermediate vermiculitic phase. Berthelin and Belgy concluded that mineral transformations were similar to field observations and that their methods could be used to simulate natural weathering processes.

Tan (1980) studied the decomposition of microcline, biotite, and muscovite by soil humic and fulvic acids at pH 7 and 2.5. Humic and fulvic acids at pH 7.0 resulted in much less mineral dissolution than had been reported by other authors (e.g., Huang and Keller 1970). Biotite was the most easily dissolved followed by microcline and muscovite. Dissolution was greatly enhanced by lowering pH values to 2.5 because of interaction between low pH and organic acid groups. X-ray diffraction analysis showed the microcline to actually be a mixture of microcline and orthoclase with the orthoclase selectively dissolved in these treatments. Generally, all treatments showed greater mineral dissolution than water solutions at the same pH.

Direct comparison of measured laboratory kinetic rates is greatly hindered by the various kinetic equations used to fit experimental data and differing experimental methods used to provide the kinetic data. Rate constants for several dissolution studies (Table 4.2) were calculated from the authors' rates where they had not been given, and units have been converted to moles, centimeters, and seconds where required.

Wollast (1967) considered the data for orthoclase feldspar dissolution to be best fit by an equation that was both first and three-halves order with respect to time. His work also suggested that the rate of Si loss was concentration- and surface area dependent, and he therefore reported dissolution rates on that basis with a solubility term. The data also showed a definite pH dependence that was not included in his rate equation. Luce et al. (1972) considered that their data for enstatite, forsterite, and serpentine dissolution showed an early period of fast dissolution followed by a period of slower dissolution that was one-half order with respect to time. The kinetic equation for their data therefore included a term, Q_0, which represented the rapid initial reaction. However, they did not report a concentration dependence and therefore expressed their rate constants in terms of moles dissolved per square centimeter surface per square root of time. Later work has generally considered the second slower reaction to be of geochemical significance. Busenberg and Clemency (1976) and Lin and Clemency (1981) considered their feldspar and muscovite dissolution data for the reaction between 100 and 1200 h to be first order with respect to time. Grandstaff (1986) also considered his olivine dissolution data to be first order with respect to time.

The methods used in the study of dissolution kinetics vary at least as much as the equations used to fit the data. Wollast (1967) used a batch-type system with pH controlled under continuous agitation. This experimental design was also used by Luce et al. (1972) with the exception that their batch reactors were stirred rather than shaken. Whereas the earlier mentioned studies were carried out with pH fixed by addition of an acid or base, the studies of Busenberg and Clemency (1976) and Lin and Clemency (1981) were carried out at basic pH values held constant by bubbled CO_2 gas with the reactors vigorously stirred. Grandstaff (1986) considered the ionic strength to be more important in the dissolution rate than did earlier researchers, and for that reason all his forsterite dissolution experiments were carried out with the ionic strength controlled by KCl, a neutral salt. He studied the effects of various organic ligands, pH,

TABLE 4.2. Measured weathering rates from laboratory studies.

A. Wollast (1967) applied $C = (3/2)k_3t^{3/2} - k_2(C - C_e)t$, to dissolution data, where C_e represents the equilibrium concentration found by graphical methods to be 5 mg/l as SiO_2. The experiments were performed in a batch type system using finely ground orthoclase in the pH range of 4–10 with continuous agitation.

pH	k_2 ($cm^{-2}s^{-1}$)	k_3 (mol $l^{-1}cm^{-2}s^{-3/2}$)
	Orthoclase	
4	1.4×10^{-11}	2.0×10^{-14}
6	1.6×10^{-11}	4.9×10^{-16}
8	2.6×10^{-10}	4.3×10^{-15}
10	6.9×10^{-11}	5.7×10^{-15}

B. Luce, Bartlett, and Parks (1972) fitted dissolution data to $Q_{Mg} = k_{Mg}t^{1/2} + Q_{Mg_0}$ and $Q_{Si} = k_{Si}t^{1/2} + Q_{Si_0}$, where Q = moles/cm² extracted, and Mg_0 and Si_0 = moles/cm² which are rapidly extracted. Experiments were conducted using stirred batch methods.

Initial pH	Q_{Mg_0}	Q_{Si_0}	k_{Mg}	k_{Si}
	(moles cm^{-2})		(moles $cm^{-2}sec^{-1/2}$)	
		Enstatite		
3.2	4.9×10^{-9}	7.3×10^{-9}	7.8×10^{-11}	2.4×10^{-11}
5.0	2.3×10^{-9}	3.0×10^{-9}	2.6×10^{-11}	2.6×10^{-11}
7.0	9.0×10^{-11}	2.0×10^{-10}	1.6×10^{-11}	1.4×10^{-11}
9.6	1.0×10^{-9}	5.7×10^{-10}	1.7×10^{-11}	2.4×10^{-11}
		Forsterite		
3.2	2.0×10^{-8}	5.0×10^{-9}	1.8×10^{-9}	2.0×10^{-10}
5.0	9.0×10^{-9}	2.0×10^{-9}	6.1×10^{-10}	3.6×10^{-11}
7.0	2.0×10^{-9}	7.0×10^{-10}	1.2×10^{-9}	2.9×10^{-11}
9.6	4.0×10^{-9}	5.8×10^{-10}	6.7×10^{-10}	3.1×10^{-12}
		Serpentine		
3.2	2.0×10^{-8}	0.0	2.7×10^{-10}	6.3×10^{-11}
5.0	5.8×10^{-9}	0.0	2.2×10^{-11}	6.8×10^{-11}
7.0	3.4×10^{-9}	2.0×10^{-10}	1.9×10^{-11}	6.8×10^{-11}
9.6	1.0×10^{-9}	2.0×10^{-10}	1.4×10^{-11}	7.0×10^{-12}

C. Busenberg and Clemency (1976) fitted data to $C = k_1t + B$, where B is the intercept. Note: All curves showed an initial parabolic region. Rates were determined for stirred batch kinetics at long time periods (most points were at times of 100–1200 h) to reduce the effects of parabolic kinetics.

	k_1 (moles $cm^{-2}sec^{-1}$)			
Feldspar	Na	Ca	K	Si
Albite	2.54×10^{-15}	†	†	4.40×10^{-16}
Oligoclase	1.67×10^{-15}	8.89×10^{-17}	†	2.60×10^{-16}
Andesine	9.75×10^{-17}	†	†	1.40×10^{-16}
Labradorite	7.16×10^{-17}	1.09×10^{-15}	†	1.26×10^{-16}
Bytownite	5.38×10^{-17}	1.29×10^{-17}	†	1.07×10^{-16}
Anorthite	3.24×10^{-17}	2.18×10^{-16}	†	1.32×10^{-16}
Orthoclase	6.64×10^{-17}	†	2.70×10^{-16}	1.67×10^{-16}
Microcline	5.46×10^{-17}	†	1.27×10^{-16}	1.52×10^{-16}

continued

TABLE 4.2. *Continued*

D. Lin and Clemency (1981) fitted data to $C = k_1t + B$, where B is the intercept. Note: All curves showed an initial parabolic region. Rates were determined for stired batch kinetics at long time periods (most points were at times of 100–1200 h) to reduce the effects of parabolic kinetics. Dissolution was incongruent in both cases.

Muscovite	k_{Si} (moles $cm^{-2}sec^{-1}$)
Green mica	2.7×10^{-17}
Ruby mica	2.4×10^{-17}

E. Grandstaff (1986) fitted data to $R_1 = dQ_1/dt = j_1k_1$, where j_1 is the stoichiometric coefficient. Dissolution experiments were carried out using batch techniques at constant ionic strength with pH held constant using pH-stat techniques.

			k_1 (moles $cm^{-2}s^{-1}$)		
pH	T (°C)	Solution composition	Mg	Fe	Si
				Forsterite	
3.48	49	KCl	3.09×10^{-14}	2.40×10^{-14}	1.41×10^{-14}
3.48	26	KCl	1.17×10^{-14}	1.05×10^{-14}	5.25×10^{-15}
3.48	1	KCl	2.75×10^{-15}	1.91×10^{-15}	1.12×10^{-15}
3.60	26	KCl + 10^{-3} M oxalate	6.76×10^{-14}	4.90×10^{-14}	2.63×10^{-14}
4.50	26	KCl + 10^{-3} M oxalate	1.54×10^{-14}	1.51×10^{-14}	6.92×10^{-15}
4.50	26	KCl + 10^{-3} M succinate	2.57×10^{-15}	2.40×10^{-15}	1.23×10^{-15}
4.50	26	KCl + 10^{-5} M phthalate	7.76×10^{-16}	4.27×10^{-16}	‡
4.50	26	KCl + 10^{-4} M phthalate	1.05×10^{-15}	‡	5.25×10^{-16}
4.50	26	KCl + 10^{-3} M phthalate	1.41×10^{-15}	7.94×10^{-16}	6.17×10^{-16}
4.50	26	KCl + 10^{-2} M phthalate	5.13×10^{-15}	‡	2.63×10^{-15}
4.50	26	KCl + $10^{-1.5}$ M phthalate	9.77×10^{-15}	6.31×10^{-15}	‡
4.50	26	KCl + 10^{-1} M phthalate	2.40×10^{-14}	1.51×10^{-14}	1.20×10^{-14}
4.50	26	KCl + 10^{-1} M acetate	9.77×10^{-16}	8.32×10^{-16}	‡
4.50	26	KCl + 10^{-2} M acetate	6.76×10^{-16}	‡	3.39×10^{-16}
4.50	26	KCl + 0.10 g fulvic acid	8.32×10^{-15}	4.68×10^{-15}	‡
4.50	26	KCl + 0.01 g fulvic acid	3.16×10^{-15}	1.51×10^{-15}	‡
4.50	26	KCl + 10^{-3} M citrate	2.00×10^{-14}	2.95×10^{-14}	9.55×10^{-15}
4.50	26	KCl + 10^{-3} M tannic acid	1.58×10^{-14}	1.12×10^{-14}	‡
4.50	26	KCl + 10^{-3} M oxalate	2.40×10^{-13}	2.14×10^{-13}	1.12×10^{-13}
4.50	26	KCl + 10^{-5} M EDTA	4.07×10^{-15}	2.51×10^{-15}	1.51×10^{-15}
4.50	26	KCl + 10^{-4} M EDTA	7.76×10^{-15}	5.50×10^{-15}	3.89×10^{-15}
4.50	26	KCl + 10^{-3} M EDTA	2.45×10^{-14}	‡	1.02×10^{-14}
4.50	26	KCl + 10^{-2} M EDTA	7.24×10^{-14}	5.25×10^{-14}	3.31×10^{-14}
3.50	26	KCl + 10^{-3} M EDTA	6.46×10^{-14}	6.61×10^{-14}	3.16×10^{-14}
3.00	26	KCl	2.40×10^{-14}	1.86×10^{-14}	1.32×10^{-14}
3.20	26	KCl	1.70×10^{-14}	1.41×10^{-14}	7.94×10^{-15}
4.00	26	KCl	2.75×10^{-15}	2.00×10^{-15}	1.23×10^{-15}
4.50	26	KCl	6.46×10^{-16}	5.01×10^{-16}	2.95×10^{-16}
5.00	26	KCl	1.78×10^{-16}	1.20×10^{-15}	8.71×10^{-17}

F. Bloom and Erich (1987) fitted data to Rate $= k_1(H^+)^n + k_0$, and with n found to be equal to 0 or 1 in all cases. The batch reactions were performed with either anion concentration or pH set constant at 298 K, 313 K, and 328 K with continuous agitation. The rate constants were found to be strongly affected by the anion present and the anion concentration.

Anion	k_0 (mol $cm^{-2}s^{-1}$)	k_1 (mol $cm^{-2}s^{-1}$)
0.1 M NO_3^-	1.4×10^{-13}	1.0×10^{-16}
0.1 M SO_4^{2-}	0.0	1.7×10^{-10}
10^{-4} M PO_4^{3-}	2.4×10^{-14}	0.0

†Kinetic rate not determined because either too low in concentration or not present in the mineral specimen.
‡Not determined.

and temperature on dissolution kinetics. He did not, however, include any of these factors in his rate equation. Bloom and Erich (1987) also noted that the ionic strength was important and so repeated the earlier work of Bloom (1983) on gibbsite dissolution with ionic strength controlled by NO_3^-, SO_4^{2-}, and PO_4^{3-}. They found that instead of obtaining odd orders for kinetic rate with respect to H^+, the reaction rate was a mixture of zero and first order. They also found that the rate was dependent on the concentration and identity of the anion controlling ionic strength.

Considering experimental variation, one might expect some differences between studies of the amount of dissolution predicted for different reaction times. Comparison of predicted dissolution vs. time for those minerals studied by more than one author shows that this is indeed the case (Table 4.3). Comparison of predicted orthoclase dissolution between the pH 10 data of Wollast (1967) and that predicted by the equation of Busenberg and Clemency (1976) indicates that the weathering rates are never within one order of magnitude of each other, and with long times differ by more than four orders of magnitude. Comparison of rates of forsterite dissolution predicted by the equations of Luce et al. (1972) and those of Grandstaff (1986) shows differences of nearly this same magnitude at short times, and dissolution over long time intervals differing by as much

TABLE 4.3. Comparison of Si and cation release predictions for selected references.

	Orthoclase Si loss (moles cm^{-2})	
Time (sec)	Wollast (1967)	Busenberg and Clemency (1976)
10^{-2}	1.00×10^{-12}	1.67×10^{-14}
10^{-4}	4.85×10^{-10}	1.67×10^{-12}
10^{-6}	4.33×10^{-7}	1.67×10^{-10}
10^{-8}	4.28×10^{-4}	1.67×10^{-8}

	Forsterite Mg and Si loss (moles cm^{-1})			
	Luce et al. (1972)		Grandstaff (1986)	
Time (sec)	Mg	Si	Mg	Si
	pH 3.48 or 3.22			
10^{-2}	3.8×10^{-8}	7.0×10^{-9}	2.47×10^{-12}	1.41×10^{-12}
10^{-4}	2.0×10^{-7}	2.5×10^{-8}	2.47×10^{-10}	1.41×10^{-10}
10^{-6}	1.8×10^{-6}	2.0×10^{-7}	2.47×10^{-8}	1.41×10^{-8}
10^{-8}	1.8×10^{-5}	2.0×10^{-6}	2.47×10^{-6}	1.41×10^{-6}
	pH 5.00			
10^{-2}	1.5×10^{-8}	2.4×10^{-9}	1.42×10^{-14}	8.71×10^{-15}
10^{-4}	7.0×10^{-8}	5.6×10^{-9}	1.42×10^{-12}	8.71×10^{-13}
10^{-6}	6.2×10^{-7}	3.8×10^{-8}	1.42×10^{-10}	8.71×10^{-11}
10^{-8}	6.1×10^{-6}	3.6×10^{-7}	1.42×10^{-8}	8.71×10^{-9}

as two-to threefold. Considering that these experiments were much closer to mimicking each other than they were to mimicking field weathering processes, the application of laboratory kinetic equations to field situations should be considered risky at best. This problem arises from not knowing the total mineral surface area per hectare, the inability to describe the relative weathering ages (for the time term in the equations) for a grain population of differing particle sizes, and differing amounts of time minerals are in contact with weathering solutions.

C. Column Studies of Mineral Weathering

Column studies of mineral dissolution are performed on reconstituted (dried, ground, and mixed) or intact soil that has been packed into a cylinder. These columns are leached either by gravity flow or by pumped leachate solutions. Gravity flow studies have the advantage that the rate of leaching is close to that observed under field conditions. In addition, most such studies include periods of reduced water content between leaching events, similar to soils under natural conditions. Problems arise in the kinetic modeling of gravity flow leaching studies because the rate of solution flow through the columns and the soil water content are not constant. These problems, however, are alleviated in columns that are leached by pumped solutions. As a result, kinetic evaluation of mineral reactions are possible using computer models (such as Parker and van Genuchten 1984).

Column leaching studies may appear to be similar to laboratory solution studies, but there are important differences. Laboratory solution studies are conducted with a high solution:soil ratio allowing for the assumption that the chemical activity of water equals one. This assumption is not valid for column leaching studies because the activity of water in a soil column is related to the pore size in which the water is flowing and to the water content of the soil during drier periods between leaching events (Tardy 1982). For these reasons, weathering rates determined in column studies often differ from results obtained with similar laboratory solution studies. The effects of wetting and drying cycles, an aspect of environmental geochemistry that is often underemphasized, can also be examined using column leaching studies.

Most of the column leaching studies published to date have not been concerned with mineral weathering as such, but were more oriented toward elucidating processes that modify leachate chemistry. Some discussion of these studies is warranted because these processes affect mineral weathering and nutrient release rates.

Adams and Boyle (1979) examined the effect of organic acids on mineral weathering rates. They used aspen (Populus grandidentata Michx.) leaf extracts to leach cations from A horizons of three Spodosols. Columns containing the soils were leached with aspen extracts until 1650 ml of leachate (approximately equal to 1 year's rainfall) had leached through the column. All soils released Mg, K, Na, Fe, and sometimes Ca. The annual rate of A horizon weathering, as indicated by the cations present in the leachate, was calculated to be as much as 1 kg ha^{-1} yr^{-1} of Ca and Mg, 1 to 3 kg ha^{-1} yr^{-1} of K, 4 to 6 kg ha^{-1} yr^{-1} of Na, and 2 to 6 kg ha^{-1} yr^{-1} of Fe. These results are generally lower than those observed under similar field conditions at the Hubbard Brook Experimental Forest by Likens et al. (1977).

Krug and Isaacson (1984) compared the leachate chemistry of acid forest soil columns that were leached with distilled- deionized water and with pH 3.0 H_2SO_4 solutions. They found that the higher release rate of Al from the organic-rich horizons when leached with acid was related to the soluble organic matter released. They concluded that chemical processes other than simple buffering and ion exchange reactions acted to consume acidity in the soil. These results agree with the column leaching data of Evans (1986), which suggested that soluble organic matter and SO_4^{2-} compete for adsorption sites on soil colloids.

Zelazny and Evans (1985) determined the influence of simulated acid precipitation with ambient or near ambient SO_4^{2-} and NO_3^- levels on the 1 M KCl exchangeable Al of a southeastern Piedmont forest soil. In this study, the soil was gravity leached with two pore volumes of solution. The results indicated that 1 M KCl exchangeable Al levels increased in all horizons when leached with the simulated acid precipitation, with the largest increase occurring at intermediate levels of SO_4^{2-} and NO_3^-. They postulated that the amount of exchangeable Al in the soil was controlled by complex interactions with SO_4^{2-} and NO_3^-, and Al chelation by the dissolved organic carbon (DOC).

Evans (1986) compared the ability of SO_4^{2-} and DOC to mobilize Al from a southeastern Piedmont forest soil in a gravity-fed column leaching study. The amount of Al mobilized by the leaching solution was of the order DOC $>$ (SO_4^{2-} + DOC) $>$ SO_4^{2-}. The number of pore volumes required for Al to appear in the eluent decreased as the DOC content of the leachate increased. Leaching the soil with SO_4^{2-} after DOC was found to reduce Al mobilization. Evans concluded that SO_4^{2-} competed with DOC for adsorption sites in the soil. In a subsequent study, Evans and Zelazny (1987) concluded from the geochemical data obtained in a gravity leaching study of this same soil, that decreases in exchangeable Al and SO_4^{2-} in the Bt1 horizon could be attributed to the precipitation of the SO_4^{2-} mineral basaluminite. Dissolved organic carbon in the leaching solution was postulated to result in complex ion formation and increased exchangeable Al levels by reducing the rate of Al precipitation.

Jardine et al. (1985a,b) leached columns of poorly crystalline kaolinite, Cheto montmorillonite, and an organic soil with acidic, dilute Al solutions to examine the interactions of Al with surfaces. In these studies, the Al concentration in the leachate was monitored continuously and modeled using the transport model of Parker and van Genuchten (1984). Aluminum in the leachate of the pumped solution was found to be best modeled by inclusion of two types of adsorption processes: one instantaneous and reversible, and the other governed by first-order kinetics. They determined that the surface of kaolinite induced increased polymerization and hydrolysis of solution Al and that the quantity of adsorbed Al was greater than the measured CEC. They concluded that the fast reaction involved exchange of Al for the Ca previously adsorbed on the cation exchange complex while the slow reaction involved Al polymerization on the kaolinite surface.

Cronan (1985) investigated the effects of soil particle size and permeability of selected northern forest soils by leaching intact soil columns. He found that the rate of nutrient release was inversely related to soil particle size and directly related to the base saturation, with large differences between adjacent soil horizons. Cronan also observed increased nutrient release rates with increasing acidity in the simulated rainfall for two soils. Soil characterization data before and after leaching were not presented to determine whether nutrient release was the result of mineral weathering or changes in ion exchange base status, but this research is a good example of the results obtainable with column leaching studies.

D. Greenhouse Studies of Mineral Weathering by Living Plants

To further approach field conditions, some researchers have studied the weathering of minerals by growing plants in experimental systems. The use of pot studies in greenhouses is advantageous because influxes and effluxes are more strictly controlled than in the field.

Spyridakis et al. (1967) grew coniferous and deciduous trees for 13 months in sand cultures with biotite as the only source of K and Mg. All seedlings altered biotite to kaolinite, except those of monterey pine (*Pinus radiata* D. Don), which altered the biotite to vermiculite. The effectiveness of kaolinite production from biotite was of the order: white cedar (*Thuga occidentalis* L.) > hemlock (*Tsuga canadensis* L.) > white pine (*Pinus strobulus* L.) > white spruce (*Picea glauca* (Moench) Voss) > red oak (*Quercus rubra* L.) > hard maple (*Acer saccharum* Marsh.). Actual amounts of nutrient uptake were not available from the published results, but the white cedar growth resulted in a transformation of the biotite to a 50:50 biotite:kaolinite mixture. If one assumes 10% K_2O in the original biotite with 2.0 g biotite per pot, then the white cedar tree used 0.1 g K_2O from biotite during the course of the experiment.

Gilkes and Young (1974a) looked at oxidized biotite as a source of K for subterranean clover. Potassium uptake was found to decrease with increasing degree of oxidation of structural Fe^{2+}. This is in agreement with their earlier leaching studies using these same biotites (Gilkes and

Young 1974b). No changes in x-ray diffraction characteristics from weathering were observed. Six harvests of the clover removed only 2% to 6% of the total K initially present. About three times as much K was removed from the least oxidized biotite than from the most oxidized sample.

A similar study was conducted by Mojallali and Weed (1978), who compared biotite with muscovite as sources for plant K by mycorrhizal soybean plants. Mycorrhizae more than doubled the rate of weathering of biotites but did not significantly increase muscovite weathering. These studies suggest that plants can grow with biotite as the only source of K.

E. Field Studies

Mass-balance studies of watersheds can provide the means to determine weathering rates over relatively large areas. In these studies, the chemistry and volume of water leaving a drainage basin and measurements of nutrient inputs by precipitation are used to quantify rates of chemical weathering in the catchment. It is assumed that the amount of dissolved load in surface water is proportional to the rate of chemical weathering. This type of information is most often used to determine rates of soil and saprolite formation and to calculate rates of surface lowering and other geomorphic parameters. However, many studies of this type have attempted to determine the contribution of mineral weathering to the nutrient supply of plants. At first glance, this would seem feasible because the original source for almost all nutrients in an unfertilized residual soil is mineral weathering and atmospheric deposition. However, the chemistry of water leaving old, highly weathered catchments in streamwater may be more indicative of weathering deep in the saprolite rather than in the zone of biological uptake (Velbel 1984, 1985, 1986).

Quantitative estimates of weathering rates are complicated by sampling and analytical errors as well as a high degree of spatial and temporal variation within all levels of biogeochemical systems (Clayton 1979). It may be impractical to adequately model all the complex interactions existing in natural weathering environments (Figure 4.2). This complexity causes published applications of mass-balance studies to vary widely in approaches, assumptions, and attention to detail. As a result, generalizations drawn from these studies may be somewhat difficult to interpret.

Detailed characterization of research areas should be an integral part of any mass-balance study. This type of information should include documentation of soil and geological spatial variability (including the nature and amount of primary and secondary soil minerals) and groundwater hydrology (including water pathways within the watershed and streamflow variability throughout the year).

Most valid models can be simplified to the mass-balance equation (modified from Clayton 1979):

$$RW = (\text{Efflux} - \text{Influx}) + \Delta(P + S)$$

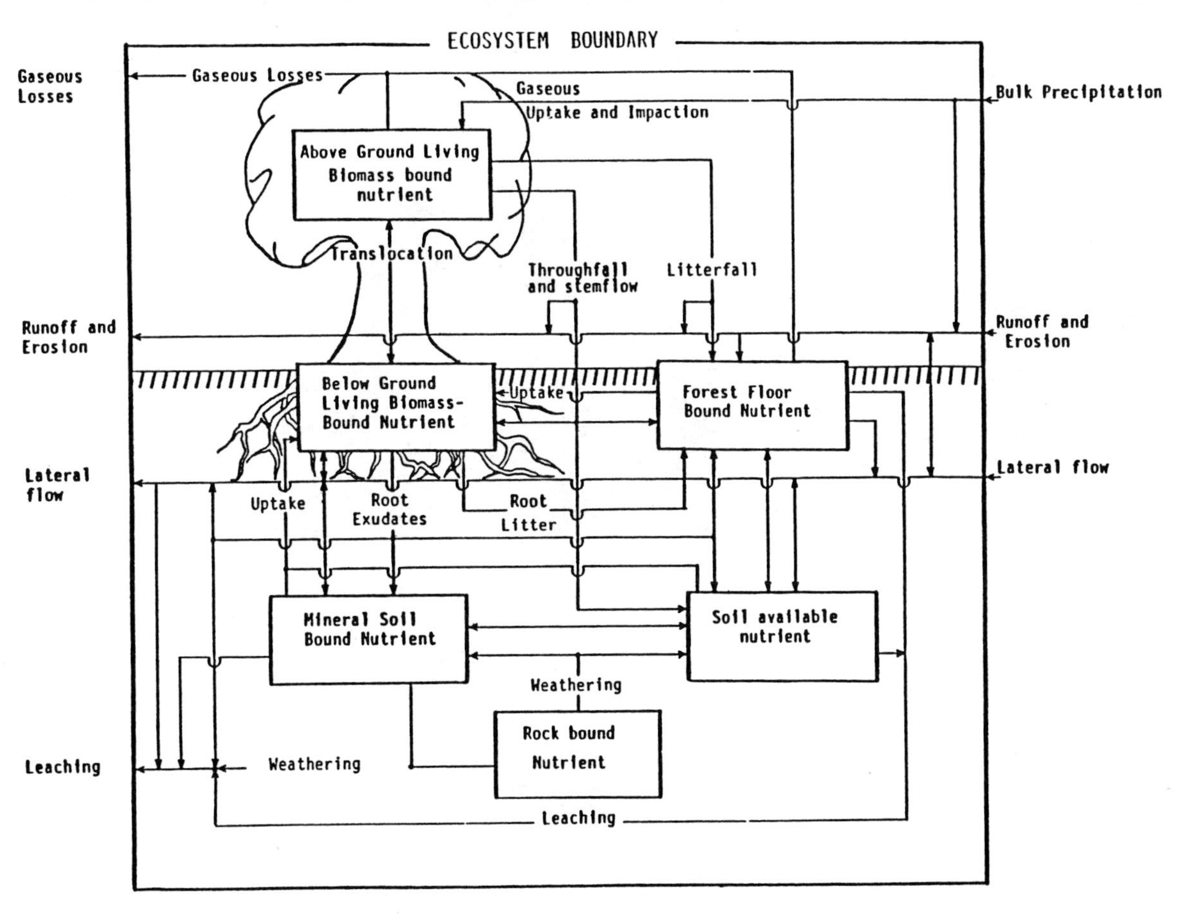

FIGURE 4.2. Mineral dynamics in weathering environments (modified after Likens et al. 1977).

where

RW = release due to weathering
Efflux = nutrient loss by leaching or erosion
Influx = nutrient input by precipitation
$\Delta(P + S)$ = net change in nutrient storage by plant and soil pools, excluding nutrients in primary minerals.

This equation represents a method for obtaining a good first approximation of actual mineral weathering rates, but confounding factors can lead to erroneous estimates. For example, the most fundamental consideration in any watershed study is that all sources of inputs and outputs are identified. It is therefore necessary to have a clearly defined catchment where streams used for measurement of effluxes are the only paths for removal of drainage water (Creasey et al. 1986). The catchment must be watertight with respect to both drainage and extraneous inputs. Losses of dissolved ionic constituents through deep seepage are commonly assumed to be negligible, for example, but may be significant in areas underlain by permeable sediments.

Basin tightness is most commonly checked through use of a Cl^- budget of the watershed (Clayton 1986). Assuming negligible Cl^- in the bedrock and steady-state biological cycling, the Cl^- inputs in precipitation should equal Cl^- outputs. With proof of watershed tightness, a drainage basin may then be used to determine weathering rates, assuming that influxes, nutrient storages, and effluxes can be measured.

Assuming no inputs by man, the most important source of nutrient influx into a watershed is precipitation. Accurate determination of precipitation inputs is therefore essential for predicting streamflow (Osbourne et al. 1982) and nutrient cycling in a watershed. Errors in the determination of precipitation inputs may arise from small intense thunderstorms of very local areal extent. Snow alters short-term water retention and release, and frost may also be a significant source of water (Osbourne et al. 1982). Dry deposition of elements is highly significant in some regions, especially in forests where the filtering action of tree canopies can substantially increase deposition rates (Mayer and Ulrich 1974).

Precipitation, although the immediate source of all water entering a tight watershed, may not be indicative of the total amount of water available for mineral weathering. Rather, the amount of water entering a soil during a precipitation event must be evaluated. It is the infiltration capacity of a soil that determines the amount of water that is available for runoff, surface storage, or leaching. Fortunately, many predictive models have been developed for predicting infiltration (Skaggs and Khaleel 1982).

That part of the rainfall which does not infiltrate into a soil immediately following a storm event becomes runoff. Part of the runoff may later infiltrate into the soil, but much of the runoff ultimately appears as flowing water in the drainage network of the watershed. Runoff in itself is not important in chemical weathering, but it affects modeling of weathering processes by acting as a diluent of the water that contains nutrients released by weathering and is essential to the estimation of erosion efflux. Like other processes in watershed modeling, it is impossible to model runoff on a physical basis alone. Many empirical models, however, are available that can approximate runoff (Huggins and Burney 1982).

Quantification of rock weathering in a watershed also requires accurate measurement of nutrient effluxes and changes in nutrient storage. The methods used for these measurements vary greatly between studies and probably account for much of the variability in reported weathering rates. A common simplification is to assume steady-state forest growth, that is, $(P + S) = 0$. In old, stable ecosystems approaching steady-state conditions, this assumption may be reasonable. However, Clayton (1979, 1984) cautions against using this assumption in rapidly changing ecosystems, for example, where rapid nutrient accumulation in new biomass is coupled with low returns by litterfall and decomposition. The assumption of $(P + S) = 0$ should be evaluated on an element-by-element basis.

Seasonal effects of temperature and precipitation should also be considered. In short-term studies, differences in rainfall and other weather patterns may cause significant changes in biomass thus affecting nutrient storage. Nutrient storages in fine roots are especially difficult to evaluate and may vary greatly over short time periods. Studies have indicated that general weight/diameter/height relations for estimation of fine root biomass are site specific (Hermann 1977). Harris et al. (1977), for example, found that the range of root size and biomass varied widely from month to month in both natural deciduous and loblolly pine (*Pinus taeda* L.) plantations. The biomass of roots less than 1 cm in diameter in the loblolly pine plantation changed from 650 to 230 g m^{-2} in 1 month. Yellow poplar (*Liriodendron tulipifera* L.) forest root biomass varied at least as much as that of loblolly pine on a monthly basis, and ranged from 230 to 1300 g m^{-2} during the 2-year course of the study. At the same time, total annual root production was estimated to be the same at both forests (9000 kg ha^{-1} yr^{-1}). No estimates were made of root sloughing or death, so estimates of annual root turnover were not possible. Other studies have shown that increases in root biomass continue after aboveground growth ceases (Hermann 1977).

Quantification of rock weathering is also hampered by problems in the determination of changes in nutrient storage in soils. Detecting significant changes in soil parameters at the precise levels needed for mass-balance studies is complicated by analytical errors, soil spatial heterogeneity (Edmonds and Lenter 1986, Riha et al. 1986a,b, Zinke 1962), and seasonal changes in soil chemistry (Nagle 1983). A simple sensitivity analysis (Table 4.4) for 20 cm of soil with bulk density = 1.5 g cm^{-3} shows that a change in exchangeable Ca as slight as 0.02 cmol kg^{-1}

TABLE 4.4. Assumptions and calculations used to develop the presented sensitivity analyses.

A. We assumed a change of 0.02 cmol exchangeable Ca kg^{-1} in a soil with an average bulk density of 1.5 g cm^{-3} to a total depth of 20 cm.

$$\frac{.02 \text{ cmol Ca}}{\text{kg soil}} \times \frac{1.5 \text{ g}}{\text{cm}^3} \times \frac{1 \text{ kg}}{1000 \text{ g}} \times \frac{10^8 \text{ cm}^2}{\text{ha}} \times 20 \text{ cm} = \frac{60{,}000 \text{ cmol Ca}}{\text{ha soil}}$$

B. This change of 60,000 cmol Ca ha^{-1} soil is equivalent to 24 kg Ca ha^{-1}, which is comparable to the average annual weathering rate value of 20 kg Ca ha^{-1} given in Table 5.

$$\frac{60{,}000 \text{ cmol Ca}}{\text{ha soil}} \times \frac{0.4 \text{ g}}{\text{cmol Ca}} \times \frac{1 \text{ kg}}{1000 \text{ g}} \times \frac{1}{\text{year}} = \frac{24 \text{ kg}}{\text{ha soil year}}$$

C. This change of 60,000 cmol Ca ha^{-1} is also equivalent to 1500 kg montmorillonite with an average CEC of 40 cmol Ca kg^{-1}.

$$\frac{60{,}000 \text{ cmol Ca}}{\text{ha soil}} \times \frac{1 \text{ kg montmorillonite}}{40 \text{ cmol Ca}} = \frac{1500 \text{ kg montmorillonite}}{\text{ha soil}}$$

D. This 1500 kg montmorillonite represents 0.05% of the total soil weight, again assuming bulk density = 1.5 g cm^{-3} and soil depth = 20 cm.

$$\frac{1500 \text{ kg}}{\text{ha soil}} \times \frac{10^3 \text{ g}}{\text{kg}} \times \frac{1 \text{ ha}}{10^8 \text{ cm}^2} \times \frac{1 \text{ cm}^3}{1.5 \text{ g}} \times \frac{1}{20 \text{ cm}} \times 100 = 0.05\%$$

(which is analytically undetectable) is equivalent to 24 kg Ca ha^{-1}. This amount is about equal to some annual estimates of Ca release by mineral weathering (Table 4.5). A change in exchangeable Ca of 0.02 cmol kg^{-1} is also equivalent to a change of about 1500 kg ha^{-1} in montmorillonite content (assuming 40 cmol Ca per kg montmorillonite), which is about 0.05% of the total soil weight (Table 4.4).

Problems also arise in field studies of weathering from the difficulty in estimating effluxes. The assumption of negligible erosion losses can be a large source of error. Dethier (1986) found, in a study of 41 watersheds in the U.S. Pacific Northwest, that physical erosion was equal to and often many times greater than the chemical denudation rate. Paces (1986) found that chemical erosion (weathering) represented about 7–17% of the total erosion in small drainage basins in the Bohemian massif. Nutrient losses through physical erosion would be highly detrimental to forest growth, because erosion losses occur at the surface where nutrients are accumulated in soil organic matter and surface horizons by nutrient cycling.

Lysimeter and streamflow data have been applied to the determination of nutrient effluxes in forest ecosystems. The lysimeters most commonly used for this purpose are porous ceramic suction samplers that are typically used to sample the pore water of unsaturated soils. Such

TABLE 4.5. Nutrient release data resulting from mass-balance studies of forested watersheds.

Location	Parent materials	Na	K	Ca	Mg	Notes	Reference
		kg/ha/hr					
White Mts., CA	Dolomite	2	4	86	52	a,b,c	Marchand 1971
	Quartz monzonite	1	8	17	2	a,b,c	Marchand 1971
Cascades, OR	Tuffs/breccias	28	1.6	47	11.6	a,b	Fredricksen 1972
Luxembourg	Metashale	9.1	0.2	8.7	15.7	b,d	Verstraten 1977
Piedmont, MD	Schist	2.6	2.3	1.3	1.7	b	Cleaves et al. 1970
	Serpentinite	Trace	Trace	Trace	34.1	b	Cleaves et al. 1974
Hubbard Brook, NH	Moraine/gneiss	5.8	7.1	21.1	3.5	b	Likens et al. 1977
Brookhaven, NY	Glacial outwash	6.7	11.1	24.2	8.4	a,d	Woodwell and Whitaker 1967
Central Wisconsin	Glacial outwash	—	6.9	22	5.0	d,e	Rockheim et al. 1983
SW Idaho	Quartz monzonite	13.5	4.1	20.1	2.4	b	Clayton 1984
Piedmont, VA	Granite	24.8	1.3	7.9	2.6	b	Pavich 1986
Bohemian Massif	Biotite-muscovite gneiss	5.0	9.4	3.7	2.9	b	Paces 1986
Czechoslovakia	Biotite gneiss	13.0	23.0	19.0	14.0	b,f	Paces 1986
	Quartzitic gneiss	6.0	13.0	8.5	6.3	b	Paces 1986
Scotland	Till	10.1	2.37	17.2	5.25	b	Creasey et al. 1986
	Gabbro	2.26	1.18	20.2	5.94	b	Creasey et al. 1986
Range							
low		Trace	Trace	Trace	1.7		
high		28	23	86	52		
Average		8.65	5.98	20.24	10.84		
Standard deviation		8.31	6.05	20.89	13.67		

[a]Calculated from original data by Clayton (1979).
[b]Streamflow data for volumes of leaching and leachate concentration.
[c]Erosion rate measured.
[d]Leachate concentration from lysimeters.
[e]Leaching volume = 98% net precipitation.
[f]Agricultural soil with uncertain biological mass balance.

lysimeters allow sampling at the base of the root zone and therefore do not sample water that is chemically controlled by deeper weathering. Under non-saturated conditions, lysimeters also sample mesopore and some micropore water to the exclusion of macropore water. In one sense, this is advantageous because micropore water has a longer residence time and is more likely to be utilized by plants. Therefore, water held under tension may be more representative of equilibrium conditions in the weathering zone than freely drained water. But, as has been previously stated, the rate of water flushing through the system is probably more important in determining streamwater chemistry than reaction conditions alone (Velbel 1985). The quantity of micropore and macropore water present at different depths would be required to determine total hydraulic flux. However, nutrient fluxes are still uncertain because of differences in micropore and macropore water chemistry (Barbee and Brown 1986). In addition, lysimeters may be subject to poor soil-to-sampler contact and to disturbance of soil water movement by the sampler itself (Barbee and Brown 1986, van der Ploeg and Beese 1977). Spatial heterogeneity in the soil and landscape also causes highly variable chemical concentrations (Barbee and Brown 1986). Therefore, the application of these results to entire watershed ecosystems may be misleading (McBratney and Webster 1983). At the same time, however, lysimeter usage may help in the study of leaching processes. For instance, Cronan and his coworkers (Cronan 1980a, Cronan and Schofield 1979, Cronan et al. 1978) used lysimeters to monitor the changes in cationic and anionic components of soil leachates as they moved through a soil following acid precipitation events. It was found that the leachates increased in Al and soluble organic anions as leaching proceeded. Such findings would not have been possible with streamflow data alone.

Streamflow data normalizes the effects of spatial variability for sampling a large area. Information is also generated for total efflux volumes and concentrations. Results generated from streamflow measurements may be valid for lithologically homogeneous watersheds with shallow soils over bedrock, but even in these cases, effects of lateral flow of soil solutions and differences in slope aspect and position limit the application to individual sites within the watershed. Streamflow data also may not be indicative of plant nutrient availability. For example, in deeply weathered watersheds in humid environments such as those in the southeastern United States, streamwater chemistry reflects deep incipient alteration of large volumes of saprolite with solution composition of percolating waters controlled by the kinetics of dissolution, precipitation, and water flux through the zone, not by chemical equilibria (Velbel 1985). Streamwater evolved from deep weathering may bear little resemblance to water percolating through highly weathered surface soils and therefore would not be representative of nutrient levels available for plant growth (Velbel 1985). On the positive side, streamflow data represents the volume-averaged nutrient release for an entire watershed. With adequate modeling of waterflow within the watershed and determination of hydrologic flow paths, valuable results may be obtained from streamflow studies.

Considering the complexity of natural ecosystems, it is not surprising that few field studies are conducted with the same underlying assumptions. However, field studies are important because the data generated represent the integration of all factors that can affect weathering in natural environments. Although the effects of such factors as acid precipitation may not be precisely evaluated because of difficulty in locating unaffected control sites, the effects of long-term additions of acid precipitation may be examined by comparing data over time. In addition, field studies can be carried on indefinitely while short-lived laboratory studies focus on details not easily measured in the field.

Despite differences in approaches and assumptions, weathering rates reported in the literature are fairly consistent (see Table 5), with differences arising from the wide variety of parent materials studied. However, it should be noted that standard deviations are often as great as the means for the nutrients indicated. Studies indicate that parent materials tend to lose soluble constituents (Ca, Mg, Na, etc.) at rates comparable to the relative concentrations available for loss. In general, more porous parent materials weather faster than less porous parent materials because there is more surface area available for attack.

Only a few studies are comparable because there are large differences in parent materials. The White Mountain, California study of Marchand (1971) and the southwestern Idaho study of Clayton (1984) both involved similar quartz monzanite parent materials. Relative amounts of common bases present in the parent rock are of the order $K > Na > Ca > Mg$. In neither of these studies was the release rate of this same order. Calcium was released faster than the other nutrients because it occurs in more easily weatherable minerals, which contain little K, Na, and Mg. This parent rock effect on nutrient release by weathering appears to be the controlling factor in relative nutrient release rates. Actual release rates are affected by reaction rates as well as flushing rates acting as a scaling factor.

Dethier (1986) determined the denudation rates for 41 watersheds in the U.S. Pacific Northwest. He found definite lithologic effects on weathering rates. The highest denudation rate in this study, 98 t km^{-2} yr^{-1}, was calculated for Franciscan rocks, a rock formation in northern California that is highly fractured with carbonate veinfillings. The intrusive and metamorphic rocks in the North Cascade Range and Klamath Mountains were found to weather least rapidly. Dethier concluded that physical denudation (erosion) is highly significant in all watersheds, varying from amounts approximately equal to chemical denudation in western Washington and Oregon to more than 20 times the chemical loss for Franciscan rocks. He also noted that chemical denudation rates for watersheds were positively correlated with mean annual runoff. This phenomenon was attributed to the larger volumes of water available for displacing pore waters and for flushing easily soluble constituents from particle surfaces.

In addition to the simplified mathematical model of Clayton (1979), Velbel (1986) and Paces (1986) have recently published more comprehensive mathematical models for the quantification of watershed nutrient cycling processes. In the approach used by Velbel (1986), nutrient fluxes in a watershed are handled as sets of simultaneous equations with each nutrient represented by one set of equations. This allowed Velbel to solve for the weathering rates of as many minerals as there were nutrients measured. This approach is a definite improvement over the method of Clayton (1979); his model required he assume that only one mineral was weathering. The rates of mineral weathering determined from watershed data by Velbel (1986) were comparable to weathering rates found in earlier laboratory kinetic studies such as those of Nickel (1973). Paces (1986) approached weathering rates in a similar manner. However, instead of determining the weathering rates of individual minerals, he solved for the total rates of chemical and physical weathering of the bedrock and saprolite. The rate of bedrock weathering of gneisses in Central Europe was found to be highly dependent on basin slope and land use. The mass of bedrock removed by weathering in representative catchments ranged from 23 to 38 g m^{-2} yr^{-1} in forested areas to 82 g m^{-2} yr^{-1} in agricultural areas. The higher weathering rates in the agricultural areas can probably be attributed to the greater temperatures under these conditions than under forest cover.

V. Conclusions and Research Needs

Mineral weathering is the initial source of most nutrients for plant growth in forest ecosystems. However, exact data on nutrient release through weathering are not easily obtained; because soil and geological heterogeneity and associated microclimatic effects make it difficult to include adequate controls in experiments, most field studies are performed with an overly simplistic view of a complex system.

The inherent variability of soils and landscapes can also lead to difficulties in predicting weathering rates for entire watersheds. Fluxes of nutrients released through weathering are well within error limits for the analytical methods used to obtain them. A negligible change in CEC resulting from either soil variability or analytical error could result in nutrient flux as great or greater than that attributable to weathering. Even for those sites in which weathering rates have been calculated, results cannot be compared to other sites because of differing experimental techniques and assumptions involved in calculations.

Data from laboratory dissolution kinetic studies give results similar to those predicted by earlier mineral persistence indices, but the data cannot easily be used in predictive field models. This results from the unrealistic methods required for data collection such as high solution-to-soil ratios, low pH values, and unrealistic concentrations of complexes. Greenhouse pot studies in which weathering rates are measured using one or more minerals as the sole nutrient source can give data on rates of mineral weathering by plants, but these studies have little relationship to field conditions because of the oversimplifications required for weathering rate determination. For example, very few minerals weather in field situations without geochemical controls on the concentrations of solution species. Another important difference between field conditions and greenhouse studies is the presence of complex plant interactions in the field. It is not possible to model the plant ecology of a mature forest with a greenhouse pot study because interactions resulting from species diversity and age differences within stands cannot be included. For these relationships, only field studies provide useful information.

To obtain a better understanding of cycling processes and to predict nutrient release by weathering, more research on mechanisms of nutrient fluxes is needed. A systems approach cannot yield a predictive model without knowledge of the mechanisms at work. Watershed studies alone cannot furnish detailed insight into the mechanistic relationships present in an entire ecosystem because of the complexity of the interactions involved. This system complexity makes it necessary to isolate factors that can be studied individually. Therefore, initial investigations should study those factors that can be quantified with reasonable accuracy in creating a model. Ideally, it would be best to start with as pure a system as possible: a mineral and water. This method would yield base rate data, assuming the reaction kinetics were favorable to obtain a measurable product. New components could be added and reevaluated using the old system as a known parameter. Complexes, chelating agents, and acids or bases could then be added to the system. With careful experimental design, this would reveal the mechanism for mineral dissolution. As previously indicated, most laboratory studies have not used realistic experimental conditions. Thus, the data often cannot be extrapolated to natural systems, although information may be gained regarding reaction mechanisms. Also, few laboratory studies have been conducted with mineral mixtures to examine potential interactions.

More information is also needed on the influence of organic matter on mineral weathering in soils. Although humic substances are ubiquitous in forest soils, the total organic fraction remains poorly characterized. Rate laws for the decomposition of organic matter in soils and sediments are not well established (Berner 1981). It is not possible at present to predict the organic chemistry of soils with any degree of precision. No currently available models can predict the type and amount of functional groups that exist on soil organic matter. Data on the rate of functional group leaching in a soil system is sparse and cannot be transferred to predictive models. Modeling of soil nutrient cycling thus is difficult because soil organics have been shown to greatly affect mineral solubility (Huang and Keller 1970, Iskander and Syers 1972, Kodama et al. 1983, Schnitzer and Kodama 1976). There are also no existing quantitative models to predict the rate of weathering of raw organics (i.e., leaves) under any given condition. Additional information in this area would help to determine how changes in soil organics accompany alterations in an ecosystem (e.g., the response of ecosystems to acidic deposition).

Additional basic chemical data is also needed on the chemistry of many soil elements. The chemistry of Al, PO_4^{3-}, and SO_4^{2-}, for example, is not adequately known for predicting their movement in the environment. Data on the chemistry of nutrients is required to accurately predict changes in their flux associated with changes in ecosystems accompanying acid precipitation. For example, the inclusion of an unknown Al-hydroxysulfate complex in chemical models would greatly increase the calculated rate of efflux of Al from soils in predictive models. Such differences in efflux would greatly affect the quantity of Al reaching aquatic systems. This type of data would also help to predict the biotoxicity of Al in aquatic ecosystems.

If simple systems are adequately characterized and modeled, it would then be possible to study a soil column which is a more complex system. The study of nutrient chemistry during water and solution movement through columns can be used to separate biological effects from soil effects. Differences between soil columns with and without plants under identical experimental conditions can be allocated directly to plant effects. In this way, column studies in the greenhouse bridge the gap between field data and laboratory data.

More carefully controlled mass-balance field experiments are also needed. Many of the studies reported in the literature did not give adequate consideration to detailed characterization of the ecosystems examined. The assumption of negligible erosion, runoff, and lateral water flow on steeply sloping forested watersheds is unjustified. The soils, geology, forest ecology, and groundwater hydrology should be delineated and documented. Soils should be chemically and physically characterized to depths beyond the root zone and field checked for variability. The cation exchange properties should be examined and the effects of expected (or observed) changes in soil solution chemistry determined. Soil water retention data should be obtained so that field measurements can be translated into water storage and combined with water inputs to calculate fluxes. Water and nutrient fluxes should be monitored at several sites within watersheds chosen for study. These samplers should be arranged to quantify nutrient fluxes from erosion and runoff both upslope and downslope of plots. The data should also be evaluated using modern hydrologic and chemical computer models to obtain fluxes and estimate flowpaths (see Smolen 1983). Data determined with careful experimental design would result in valid weathering estimations and would lead to better predictive models.

It is also important that watershed studies be expanded to include several types of ecosystems in addition to steeply sloping forested areas. The pioneering study of Paces (1986), which included agricultural soils, is an important step in this direction. Nutrient fluxes also may be important in areas with high water tables where groundwater contamination is possible. Results may show that other ecosystems are more or less sensitive to changes in nutrient fluxes than forests. This type of data would be of critical importance in evaluating proposed pollution regulations as well as to land and ecosystem management.

The release of nutrients during weathering largely determines the capacity of soils to be buffered against long-term chronic nutrient depletion that may be attributed to acid deposition. However, current information is inadequate to determine the effects of acidic deposition on mineral weathering rates, and more research is essential.

Acknowledgment. The authors are indebted to James A. Burger, Alan A. Lucier, and Stephen H. Schoenholtz for initially reviewing this manuscript.

References

Abudelgawad, G., A.L. Page, and L.J Lund. 1975. Chemical weathering of glauconite. Soil Sci Soc Am Proc 39:567–571.

Adams, P.W. and J.R. Boyle. 1979. Cation release from Michigan spodosols leached with aspen leaf extracts. Soil Sci Soc Am J 43:593–596.

Allen, B.F. and D.S. Fanning. 1983. Composition and soil genesis. *In*: Pedogenesis and soil taxonomy. I. Concepts and Interactions, L.P. Wilding, N.E. Smeck, and G.F. Hall (eds.), pp. 141–192. Amsterdam: Elsevier.

Amonette, J., F.T. Ismail, and A.D. Scott. 1985. Oxidation of iron in biotite by different oxidizing solutions at room temperature. Soil Sci Soc Am J 49:772–777.

Barbee, G.C. and K.W. Brown. 1986. Comparison between suction and free-drainage soil solution samplers. Soil Sci 141:149–154.

Barshad, I. 1966. The effect of variation of precipitation on the nature of clay mineral formation in soils from acid and basic igneous rocks. *In*: Proceedings of International Clay Conference, Jerusalem, Israel, Vol. 1, pp. 167–173.

Berner, R.A. 1971. Principles of Chemical Sedimentology. New York: McGraw-Hill.

Berner, R.A. 1978. Rate control of mineral dissolution under earth surface conditions. Am J Sci 278:1235–1252.

Berner, R.A. 1981. Kinetics of weathering and digenesis. *In*: Kinetics of geochemical processes, A.C. Lasaga and R.J Kirkpatrick (eds.), pp. 111–134. Reviews in Mineralogy, Vol. 8. Mineralogical Society of America.

Berner, R.A., E.L. Sjoberg, M.A. Velbel, and M.D. Krom. 1980. Dissolution of pyroxenes and amphiboles during weathering. Science 207:1205–1206.

Berthelin, J. and G. Belgy. 1979. Microbial degradation of phyllosilicates during simulated podzolization. Geoderma 21:297–310.

Birkeland, P.W. 1974. Pedology, Weathering, and Geomorphological Research. London: Oxford University Press.

Birkeland, P.W. 1984. Soils and Geomorphology. London: Oxford University Press.

Bloom, P.R. 1983. The kinetics of gibbsite dissolution in nitric acid. Soil Sci Soc Am J 47:164–168.

Bloom, P.R. and M.S. Erich. 1987. Effect of solution composition on the rate and mechanism of gibbsite dissolution in acid solutions. Soil Sci Soc Am J 51:1131–1136.

Brewer, R. 1976. Fabric and Mineral Analysis of Soils. Huntington, New York: Robert E. Krieger.

Busenberg, E. and C.V. Clemency. 1976. The dissolution kinetics of feldspars at 25°C and 1 atm CO_2 partial pressure. Geochim Cosmochim Acta 40:41–19.

Carroll, D. 1970. Rock Weathering. New York: Plenum Press.

Chesworth, W. 1972. The stability of gibbsite and boehmite at the surface of the earth. Clays Clay Miner 20:369–374.

Chesworth, W. 1973. The parent rock effect in the genesis of soil. Geoderma 10:215–225.

Clayton, J.L. 1979. Nutrient supply to soil by rock weathering. *In*: Proceedings, Impact of Intensive Harvesting on Forest Nutrition, pp. 75–96. Syracuse, New York: State University of New York.

Clayton, J.L. 1984. A rational basis for estimating elemental supply rate from weathering. *In*: Proceedings Sixth North American Forest Soils Conference, E.L. Stone (ed.), pp. 405–419. Knoxville: University of Tennessee.

Clayton, J.L. 1986. An estimate of plagioclase weathering rate in the Idaho batholith based upon geochemical transport rates. *In*: Rates of Chemical Weathering of Rocks and Minerals, S.M. Colman and D.P. Dethier (eds.), pp. 453–467. Orlando: Academic Press.

Cleaves, E.T., D.W. Fisher, and O.P. Bricker. 1974. Chemical weathering of serpentinite in the eastern Piedmont of Maryland. Geol Soc Am Bull 85:437–444.

Cleaves, E.T., A.E. Godfrey, and O.P. Bricker. 1970. Geochemical balance of a small watershed and its geomorphic implications. Geol Soc Am Bull 81:3015–3032.

Courbe, C., B. Velde, and A. Meunier. 1981. Weathering of glauconite: Reversal of the glauconization process in a soil profile in western France. Clay Miner 16:231–243.

Creasey, J., A.C. Edwards, J.M. Reid, D.A. MacLeod, and M.S. Cresser. 1986. The use of catchment studies for assessing chemical weathering rates in two contrasting upland areas in Northeast Scotland. *In*: Rates of Chemical Weathering of Rocks and Minerals, S.M. Colman and D.P. Dethier (eds.) pp. 468–502. Orlando: Academic Press.

Cronan, C.S. 1980a. Solution chemistry of a New Hampshire subalpine ecosystem: A biogeochemical analysis. Oikos 34:272–281.

Cronan, C.S. 1980b. Consequences of sulfuric acid input to a forest soil. *In*: Atmospheric Sulfur Deposition—Environmental Impact and Health Effects, D.S. Shriner, C.R. Richmond, and S.E. Lindberg (eds.), pp. 335–343. Ann Arbor, Michigan: Ann Arbor Science.

Cronan, C.S. 1985. Chemical weathering and solution chemistry in acid forest soils: Differential influence of soil type, biotic processes, and H^+ deposition. *In*: The Chemistry of Weathering, J.I. Drever (ed.), pp. 175–196. Dordrecht, Holland: D. Reidel.

Cronan, C.S. and C.L. Schofield. 1979. Aluminum leaching response to acid precipitation: Effects on high-elevation watersheds in the Northwest. Science 204:304–306.

Cronan, C.S., W.A. Reiner, R.C. Reynolds, and G.E. Lang. 1978. Forest floor leaching: Contributions from mineral, organic, and carbonic acids in New Hampshire subalpine forests. Science 200:309–311.

Daniels, W.L., C.J. Everett, and L.W. Zelazny. 1987. Virgin hardwood forest soils of the southern Appalachian mountains: I. Soil morphology and geomorphology. Soil Sci Soc Am J 51:722–729.

Dethier, D.P. 1986. Weathering rates and chemical flux from catchments in the Pacific Northwest, U.S.A. *In*: Rates of Chemical Weathering of Rocks and Minerals, S.M. Colman and D.P. Dethier (eds.), pp. 503–530. Orlando: Academic Press.

Eckhardt, F.E.W. 1985. Solubilization, transport, and deposition of mineral cations by microorganisms—Efficient rock weathering agents. *In*: The Chemistry of Weathering, J.I. Drever (ed.), pp. 161–174. Dordrecht, Holland: D. Reidel.

Edmonds, W.G. and M. Lentner. 1986. Statistical evaluation of the taxonomic composition of three soil map units in Virginia. Soil Sci Soc Am J 50:997–1001.

Eggleton, R.A. 1986. The relation between crystal structure and silicate weathering rates. *In*: Rates of Chemical Weathering of Rocks and Minerals, S.M. Colman and D.P. Dethier (eds.), pp. 21–40. Orlando: Academic Press.

El-Amamy, M. M., A.L. Page, and A. Abudelgawad. 1982. Chemical and mineralogical properties of glauconitic soil as related to potassium depletion. Soil Sci Soc Am J 46:426–430.

Evans, A., Jr. 1986. Effects of dissolved organic carbon and sulfate on aluminum mobilization in forest soil columns. Soil Sci Soc Am J 50:1576–1578.

Evans, A., Jr. and L.W. Zelazny. 1987. Effects of sulfate addition on the status of exchangeable Al in a Cecil soil. Soil Sci 143:410–417.

Franzmeier, D.P., E.J. Pederson, T.J. Longwell, J.G. Byrne, and C.K. Losche. 1969. Properties of some soils of the Cumberland Plateau as related to slope aspect and position. Soil Sci Soc Am Proc 33:755–761.

Fredricksen, R.L. 1972. Nutrient budget of a Douglas-fir forest on an experimental watershed in Oregon. *In*: Proceedings, Research on Coniferous Ecosystems Symposium, pp. 115–131, Bellingham, Washington.

Frenkel, H., C. Armhein, and J.J. Jurinak. 1983. The effect of exchangeable cations on soil mineral weathering. Soil Sci Soc Am J 47:649–653.

Frenkel, H. and D. L. Suarez. 1977. Hydrolysis and decomposition of calcium montmorillonite. Soil Sci Soc Am J 41:887–891.

Gersper, P.L. and N. Holowaychuk. 1970a. Effects of stemflow water on the Miami soil under a beech tree: I. Morphological and physical properties. Soil Sci Soc Am J 34:779–786.

Gersper, P.L. and N. Holowaychuk. 1970b. Effects of stemflow water on the Miami soil under a beech tree: II. Chemical properties. Soil Sci Soc Am J 34:786–794.

Gersper, P.L. and N. Holowaychuk. 1971. Effects of stemflow from canopy trees on chemical properties of soil. Ecology 52:691–702.

Gilkes, R.J. and R.J. Young. 1974a. Artificial weathering of oxidized biotite: III. Potassium uptake by subterranean clover. Soil Sci Soc Am Proc 38:41–43.

Gilkes, R.J. and R.J. Young. 1974b. Artificial weathering of oxidized biotite: IV. The inhibitory effect of potassium on dissolution rate. Soil Sci Soc Am J 38:529–532.

Gilkes, R.J., R.C. Young, and J.P. Quirk. 1973a. Artificial weathering of oxidized biotite: I. Potassium removal by sodium chloride and sodium tetraphenylboron solutions. Soil Sci Soc Am J 37:25–28.

Gilkes, R.J., R.C. Young, and J.P. Quirk. 1973b. Artificial weathering of oxidized biotite: II. Rates of dissolution in 0.1, 0.01, 0.001M HCl. Soil Sci Soc Am J 37:29–33.

Goldich, S.S. 1938. A study in rock weathering. J Geol 46:17–58.

Grandstaff, D.E. 1986. The dissolution rate of forsteritic olivine from Hawaiian beach sand. *In*: Rates of Chemical Weathering of Rocks and Minerals, S.M. Colman and D.P. Dethier (eds.), pp. 41–60. Orlando: Academic Press.

Grim, R.E. 1968. Clay Mineralogy. New York: McGraw-Hill.

Harris, W.F., R.S. Kinerson, Jr., and N.T. Edwards. 1977. Comparison of belowground biomass of natural deciduous forests and loblolly pine plantations. *In*: The Belowground Ecosystem: A Synthesis of Plant-Associated Processes, J.K. Marshall (ed.), pp. 29–37. Range Science Dept. Science Series No. 26, Colorado State University, Ft. Collins, Colorado.

Helgeson, H.C. 1971. Kinetics of mass transfer among silicates and aqueous solutions. Geochim Cosmochim Acta 35:421–469.

Helgeson, H.C., R.M. Garrels, and F.T. MacKenzie. 1969. Evaluation of irreversible reactions in geochemical processes involving minerals and aqueous solutions—II. Application. Geochim Cosmochim Acta 33:455–481.

Hermann, R.K. 1977. Growth and production of tree roots: A review. *In*: The Belowground Ecosystem: A Synthesis of Plant-Associated Processes, J.K. Marshall (ed.), pp. 7–28. Range Science Dept. Science Series No. 26, Colorado State University, Ft. Collins, Colorado.

Holdren, G.R., Jr. and R.A. Berner. 1979. Mechanism of feldspar weathering. I. Experimental studies. Geochim Cosmochim Acta 43:1161–1171.

Huang, W.H. and W.D. Keller. 1970. Dissolution of rock-forming silicate minerals in organic acids: Simulated first stage weathering of fresh mineral surfaces. Am Miner 55:2076–2094.

Huggins, L.F. and J.R. Burney. 1982. Surface runoff, storage, and routing. *In*: Hydrologic Modelling of Small Watersheds, C.T. Haan, H.P. Johnson, and D.L. Brakensick (eds.), pp. 171–225. St. Joseph, Missouri: American Society of Agricultural Engineers.

Iskandar, I.K. and J.K. Syers. 1972. Metal-complex formation by lichen compounds. J Soil Sci 23:255–265.

Jackson, M.L. and W.D. Keller. 1970. A comparative study of the role of lichens and "inorganic" processes in the chemical weathering of recent Hawaiian lava flows. Am J Sci 269:446–466.

Jackson, M.L. and G.D. Sherman. 1953. Chemical weathering of minerals in soils. Adv Agron 5:219–318.

Jardine, P.M., J.C. Parker, and L.W. Zelazny. 1985a. Kinetics and mechanisms of aluminum transport on kaolinite using a two-site nonequilibrium transport model. Soil Sci Soc Am J 49:867–873.

Jardine, P.M., L.W. Zelazny, and J.C. Parker. 1985b. Mechanisms of aluminum adsorption on clay minerals and peat. Soil Sci Soc Am J 49:862–867.

Jenny, H. 1941. Factors of soil formation. New York: McGraw-Hill.

Keller, W.D. 1955. The Principles of Chemical Weathering. Columbia, Missouri: Lucas Brothers.

Kellogg, C.E. 1941. The Soils That Support Us: An Introduction to the Study of Soils and Their Use by Men. New York: Macmillan.

Kittrick, J.A. 1969. Soil minerals in the Al_2O_3-SiO_2-H_2O system and a theory of their formation. Clays Clay Miner 17:157–167.

Kodama, H. and A.E. Foscolos. 1981. Occurrence of bertherine in Canadian arctic desert soils. Can Mineralogist 19:279–283.

Kodama, H., M. Schnitzer, and M. Jaakkimainen. 1983. Chlorite and biotite weathering by fulvic acid solutions in closed and open systems. Can J Soil Sci 63:619–629.

Krug, E.C. and P.J Isaacson. 1984. Comparison of water and dilute acid treatment on organic and inorganic chemistry of leachate from organic-rich horizons of an acid forest soil. Soil Sci 137:370–378.

Lanyon, L.E. and G.F. Hall. 1979. Dissolution of selected rocks and minerals in dilute salt solution as influenced by temperature regime. Soil Sci Soc Am J 43:192–195.

Likens, G.E., F.H. Bormann, R.S. Pierce, J.S. Eaton, and N.M. Johnson. 1977. Biogeochemistry of a Forested Ecosystem. New York: Springer-Verlag.

Lin, F.C. and C.V. Clemency. 1981. Dissolution kinetics of phlogopite. I. Closed system: Clays Clay Miner 29:101–106.

Losche, C.K., R.J. McCracken, and C.B. Davey. 1970. Soils of the steeply sloping landscapes in the southern Appalachian Mountains. Soil Sci Soc Am Proc 34:473–478.

Loughnan, F.C. 1969. Chemical Weathering of the Silicate Minerals. New York: American Elsevier.

Luce, R.W., R.W. Bartlett, and G.A. Parks. 1972. Dissolution kinetics of magnesium silicates. Geochim Cosmochim Acta 36:36–50.

Marchand, D.E. 1971. Rates and modes of denudation, White Mountains, eastern California. Am J Sci 270:109–135.

Mattigod, S.V. and J.A. Kittrick. 1980. Temperature and water activity as variables in soil mineral activity diagrams. Soil Sci Soc Am J 44:149–154.

Mayer, R. and B. Ulrich. 1974. Conclusions on the filtering action of forests from ecosystem analysis. Oecol Plant 9:157–168.

McBratney, A.B. and R. Webster. 1983. How many observations are needed for regional estimation of soil properties? Soil Sci 135:177–183.

Millot, G. 1970. Geology of Clays. New York: Springer-Verlag.
Mojallali, H. and S.B. Weed. 1978. Weathering of micas by mycorrhizal soybean plants. Soil Sci Soc Am J 42:367–372.
Mortland, M.M. 1958. Kinetics of potassium release from biotite. Soil Sci Soc Am Proc 22:503–508.
Nagle, S.M. 1983. Evaluation of selected lime requirement tests for Virginia soils developed through field response of soil pH and crop yields. M.S. Thesis, Virginia Polytechnic Institute and State University, Blacksburg, Virginia.
Neary, D.G., L.A. Morris, and B.F. Swindel. 1984. Site preparation and nutrient management in southern pine forests. *In*: Forest Soils and Treatment Impacts, E.A. Stone (ed.), pp. 121–144. Proceedings, Sixth North American Forest Soils Conference, University of Tennessee, Knoxville.
Nickel, E. 1973. Experimental dissolution of light and heavy minerals in comparison with intrastratal solution. Contrib Sediment 1:1–68.
Ollier, C. 1984. Weathering. London: Longman.
Osborne, H.B., L.J. Lane, C.W. Richardson, and M.L. Molnau. 1982. Precipitation. *In*: Hydrologic Modeling of Small Watersheds, C.T. Haan, H.P. Johnson, and D.L. Brakensick (eds.), pp. 81–118. St. Joseph, Missouri: American Society of Agricultural Engineers.
Oster, J.D. and I. Shainberg. 1979. Exchange cation hydrolysis and soil weathering as affected by exchangeable sodium. Soil Sci Soc Am J 43:70–75.
Ovington, J.D. 1965. Nutrient cycling in woodlands. *In*: Experimental Pedology, E.G. Hallsworth and D.V. Crawford (eds.), pp. 208–218. London: Butterworth.
Paces, T. 1986. Rates of weathering and erosion derived from mass balance in small drainage basins. *In*: Rates of Chemical Weathering of Rocks and Minerals, S.M. Colman and D.P. Dethier (eds.), pp. 531–551. Orlando: Academic Press.
Parker, J.C. and M. Th. van Genuchten. 1984. Determining transport parameters from laboratory and field tracer experiments. Virginia Agric Exp Stat Bull 84-3.
Pavich, M.J. 1986. Processes and rates of saprolite production and erosion on a foliated granitic rock of the Virginia Piedmont. *In*: Rates of Chemical Weathering of Rocks and Minerals, S.M. Colman and D.P. Dethier (eds.), pp. 552–590. Orlando: Academic Press.
Pettijohn, F.J. 1941. Persistence of heavy minerals and geologic ages. J Geol 49:610–625.
Petrovic, R. 1976. Rate control in feldspar dissolution. II. The protective effect of precipitates. Geochim Cosmochim Acta 40:1509–1522.
Petrovic, R., R.A. Berner, and M.B. Goldhaber. 1976. Rate control in dissolution of alkali feldspars. I. Studies of residual feldspar grains by x-ray photoelectron spectroscopy. Geochim Cosmochim Acta 40:537–548.
Reed, M.G. and A.D. Scott. 1962. Kinetics of potassium release from biotite and muscovite in sodium tetraphenylborate solutions. Soil Sci Soc Am Proc 26:437–440.
Reiche, P. 1943. Graphic representation of chemical weathering. J Sediment Petrol 13:58–68.
Reiche, P. 1950. A survey of weathering processes and products. Univ New Mexico Publ in Geol No 3. 95pp.

Reuss, J.O. and D.W. Johnson. 1986. Acid Deposition and the Acidification of Soils and Waters. New York: Springer-Verlag.

Reynolds, R.C. 1971. Clay mineral formation in an alpine environment. Clays Clay Miner 19:361–374.

Riha, S.J., B.R. James, G.P. Senesac, and E. Pallant. 1986a. Spatial variability of soil pH and organic matter in forest plantations. Soil Sci Soc Am J 50:1347–1352.

Riha, S.J., G.P. Senesac, and E. Pallant. 1986b. Effects of forest vegetation on spatial variability of surface mineral soil pH, soluble aluminum and carbon. Water Air Soil Pollut 31:929–940.

Robert, M. 1973. The experimental transformation of mica toward smectite: Relative importance of total charge and tetrahedral substitution. Clays Clay Miner 21:167–174.

Schnitzer, M. 1969. Reaction between fulvic acid, a soil humic compound and inorganic soil constituents. Soil Sci Soc Am Proc 33:75–81.

Schnitzer, M. and H. Kodama. 1976. The dissolution of micas by fulvic acid. Geoderma 15:381–391.

Schott, J. and R.A. Berner. 1985. Dissolution mechanisms of pyroxenes and olivines during weathering. *In*: The Chemistry of Weathering, J.I. Drever (ed.), pp. 35–54. Dordrecht, Holland: D. Reidel.

Skaggs, R.W. and R. Khaleel. 1982. Infiltration. *In*: Hydrologic Modeling of Small Watersheds, C.T. Haan, H.P. Johnson, and D.L. Brakensick (eds.), pp. 122–166. St. Joseph, Missouri: American Society of Agricultural Engineers.

Skopp, J. 1986. Analysis of time-dependent chemical processes in soils. J Environ Qual 15:205–213.

Smeck, N.E., E.C.A. Runge, and E.E. Mackintosh. 1983. Dynamics and genetic modelling of soil systems. *In*: Pedogenesis and Soil Taxonomy. I. Concepts and Interactions, L.P. Wilding, N.E. Smeck, and G.F. Hall (eds.), pp. 51–82. Amsterdam: Elsevier.

Smolen, M.D. 1983. Hydrologic/water quality models. Southern Cooperative Series Bulletin No. 291, Virginia Agricultural Experimental Station, Blacksburg, Virginia.

Sparks, D.L. 1989. Kinetics of soil chemical processes. San Diego: Academic Press.

Spyridakis, D.E., G. Chesters, and S.A. Wilde. 1967. Kaolinization of biotite as a result of coniferous and deciduous seedling growth. Soil Sci Soc Am Proc 31:203–210.

Stumm, W. and J.J. Morgan. 1981. Aquatic Chemistry—An Introduction Emphasizing Chemical Equilibria in Natural Waters. New York: Wiley.

Suarez, D.L. and H. Frenkel. 1981. Cation release from Na- and Ca-saturated clay-sized soil fractions. Soil Sci Soc Am J 45:716–721.

Tan, K.H. 1980. The release of silicon, potassium, and aluminum during the decomposition of soil minerals by humic acid. Soil Sci 129:5–11.

Tan, K.H. 1986. Degradation of soil minerals by organic acids. *In*: Interactions of Soil Minerals with Natural Organics and Microbes, P.M. Huang and M. Schnitzer (eds.), pp. 1–28. Spec. Pub. No. 17, Soil Society of America, Madison, Wisconsin.

Tardy, Y. 1982. Kaolinite and smectite stability in weathering conditions. Estud Geol 38:295–312.

Twidale, C.R. 1976. Analysis of Landforms. Sydney, Australia: Wiley.
van der Ploeg, R. R. and F. Beese. 1977. Model calculations for the extraction of soil water by ceramic cups and plates. Soil Sci Soc Am J 41:466–470.
Velbel, M.A. 1984. Natural weathering mechanisms of almadine garnet. Geology 12:631–634.
Velbel, M.A. 1985. Hydrogeochemical constraints on mass balances in forested watersheds of the southern Appalachians. *In*: The Chemistry of Weathering, J.I. Drever (ed.), pp. 231–247. Dordrecht, Holland: D. Reidel.
Velbel, M.A. 1986. Influence of surface area, surface characteristics, and solution composition on feldspar weathering rates. *In*: Geochemical Processes at Mineral Surfaces, M. Joan Comstock (ed.), pp. 615–634. ACS Symposium Series 323, American Chemical Society, Washington, D.C.
Velbel, M.A. 1986. The mathematical basis for determining rates of geochemical and geomorphic processes in small forested watersheds by mass balance: Examples and implications. *In*: Rates of Chemical Weathering of Rocks and Minerals, S.M. Colman and D.P. Dethier (eds.), pp. 439–452. Orlando: Academic Press.
Verstraten, J.M. 1977. Chemical erosion in a forested watershed in the Oesling, Luxembourg. Earth Surf Processes 2:175–184.
von Reichenbach, H.G. 1972. Factors of mica transformation. *In*: Proceedings, Ninth Colloquium of the International Potash Institute, pp. 33–42, Landshut, Federal Republic of Germany.
von Reichenbach, H.G. and C.I. Rich. 1969. Potassium release from muscovite as influenced by particle size. Clays Clay Miner 17:23–29.
Wilding, L.P., N.E. Smeck, and L.R. Drees. 1977. Silica in soils: Quartz, cristobalite, tridymite, and opal. *In*: Minerals in Soil Environments, J.B. Dixon and S.B. Weed (eds.), pp. 471–552. Madison, Wisconsin: Soil Science Society of America.
Wollast, R. 1967. Kinetics of alteration of K-feldspar in buffered solutions. Geochim Cosmochim Acta 31:635–648
Wollast, R. and L. Chou. 1985. Kinetic study of the dissolution of albite with a continuous flow-through fluidized bed reactor. *In*: The Chemistry of Weathering, J.I. Drever (ed.), pp. 75–96. Dordrecht, Holland: D. Reidel.
Woodwell, G.M. and R.H. Whittaker. 1967. Primary production and the cation budget of the Brookhaven forest. *In*: Proceedings, Primary Productivity and Mineral Cycling in Natural Ecosystems Symposium, pp. 151–166. University of Maine, Orono, Maine.
Yaalon, D.H. 1983. Climate, time and soil development. *In*:Pedogenesis and Soil Taxonomy. I. Concepts and Interactions, L.P. Wilding, N.E. Smeck, and G.F. Hall (eds.), pp. 233–252. Amsterdam: Elsevier.
Zelazny, L.W. and A. Evans, Jr. 1985. The release of aluminum by acid rain from soil materials. Am Water Works Assoc Proc Acid Rain 20187:31–42.
Zinke, P.J. 1962. The pattern of individual forest trees on soil properties. Ecology 43:130–133.

5
Effects of Acidic Deposition on Soil Organisms

David D. Myrold

I. Introduction

The first studies of atmospheric deposition effects on soil organisms and biological processes were conducted near industrial sources of heavy metals and SO_2. Subsequently, the emphasis has shifted to regional-scale pollution problems and the possible influence of acidic deposition on soil biology.

Numerous reviews on acidic deposition and its effects on forests and soils have appeared over the past decade (Abrahamsen 1980, 1984, Abrahamsen et al. 1977, Arthur and Wagner 1983, Evans 1984, Fielder and Thakur 1985, Johnson and Siccama 1983, Kennedy 1986, McFee 1983, McLaughlin 1985, Miller 1984, Morrison 1984, Rechcigl and Sparks 1985, Reuss and Johnson 1986, Tabatabai 1985, Tamm 1977, vanLoon 1984), as well as several bibliographies (Anonymous 1987, Ridout 1982, Stopp 1985). In addition to these general reviews, several reviews have examined the effects of acidic deposition on soil organisms (Alexander 1980a,b, Hågvar 1986) or nutrient cycling (Cole and Stewart 1983, Coleman 1983, Cook 1983, De Vries and Breeuwsma 1987, Olson 1983).

This review incorporates many recent investigations and serves as an update of earlier reviews on acidic deposition impacts on soil organisms. It also attempts to present information in a format that permits evaluation of potential effects of current deposition levels. Emphasis is given to effects of acidity and ionic oxides of sulfur and nitrogen. Effects of gaseous pollutants commonly associated with acidic deposition (i.e., SO_2, NO_x, O_3) are considered where appropriate.

This review has three main sections: first, an introduction to the components of acidic deposition and the direct and indirect mechanisms of their effects on soil organisms; second, an in-depth look at the research findings on the effects of acidic deposition on populations of soil organisms; and third, an examination of research on the effects of atmospheric inputs on soil biological activity. A conclusion and listing of projected research needs summarize the review.

II. Mechanisms of Acidic Deposition Effects on Soil Organisms

A. Direct Effects

The component of acidic deposition that has received greatest attention in recent years is H^+, that is, acidity. Although the effect of acidity is complicated by the association of H^+ with various anions, especially NO_3^- and SO_4^{2-}, the effect of these ions on soil organisms is considered separately.

Most soil organisms grow best at pH values near neutrality, with activity, growth, and reproduction curtailed at lower pH. Decreased pH can adversely affect membrane permeability, enzyme activity, and cellular bioenergetics (Padin 1984), causing slower growth and lower activity. Direct pH effects on soil animals are likely to be of greatest consequence with soft-bodied microfauna (e.g., Enchytraeidae), which have no external protection from H^+. Direct effects on microorganisms are most likely for bacteria, which are generally less tolerant of acidic conditions than fungi.

Not all bacteria, however, react negatively to low pH conditions, and many acidophilic species are known. Even presumed acid-intolerant bacteria, like the autotrophic nitrifiers, can apparently exist in acidic environments (Walker and Wickramasinghe 1979). In addition, microbial species often have the capability to adapt to altered soil pH (Parkin et al. 1985).

Deposition of NO_x (mainly NO_2) and its reaction products from the atmosphere adds NO_3^-, and perhaps some NO_2^-, to the soil system. Because many soil systems are limited by nitrogen, inputs of NO_3^- may stimulate the activity of soil organisms and the cycling of nitrogen (Aber et al. 1982). Decreased microbial activity is also possible, if inputs of NO_2^- are substantial, because even relatively low concentrations of NO_2^- are known to inhibit microorganisms (Grant et al. 1979). However, NO_2^- inputs are probably not large enough to affect soil NO_2^- concentrations.

Sulfur dioxide emissions ultimately enter the soil system as HSO_3^-/SO_3^{2-} or HSO_4^-/SO_4^{2-}. HSO_3^-/SO_3^{2-} is toxic to a wide variety of microorganisms and consequently inhibits carbon and nitrogen cycling (Babich and Stotzky 1978, Labeda and Alexander 1978). The toxicity of HSO_3^-/SO_3^{2-} is exacerbated as pH decreases. Inputs of SO_4^{2-} could be beneficial to soil organisms because SO_4^{2-} is a readily available form of sulfur.

Other components of atmospheric deposition may also directly affect soil organisms. Ozone can have detrimental effects on the viability of fungal spores (Babich and Stotzky 1980), but direct effects of O_3 in soils are unlikely because O_3 is rapidly degraded in soil (Turner et al. 1973). Organic compounds form another component of atmospheric deposition (Ligocki et al. 1985), but their effect on soil organisms has not been assessed.

B. Indirect Effects

Acidic deposition can indirectly affect biological activity in the soil. In some instances, indirect effects may be more significant than direct effects. Indirect effects of atmospheric inputs on soil organisms may occur through two major mechanisms: an altered soil chemical environment and altered organic matter cycling.

Acidic deposition can affect soil chemical properties in a variety of ways. Most apparent is the lowering of the soil solution pH. Decreased pH generally decreases the availability of soil nutrients such as phosphorus and the basic cations, but increases soil solution concentrations of potentially toxic elements such as aluminum and heavy trace metals. Atmospheric inputs of H^+ to soil solutions can cause displacement of other cations from soil cation exchange sites, resulting in higher soil solution osmotic strength—the so-called salt effect. Because the hydrogen ions in acidic rain are accompanied by relatively mobile anions, NO_3^- and SO_4^{2-}, the cations displaced from exchange sites by H^+ have a tendency to be leached from the soil profile. Acidic inputs may also increase weathering rates, which may compensate for cation leaching losses to some degree (Johnson et al. 1982, vanLoon 1984).

Changes in organic matter cycling as a result of pollutants can occur in several ways. Soil organic matter leaching is known to be influenced by the soil solution pH (Chang and Alexander 1984, Cronan 1985). In addition, direct damage to foliage from acidic deposition and associated pollutants could affect organic matter inputs by litter and throughfall (Evans 1984). Such damage might increase susceptibility to disease, which could further alter litterfall inputs (McLaughlin 1985). Belowground inputs of carbon from root exudation or fine-root turnover might be altered by acidic deposition or ozone, although this has not been extensively studied (Smith, this volume).

In practice, effects of acidic deposition on soil chemistry and organic matter cycling are difficult to separate. For example, nitrogen and sulfur in deposition could alter the chemical composition of foliage and litterfall, thus influencing tree nutrition, forest productivity, total litterfall, and litter decomposition rate (Johnson et al. 1982). Increased nitrogen availability could also increase internal H^+ production by favoring higher rates of nitrification (van Miegroet and Cole 1984). Similarly, changes in soil pH and associated changes in nutrient availability resulting from H^+ deposition and leaching could affect litterfall quantity and quality.

C. Input Amounts and Composition

Responses of soil organisms and biological processes to acidic inputs are dose-dependent phenomena. In studies reported in the literature, many different doses have been used. To aid in interpreting the experiments described here, I have divided the responses into low (<1kmol H^+ ha^{-1} yr^{-1}), moderate (1–10 kmol H^+ ha^{-1} yr^{-1}), and high (>10kmol H^+ ha^{-1} yr^{-1}) levels of H^+ input. The lower and upper boundaries of the moderate-level category correspond to 100 cm of rain at pH 4 or pH 3 per year, respectively. Most forest ecosystems currently experience low-level inputs, although some forests in Western Europe experience H^+ inputs between 1 and 3.7 kmol ha^{-1} yr^{-1} (van Breemen et al. 1984).

The composition of the simulated acidic rain used in artificial acidification experiments has varied considerably: for example, pH adjusted with only H_2SO_4; with H_2SO_4 and HNO_3 in the same proportions as in naturally occurring precipitation; or with H_2SO_4 and HNO_3 plus all other cations and anions at concentrations similar to natural rain. Although the latter amendment undoubtedly best simulates the field situation, no consistent effects attributable to minor cations and ions in artificial rain mixtures have been noted. Attempts to simulate dry acidic deposition inputs have generally not been included in experimental designs.

III. Effects on Soil Organism Populations

Soil is the home for a wide variety of organisms, ranging in size from 1 mm to several cm. The types of organisms present in a given soil are largely determined by the physical and chemical environment of the soil.

Soil organisms are essential to the cycling of nutrients, acting both as catalytic agents and as sinks for nutrients. They help to build and stabilize soil structure. Some microorganisms form symbiotic relationships with plants; others are pathogenic in nature. Because of the diversity and importance of these different functions, several studies have attempted to determine the influence of atmospheric deposition on soil organism populations.

A. Soil Animals

Studies of acidic deposition effects on the soil mesofauna have concentrated mainly on the more numerous soil animals: Enchytraeidae (potworms), Collembola (springtails), and Acarina (mites). Less information is available on the Lumbricidae (earthworms), Nematoda (nematodes), Protura, Araneida (spiders), and Chilopoda (centipedes).

The most detailed studies of mesofaunal responses to acidic inputs were done in the coniferous forests of Scandinavia by Bååth and coworkers in Sweden and Hågvar and co-workers in Norway. The results of several of their studies are summarized in Table 5.1. The Enchytraeidae seem to be most affected by acidic inputs, with both groups of workers showing a significant decrease in Enchytraeidae numbers at moderate to

TABLE 5.1. Effects of acidic inputs on soil animal populations.[a]

	H^+ Load (kmol ha^{-1} yr^{-1})[b]		
Organism	<1	1–10	>10
Enchytraeidae			
Cognettia sphagnetorum	o	--	--
TOTAL	o	--	--
Collembola			
Anurida pygmaea	+	o	o
Anurophorus binoculatus	o	−	
Anurophorus septentrionalis	o	o	
Folsomia litsteri	o	−	
Isotoma notabilis	+	−	--
Isotomiella minor	−	−	--
Mesaphorura yosii	+	+	++
Neanura muscorum	o	o	+
Neelus minimus	−	o	--
Onychiurus absoloni	+	o	
Onychiurus armatus	o	o	--
Tullgergia krausbaueri	+	++	
Willemia anophthalma	−	−	−
Xenylla borneri	−	−	
TOTAL	+	+	+/−
Acarina			
Astigmata	o	+/−	+
Cryptostigmata (Oribatei)			
Brachychthoniidae spp.	−	+/−	+/−
Carabodes spp	o	−	
Oppia nova	o	+/−	+/−
Oppia obsoleta	+/−	+/−	++
Oppia ornata	o	o	o
Oppioidea spp	o	+	
Steganacarus spp	o	+	++
Suctobelba spp	o	−	−
Tectocepheus velatus	+	++	++
TOTAL	+	++	++
Mesostigmata	o	o	o
Prostigmata	o	+/−	+/−
TOTAL	+	+	+
Protura			
TOTAL	o	o	o

[a]Data taken from Abrahamsen et al. 1980, Bååth et al. 1980a,b, Hågvar 1980, 1984, Hågvar and Amundsen 1981, and Hågvar and Kjøndal 1981b.

[b]++, significant increase; +, increase; o, no difference; −, decrease; --, significant decrease; +/−, mixed results in different studies.

high H^+ loads. Total numbers of Collembola generally increase in response to artificial acid rain. However, there is considerable variability in responses among Collembola species. *Anurida pygmaea*, *Mesaphorura yosii*, and *Tullgergia krausbaueri* appear to be acidophilic. Populations of all three species increased as the intensity of acidic input increased, and liming decreased the numbers of *A. pygmaea* and *M. yosii* (Hågvar 1984). Other Collembola species, such as *Isotoma notabilis*, *Isotomiella minor*, and *Neelus minimus* seem to prefer less acidic conditions. Acarina responded similarly to the Collembola, primarily because the Oribatei mites generally increased in number in the high-acidity treatments. This positive response to applied acidity was especially marked for *Oppia obsoleta*, *Steganacarus* spp., and *Tectocepheus velatus*.

A study in a deciduous forest in Tennessee, United States, involving artificial H^+ inputs of 1.2 to 6 kmol ha^{-1} yr^{-1} showed a similar positive response of microarthropods (Collembola and Acarina) to acidic inputs (Craft and Webb 1984). The same study showed no effect of acidic inputs on total litter macroarthropods.

Few studies have been reported on the effects of atmospheric deposition on other groups of soil animals. Moderate to high levels of acidic inputs reduced numbers of earthworms (Hågvar 1980), had mixed effects on populations of soil nematodes (Timans 1986) or plant parasitic nematodes (Heagle et al. 1983, Shriner 1977), and had little effect on Protura populations (Bååth et al. 1980b, Hågvar 1984). Plant parasitic nematodes were inhibited by 0.25 ppm O_3 or SO_2 (Weber et al. 1979).

Studies of mesofauna populations along point-source pollution gradients have found results somewhat different than those just described. Microarthropod populations (spiders and centipedes) were lower downwind of a coking plant (Killham and Wainwright 1981). Decreased numbers of total arthropods and Acarina, but no difference in Collembola, were found near a coal-fired power plant in Tennessee (Larkin and Kelly 1987). Whether the reductions in arthropod populations were caused by SO_2, heavy metals, or long-term chronic acidic inputs is uncertain.

B. Microorganisms

Some artificial acidification field experiments have shown reductions in total bacterial populations in the F and H horizons of the forest floor (Bååth et al. 1980a, Kreutzer and Zelles 1986, Zelles et al. 1987) or upper portion of mineral soil (Mancinelli 1986) in response to moderate H^+ inputs. Others have shown no significant change in total numbers (Bååth et al. 1979) or total microbial biomass (McColl and Firestone 1987). Contrasting results have also been observed for metabolically active (respiring) bacteria (Bååth et al. 1979, Zelles et al. 1987). One study of rhizosphere bacterial populations found increases in total bacteria, but decreases in certain physiological groups, as acidity of simulated rain increased (Shafer 1988). Stimulation of bacterial populations by ozone (0.15 ppm) was also observed by Shafer (1988).

Studies along point-source gradients of SO_2 pollution have often shown a significant detrimental effect of SO_2 on bacterial numbers in the F and H litter layers (Bewley and Parkinson 1984, Lettl 1984, Prescott et al. 1984), although Larkin and Kelly (1987) observed no differences. Both negative effects (Bryant et al. 1979, Lettl 1984, Wainwright 1979) or no effect (Bewley and Parkinson 1984, Wainwright 1980) have been found in mineral soil.

Along with changes in total bacterial populations, the bacterial species composition can also be altered. Increased SO_2 levels have been associated with increased populations of sulfur oxidizers (Lettl 1984, 1985, Wainwright 1979, but see Wainwright 1980), starch hydrolyzers (Bewley and Parkinson 1984), and spore formers (Prescott et al. 1984), but lower numbers of ammonifiers (Lettl 1984). Low to moderate acidic inputs have increased actinomycetes, proteolytic bacteria (Mancinelli 1986), and spore formers (Bååth et al. 1980b); had no effect (Mancinelli 1986) or decreased (Francis 1982) denitrifier populations; and decreased ammonium and nitrite oxidizers (Francis 1982, Mancinelli 1986), lipolytic bacteria (Mancinelli 1986), and gram-negative bacteria (Bååth et al. 1980b). Acidic inputs have been found to increase (Bååth et al. 1980b), have no effect (Mancinelli 1986), or decrease (Shafer 1988) numbers of starch-hydrolyzing bacteria. Bacteria in acid-treated soil may be smaller in size than those in control soils (Bååth et al. 1979 and 1980b).

Many of the studies that have monitored the effect of atmospheric deposition on bacterial populations have also measured fungal populations. Generally, moderate to high acidic inputs or increased SO_2 levels have not been found to affect fungal populations in organic or mineral soil horizons (Bååth et al. 1979, 1980a, 1984, Bewley and Parkinson 1984, Larkin and Kelly 1987, Wainwright 1979 and 1980), although Mancinelli (1986) observed an increase in fungal populations. An exception is a study by Fritze (1987), which found significantly less total hyphal length but no difference in fluoresein diacetate-active mycelium at sites exposed to elevated SO_2 and NO_x. The effect of moderate to high acidic input levels on metabolically active fungi is mixed, with either no difference (Bååth et al. 1979) or a decrease in active mycelium (Bååth et al. 1980a, and 1984). Bååth et al. (1984) observed some shifts in the fungal species composition; the abundance of *Penicillium spinulosum* and *Oidiodendron* cf. *echinulatum* II increased with increasing H^+ inputs of 1 to 3 kmol ha^{-1} yr^{-1}. Wainwright (1979) found higher numbers of S-oxidizing fungi in SO_2 polluted soils.

Research on the effect of acidic deposition on soil-borne plant pathogens is scarce (Shriner and Cowling 1980). There is some evidence that increased acidity can predispose trees to bacterial (Raynal et al. 1982, Ulrich 1981) or fungal root pathogens (Carey et al. 1984, Skelly 1980).

However, work with *Phytophthora cinnamomi*, a fungal root-rot pathogen, on lupine suggests that low pH may inhibit sporangia formation and decrease the number of infection sites (Shafer et al. 1985b). Clearly there is not yet enough information to draw conclusions on the effects of acidic deposition on pathogenic microorganisms.

Considerably more work has been done on beneficial fungi, the ectomycorrhizae and vesicular-arbuscular mycorrhizae (VAM). It appears that many ectomycorrhizal fungi are capable of good growth in culture at pH 3 (Hung and Trappe 1983), and ectomycorrhizal tree seedlings have been successfully used to revegetate acid mine spoils (Marx and Bryan 1975). However, ectomycorrhizal infection of *Abies balsamea* decreased as the pH of potting mixture decreased from 5 to 3 (Entry et al. 1987).

Laboratory studies suggest that mycorrhizal infection of pine (Dighton and Skeffington 1987, Shafer et al. 1985a, Stroo and Alexander 1985, Stroo et al. 1988), birch (Keane and Manning 1988), and oak seedlings (Reich et al. 1985, 1986) is decreased by artificial rain inputs of pH 3.0–3.5 compared to inputs of pH 5.6. However, Shafer et al. (1985a) found increased ectomycorrhizal infection with pH 2.4 artificial rain. Meier et al. (1989) found no effect of simulated rain at pH 3.5 or 2.5 on numbers of ectomycorrhizal root tips of red spruce, although qualitative differences in mycorrhizal types were observed.

The effects of acidic inputs in the field are not well studied but Blaschke (1986) observed that numbers of *Picea abies* mycorrhizal root tips decreased in plots exposed to moderate rates of pH 2.7 simulated acid rain. In addition to the effect of H^+, decreased ectomycorrhizal colonization of spruce as a result of nitrogen inputs has also been observed (Alexander and Fairley 1983, Meyer 1985).

Mixed results have been reported concerning effects of SO_2 and O_3 on ectomycorrhizae. Reich et al. (1985) found decreased mycorrhizal infection in SO_2-exposed red oak seedlings in the field, whereas seedlings exposed to O_3 showed an increase in mycorrhizal roots. These results were confirmed in controlled laboratory conditions at SO_2 and O_3 concentrations from 0.02 to 0.12 ppm. However, Mahoney et al. (1985) found no effect of 0.06 ppm SO_2 or 0.07 ppm O_3 on ectomycorrhizal formation on loblolly pine seedlings. Similarly, O_3 at 0.07 ppm alone, or in combination with low levels of simulated acid rain, had no effect on formation of birch ectomycorrhizae (Keane and Manning 1988). These results were confirmed by Stroo et al. (1988), who found no interaction between O_3 concentrations and simulated acid rain on ectomycorrhizal infection of white pine roots. The effects of these gaseous pollutants are likely to be indirect, through altered plant host metabolism. There is also evidence that ectomycorrhizae may provide some degree of protection to the tree from these pollutant gases (Carney et al. 1978, Garrett et al. 1982, Mahoney et al. 1985).

Vesicular-arbuscular mycorrhizae may be adversely affected by exposure to both acidic inputs and O_3 (Brewer and Heagle 1983, Ho and Trappe 1984, McCool et al. 1979). However, Killham and Firestone (1983) showed no effect of simulated rain at pH 3 or 4. Like their ectomycorrhizal counterparts, VAM seem to protect the host from O_3 damage (Ho and Trappe 1984).

Direct effects of acidic rain on other microorganisms, like algae and protozoa, are less well studied. However, laboratory work has shown HSO_3^-/SO_3^{2-} and NO_2^- to be more inhibitory to cyanobacteria than to eukaryotic algae, whereas only HSO_3^-/SO_3^{2-} inhibited protozoan respiration (Wodzinski et al. 1977, 1978).

IV. Effects on Soil Biological Processes

Although changes in organism populations as a result of atmospheric deposition may be good indicators of ecosystem stress, it is the activity of these organisms that is most important in terms of ecosystem response. Because of this, most research on the effects of atmospheric inputs has concentrated on measuring changes in the various nutrient cycling processes, especially litter decomposition and soil respiration.

A. Carbon Cycle

Many field decomposition studies have found no impact of acidic inputs on leaf litter decomposition, even under H^+ loading rates as high as 25 kmol ha^{-1} yr^{-1} (Berg 1986a, Gray and Ineson 1981, Hågvar and Kjøndal 1981a). Studies reporting decreased litter decomposition have either involved acute doses of acid (0.18M H_2SO_4) applied yearly (Berg 1986b) or found decreases over only some periods of decomposition (Bååth et al. 1980b). Increased pine needle decomposition in response to moderate H^+ inputs was found in another study (Roberts et al. 1980).

Similarly, laboratory studies have usually found that low to moderate acidic inputs either do not affect or tend to stimulate leaf litter decomposition (Abrahamsen et al. 1980, Hågvar and Kjøndal 1981a, Lee and Weber 1983) and the degradation of model substrates, like cellulose (Biénkowski et al. 1986). Hågvar and Kjøndal (1981a) observed decreased leaf decomposition only at extremely high H^+ inputs of $>$100 kmol ha^{-1} yr^{-1}. Field decomposition studies along gradients of SO_2, on the other hand, have often shown decreased organic matter decomposition closer to the source of emission (Bewley and Parkinson 1986a, Killham and Wainwright 1981, Prescott and Parkinson 1985, Prescott et al. 1984). However, exceptions to this generalization can be found (Larkin and Kelly 1987).

Research on the effects of atmospheric deposition on soil and litter respiration is extensive. Therefore, it is convenient to group and examine these studies according to: (1) measurement in the field or lab, (2) type of deposition, and (3) presence of any organic amendments.

Field studies of CO_2 evolution from soils under deciduous (Johnson and Todd 1984, Kelly and Strickland 1984), mixed (Bitton et al. 1985), or pine forests (Roberts et al. 1980, Nohrstedt 1987) have usually shown no effects with H^+ inputs of 0.25–12 kmol ha^{-1} yr^{-1}. An exception is the study by Zelles et al. (1987), which found depressed laboratory respiration rates of F horizon material exposed to 3 kmol H^+ ha^{-1} yr^{-1}; however, these differences disappeared when the samples were amended with glucose. No differences were found in soil respiration rates along gradients of SO_2 (Larkin and Kelly 1987, Wainwright 1980).

The effects of acidic inputs on CO_2 evolution in laboratory incubations of unamended soils and litter are summarized in Table 5.2. Low-level additions of acid had no effect on respiration from litter layers and occasionally increased CO_2 evolution in mineral soils. Moderate H^+ inputs generally had no effect or decreased respiration in incubated litter layers or organic soils, and usually had either no effect or increased CO_2 evolution in samples with a mineral soil component. High acidification rates generally had no effect or tended to decrease respiration from all samples. Positive responses to high acidic inputs in litter layers were restricted to one organic horizon that had a relatively high pH (Chang and Alexander

Table 5.2. Effects of acidic inputs on laboratory soil respiration.

	H^+ Load (kmol ha^{-1} yr^{-1})[a]			
Soil sample type	<1	1–10	>10	Reference
Litter or organic soil		o	−	Bååth et al. 1979
		−		Bååth et al. 1980a, 1984
		−		Bitton and Boylan 1985
		−/o	−/+	Chang and Alexander 1984
	o	o	o	Cronan 1985
		o		Hovland 1981
		o	−/+	Hovland et al. 1980
		−		Lohm et al. 1984
		o	−	Persson et al. 1989
		o		Popovic 1984
Organic and mineral soil		+/o	+/o	Bitton and Boylan 1985
		o		Cronan 1985
			−	Klein et al. 1984
	o	o		Will et al. 1986
Mineral soil		+/o	−/o	Bieńkowski et al. 1986
		+/o	+/o	Killham and Firestone 1982
	+/o	+	−	Killham et al. 1983
		o		Lohm et al. 1984
	o	o		McColl and Firestone 1987

[a] +, significant increase; o, no significant difference; −, significant decrease; +/o, +/−, and −/o, mixed results.

1984) and to the very early stages of spruce needle decomposition (Hovland et al. 1980). Positive responses to high acidic input levels were obtained with only two mineral soils, both of which had relatively high organic matter levels (Bitton and Boylan 1985).

Some of the studies cited in Table 5.2 also looked at the effect of acidic inputs on organic carbon leaching (Chang and Alexander 1984, Cronan 1985) or available carbon (Killham and Firestone 1982, Killham et al. 1983). High levels of H^+ loading decreased both the amount of available carbon and the amount of organic carbon leached (Chang and Alexander 1984, Hay et al. 1985, Killham et al. 1983), with corresponding inhibition of CO_2 evolution in most cases. Low or moderate acidic input levels, on the other hand, increased available carbon (Killham et al. 1983), corresponding to increased respiration, or had mixed (Cronan 1985) or negative (Chang and Alexander 1984) effects on organic carbon leaching and respiration.

Additional studies in which the pH of the soil was directly adjusted with additions of mineral acids have shown results similar to those outlined in Table 5.2, that is, high levels of acidic inputs have resulted in decreased respiration (Francis 1982, Francis et al. 1980, Hendrickson 1985), whereas lower level inputs have had no effect on heat output, another general measure of microbial activity (Ljungholm 1979).

Soils amended with various carbon sources have generally responded in a manner similar to unamended soils when subjected to various levels of acid inputs. Glucose-amended soils have shown decreased respiration once the soil pH was lowered below 3 (Bewley and Stotzky 1983a and 1983b), to 3.8 (Lohm 1980), or 4 (Ljungholm et al. 1979, Lohm et al. 1984), and also after high acute or chronic levels of H^+ input (Strayer and Alexander 1981). These studies might represent the effect of acidic inputs on the degradation of readily available carbonaceous compounds in newly fallen leaves, recently dead roots, or dissolved soil organic compounds. Soils amended with more complex carbon compounds, such as cellulose (Ljungholm et al. 1979, Moloney et al. 1983), vanillin (Bewley and Stotzky 1984), oak leaves (Francis 1982, Francis et al. 1980), or spruce and fir needles (Moloney et al. 1983), have also shown depressed microbial activity when the soil pH was lowered to 3.2–4.0. At higher pH values, no effect was observed. The addition of more complex carbon substrates may simulate the decomposition of less readily available carbon compounds; for example, vanillin is quite similar to organic compounds isolated from soil humus.

Responses of soil respiration to point-source gradients of SO_2 have been mixed. Some sites have shown no effect of SO_2 pollution on respiration of the forest floor (Lettl 1984, Nohrstedt 1985) or mineral soil (Lettl 1984, Wainwright 1980); others have demonstrated lower CO_2 evolution rates from the forest floor (Lettl 1984) or mineral soil (Bryant et al. 1979) at polluted sites. Evolution of CO_2 was depressed in soils collected near the pollution source after the addition of starch, cellulose, or vanillin, whereas adding glucose stimulated CO_2 evolution (Bewley and Parkinson 1986a, Bryant et al. 1979).

B. Nitrogen Cycle

Microorganisms are the primary agents responsible for nitrogen cycling in soil systems. Mineralization of organic nitrogen and immobilization of inorganic nitrogen are accomplished by all types of microorganisms; nitrification and denitrification are predominantly bacterial processes; dinitrogen fixation is by bacteria, actinomycetes, and cyanobacteria.

Field studies have shown no effects of acidic inputs on NH_4^+ or NO_3^- concentrations in the soil solution (Hern et al. 1985) or in soil extracts (Johnson and Todd 1984), even under high H^+ inputs. Laboratory studies, however, have shown that high levels of artificial acidic rain increased NH_4^+ concentrations, presumably through chemical deamination (Biénkowski et al. 1986, Killham and Firestone 1982, Killham et al. 1983), but there was no plant uptake in these systems.

The results of laboratory incubation studies on net nitrogen mineralization and nitrification are summarized in Table 5.3. Generally, net nitrogen mineralization is unaffected by acidic inputs, although there

TABLE 5.3. Effects of acidic inputs on laboratory net nitrogen mineralization and nitrification.

N Mineralization[a]			Nitrification			
H^+ Load ($kmol\ ha^{-1}\ yr^{-1}$)						
<1	1–10	>10	<1	1–10	>10	Reference
	+			+/−		Bieńkowski et al. 1986
−	o		−	−		Bitton et al. 1985
−/o	−/o	+/−	o	o	+/−	Firestone et al. 1984
		o			−	Hern et al. 1985
o	o		o	o		Johnson and Todd 1984
	−	+/−		−	+/−	Klein and Alexander 1986
	+/−			+/−		Klein et al. 1983
		−			−	Klein et al. 1984
	+/−			o		Like and Klein 1985
	o					Lohm et al. 1984
o	+/o		o	−/o		McColl and Firestone 1987
		+/−			−/o	Novick et al. 1983
	o	+		o	o	Persson et al. 1989
	o			o		Popovic 1984
		+/o			+/o	Strayer et al. 1981
	+/−	+/−		−	−	Stroo and Alexander 1986
	o	−				Will et al. 1986

[a]+, significant increase; o, no significant difference; −, significant decrease; +/o, +/−, and −/o, mixed results.

appears to be a trend toward greater inhibition at the highest H^+ loading rates. Acidifying soils to low pH values has been shown to inhibit nitrogen mineralization (Francis 1982, Francis et al. 1980, Hendrickson 1985). Instances of increased nitrogen mineralization after acidification could be the result of increased nitrogen availability, because acidic inputs have been found to increase leaching of amino acids from soil (Hay et al. 1985).

Nitrification, a process known to be inhibited in most acidic soils, is usually depressed by acidic inputs. However, some studies showed no effect or even a stimulation of NO_3^- production with acidification. Two of the studies showing a positive nitrification response to acidic inputs also contained results suggesting that heterotrophic nitrification (a process not as affected by low pH) was a dominant contributor to NO_3^- production (Klein et al. 1983, Strayer and Alexander 1981).

Inputs of SO_2 have been found to either increase (Wainwright 1980) or decrease (Nohrstedt 1985) net nitrogen mineralization. Nevell and Wainwright (1987) found no difference in nitrification rates between SO_2- polluted and unpolluted sites.

Considerably less work has been done on other aspects of nitrogen cycling. It is known that there are denitrifying organisms adapted to low soil pH that function at rates comparable to those found in neutral soils (Parkin et al. 1985), and that the composition of denitrification gases shifts toward N_2O at low pH levels (Firestone et al. 1980). Francis (1982) and Francis et al. (1980) reported that rapidly acidified soils had decreased N_2O production (actual denitrification, i.e., N_2O production in the presence of the C_2H_2 block, was not measured).

Denitrification potential measurements have shown a complex response to acid additions: low to moderate H^+ inputs either depress or have no effect, whereas high inputs tend to stimulate denitrification potentials (Firestone et al. 1984, McColl and Firestone 1987). The enhancement at high H^+ loading rates may result from increased NO_3^- availability from the mixed acid treatment. To date, no studies have examined the effects of actual or simulated acid rain on actual denitrification rates.

Acidic rain appears to have little effect on nonsymbiotic nitrogen fixation in soils (Bitton and Boylan 1985, Francis 1982, Francis et al. 1980), although heavy H^+ inputs significantly decreased acetylene reduction by cyanobacteria in several forest soils (Chang and Alexander 1983b). Decreased pH adversely affected nitrogen fixation by the cyanobacterial symbiont in epiphytic lichens (Denison et al. 1977). Sulfur dioxide pollution had no effect on soil nitrogenase activity (Nohrstedt, 1985).

Acidic inputs appear to inhibit symbiotic nitrogen fixation, at least in the *Rhizobium*-legume symbiosis, by decreasing nodulation (Chang and Alexander 1983a). Effects of acidic inputs on the *Frankia* symbiosis are unknown, although numbers of *Frankia* inoculated into low pH(<4.2) soils often decrease (Griffiths and McCormick 1984, Smolander and Sundman 1987).

C. Sulfur Cycle

Because sulfur is one of the main components of atmospheric deposition, the impacts of atmospheric deposition on sulfur cycling are both direct and indirect. However, despite the relatively large inputs of sulfur (1–60 kg S ha^{-1} yr^{-1}) to some ecosystems (Johnson 1984), few studies have been done on the effect of these inputs on sulfur cycling. Increases in S-oxidizers have been measured, but effects on sulfur oxidation have been mixed (Wainwright 1979, 1980). Enhanced microbial immobilization of sulfur after simulated acidic rain inputs has been measured (Hern et al. 1985).

D. Phosphorus Cycle

Microorganisms are active in the mineralization of organic phosphorus and immobilization of inorganic phosphate into organic forms. The few studies that have been done have found no changes in extractable phosphorus (Johnson and Todd 1984) or phosphorus solubility (Wainwright 1980) as the result of acidic inputs.

E. Soil Enzyme Activity

Soil enzymes play an important role in the cycling of many nutrients and also serve as an indicator of microbial activity. Because enzyme production and activity are affected by soil chemical properties, atmospheric inputs could have an impact.

The effects of acidic inputs on several soil enzymes are presented in Table 5.4. Dehydrogenase, an index of overall microbial activity, showed results similar to those found for soil respiration: possible stimulation at low input rates, mixed effects at moderate input rates, and generally a depression of activity under high H^+ loads. Cellulase activity showed similar trends: behavior that is in agreement with studies on carbon cycling processes. Low to moderate acidic inputs either stimulated or did not affect urease and protease activity, whereas high input rates decreased activity. These observations are consistent with observed effects of acidic inputs on nitrogen mineralization. Arylsulfatase, an enzyme involved in ester sulfate bond cleavage, generally responded in the same manner as urease and protease. Phosphatase, an enzyme that cleaves ester phosphate bonds, showed mixed responses to low and moderate H^+ loads, and was generally inhibited at high acidic input levels.

Enzyme assays of soils occurring along SO_2 gradients have generally shown no response (Bewley and Parkinson 1986b, Nevell and Wainwright 1987, Nohrstedt 1985, Wainwright 1980), except for higher rhodanase activity, an enzyme that is responsible for the cleavage of thiosulfate to sulfate and elemental sulfur (Wainwright 1980).

TABLE 5.4. Effects of acidic inputs on soil enzyme activities.

Enzyme	H^+ Load (kmol ha^{-1} yr^{-1})[a] <1	1–10	>10	Reference
Dehydrogenase	o	o		Bitton et al. 1985
		−/o	o	Bitton and Boylan 1985
			−	Francis et al. 1980, Francis 1982
	+/o	+	−	Killham et al. 1983
	+	o		Will et al. 1986
Phosphatase	o	o		Bitton et al. 1985
		−/o	o	Bitton and Boylan 1985
		+/o	−/o	Killham and Firestone 1982
	+/o	−/o	−/o	Killham et al. 1983
	o	−		Will et al. 1986
Urease	+	+		Bitton et al. 1985
			−	Francis et al. 1980, Francis 1982
		o	−/o	Killham and Firestone 1982
	+/o	+/o	−	Killham et al. 1983
	o	+		Will et al. 1986
Arylsulfatase		+/o	−/o	Killham and Firestone 1982
	+/o	+	−/o	Killham et al. 1983
		−		Press et al. 1985
	o	o		Will et al. 1986
Protease	o	+		Bitton et al. 1985
		o	−/o	Bitton and Boylan 1985
Cellulase		o	−	Hovland 1981

[a] +, significant increase; o, no significant difference; −, significant decrease; +/o, +/−, and −/o, mixed results.

V. Conclusions and Research Needs

Acidic deposition, especially at high loading rates, definitely affects soil organisms and biological processes. These effects may be positive or negative, depending upon environmental conditions and the organism or process involved.

High inputs of acidity (>10 kmol H^+ ha^{-1} yr^{-1}) usually decrease populations of soil organisms and decrease biological activity. Lower inputs of H^+ have more mixed effects. For example, there may be no change in total soil microfauna populations, but species composition may be altered. Respiration from the forest floor is sometimes decreased, but there is often little negative effect, or even a positive effect on CO_2 evolution from mineral soil. Because most research on acid rain effects has been short term, even these tentative conclusions are subject to change.

The effects of gaseous components, like O_3, SO_x, and NO_x, may be both direct (increased nitrogen and sulfur fertility) and indirect, through their effect on plants. Consequently, their impact is more difficult to assess. Certainly, high levels of these inputs would be expected to be harmful, but the balance between potentially beneficial fertilizer effects and detrimental toxic effects is not well defined.

Several areas for further investigation are apparent after reviewing the results obtained from experimentation on the effects of atmospheric deposition.

1. Establish long-term studies of acidic deposition impacts on soil organisms and processes. Baseline data are needed for currently nonimpacted sites that may serve as controls and also as sites for manipulative experiments. Long-term studies with realistic levels of atmospheric inputs are crucial because short-term, acute loading is not a representative model system. Finally, attention must be given to quantifying input intensities so that studies can be compared.
2. Determine the effects of organic inputs from the atmosphere on soil organisms and biological processes. Levels of organics may be too low to have any effect, but no data are currently available to address this issue.
3. Determine the relative importance of natural acidification processes, like nitrification and litter decomposition, versus H^+ inputs on soil biology. Although air pollution effects are an important environmental concern, many other management inputs (e.g., fertilization, shortened rotations, etc.) or natural phenomena (e.g., presence of nitrogen-fixing plants) may make a greater contribution to soil acidification than anthropogenic sources. Before this question can be evaluated, we need to know what the natural rates of acidification are and what processes are involved.
4. Investigate how acidic deposition affects sulfur cycling. There is a growing body of literature on sulfur budgets, but less is known about the biological aspects of sulfur cycling. The work of Fitzgerald et al. (1988) and Mitchell et al. (in press) are notable exceptions. Information is needed concerning possible acidic deposition effects on rates of sulfur mineralization, immobilization, oxidation, and reduction.
5. Investigate how acidic deposition affects phosphorus availability and cycling. It is generally accepted that the availability of inorganic phosphorus decreases as pH is lowered in acidic soils. Consequently, the cycling of organic phosphorus is likely to be more important as soil pH decreases. Because phosphorus is often the second most limiting plant nutrient after nitrogen, the need for information about phosphorus is apparent.
6. Determine the mechanisms of response of soil organisms to pollutant inputs. By acquiring a more basic understanding of how organisms respond to environmental stresses, it should be possible to better predict or model the response of soil biological processes to atmospheric deposition over a wide range of soils and forests. If possible, the independent

effects of increased nitrogen and sulfur inputs should be separated from H^+ inputs. The synergistic effects of these inputs need to be quantified. This will likely require the cooperative research of soil chemists, foresters, and soil biologists.

7. Determine effects of acidic deposition on specific physiological groups of microorganisms. Work on impacts of various components of atmospheric deposition on mycorrhizal fungi and pathogenic microorganisms has begun, but more research remains to be done with these organisms and others. *Frankia*, the nitrogen-fixing symbiont of a variety of plant genera common in forests, is an example of an important microorganism for which no information exists regarding sensitivity to atmospheric inputs. Information about the effect of acidic deposition on mycorrhizae and *Frankia* symbioses will contribute not only to a better understanding of these microorganisms, but will also provide insight into possible effects of acidity on phosphorus and nitrogen cycling.

Acknowledgement. I would like to express my appreciation to Nanci Pascoe, who did much of the library work associated with this review, and to Frank Wildensee, who translated some of the articles in German. This project was supported by The National Council of the Paper Industry for Air and Stream Improvement.

References

Aber, J.D., G.R. Hendrey, D.B. Botkin, A.J. Francis, and J.M. Melillo. 1982. Potential effects of acid precipitation on soil nitrogen and productivity of forest ecosystems. Water Air Soil Pollut 18:405–412.

Abrahamsen, G. 1980. Acid precipitation, plant nutrients and forest growth. *In*: Ecological Impact of Acid Precipitation, Proceedings of an International Conference, D. Drabloes and A. Tollan, eds., pp. 58–63. SNSF, Oslo, Norway.

Abrahamsen, G. 1984. Effects of acidic deposition on forest soil and vegetation. Philos Trans Royal Soc London B 305:369–381.

Abrahamsen, G., R. Horntvedt, and B. Tveite. 1977. Impacts of acid precipitation on coniferous forest ecosystems. Water Air Soil Pollut 8:57–73.

Abrahamsen, G., J. Hovland, and S. Hågvar. 1980. Effects of artificial acid rain and liming on soil organisms and the decomposition of organic matter. *In*: Effects of Acid Precipitation on Terrestrial Ecosystems, T.C. Hutchinson and M. Havas, eds., pp. 341–362. New York: Plenum Press.

Alexander, I.J. and R.I. Fairley. 1983. Effects of N fertilization on populations of fine roots and mycorrhizas in spruce humus. Plant Soil 72:49–53.

Alexander, M. 1980a. Effects of acidity on microorganisms and microbial processes in soil. *In*: Effects of Acid Precipitation on Terrestrial Ecosystems, T.C. Hutchinson and M. Havas (eds.),pp. 363–374. New York: Plenum Press.

Alexander, M. 1980b. Effects of acid precipitation on biochemical activities in soil. *In*: Ecological Impact of Acid Precipitation, Proceedings of an International Conference, D. Drabloes and A. Tollan (eds.), pp. 40–52. SNSF, Oslo, Norway.

Anonymous. 1987. International bibliography of acid rain,1977–1986. Philadelphia: BIOSIS.

Arthur, M.F. and C.K. Wagner. 1983. Supplemental literature review. Environ Exp Bot 23:259–279.

Bååth, E., B. Lundgren, and B. Söderström. 1979. Effects of artificial acid rain on microbial activity and biomass. Bull Environ Contam Toxicol 23:737–740.

Bååth, E., B. Lundgren, and B. Söderström. 1984. Fungal populations in podzolic soil experimentally acidified to simulate acid rain. Microb Ecol 10:197–203.

Bååth, E., B. Berg, U. Lohm, B. Lundgren, H. Lundkvist, T. Rosswall, B. Söderström, and A. Wiren. 1980a. Soil organisms and litter decomposition in a Scots pine forest - Effects of experimental acidification. *In*: Effects of Acid Precipitation on Terrestrial Ecosystems, T.C. Hutchinson and M. Havas, (eds.), pp. 375–380. New York: Plenum Press.

Bååth, E., B. Berg, U. Lohm, B. Lundgren, H. Lundkvist, T. Rosswall, B. Söderström, and A. Wiren. 1980b. Effects of experimental acidification and liming on soil organisms and decomposition in a Scots pine forest. Pedobiologia 20:85–100.

Babich, H. and G. Stotzky. 1978. Atmospheric sulfur compounds and microbes. Environ Res 15:513–531.

Babich, H. and G. Stotzky. 1980. Environmental factors that influence the toxicity of heavy metal and gaseous pollutants to microorganisms. CRC Crit Rev Microbiol 8:99–145.

Berg, B. 1986a. The influence of experimental acidification on nutrient release and decomposition rates of needle and root litter in the forest floor. For Ecol Manage 15:195–213.

Berg, B. 1986b. The influence of experimental acidification on needle litter decomposition in a *Picea abies* L. forest. Scand J For Res 1:317–322.

Bewley, R.J. and D. Parkinson. 1984. Effects of sulphur dioxide pollution on forest soil microorganisms. Can J Microbiol 30:179–185.

Bewley, R.J. and D. Parkinson. 1986a. Monitoring the impact of acid deposition on the soil microbiota, using glucose and vanillin decomposition. Water Air Soil Pollut 27:57–68.

Bewley, R.J.F. and D. Parkinson. 1986b. Sensitivity of certain soil microbial processes to acid deposition. Pedobiologia 29:73–84.

Bewley, R.J. and G. Stotzky. 1983a. Simulated acid rain (H_2SO_4) and microbial activity in soil. Soil Biol Biochem 15:431–437.

Bewley, R.J. and G. Stotzky. 1983b. Anionic constituents of acid rain and microbial activity in soil. Soil Biol Biochem 15:431–437.

Bewley, R.J. and G. Stotzky. 1984. Degradation of vanillin in soil-clay mixtures treated with simulated acid rain. Soil Sci 137:415–418.

Bieńkowski, P., Z. Fischer, and J. Goździewicz. 1986. Reaction of soil system to "acid rain" under conditions of a laboratory experiment. Ekol Pol 34:75–86.

Bitton, B. and R.A. Boylan. 1985. Effect of acid precipitation on soil microbial activity: I. Soil core studies. J Environ Qual 14:66–69.

Bitton, G., B.G. Volk, D.A. Graetz, J.M. Bossart, R.A. Boylan, and G.E. Byers. 1985. Effect of acid precipitation on soil microbial activity. II. Field studies. J Environ Qual 14:69–71.

Blaschke, H. 1986. Effect of artificial acid rain on the development of fine-roots and mycorrhizae of Norway spruce. Forstwiss Centralbl (Hamburg) 105:324–329.

Brewer, P.F. and A.S. Heagle. 1983. Interactions between *Glomus geosporum* and exposure of soybeans to ozone or simulated acid rain in the field. Phytopathology 73:1035–1040.

Bryant, R.D., E.A. Gordy, and E.J. Laishley. 1979. Effect of soil acidification on the soil microflora. Water Air Soil Pollut 11:437–445.

Carey, A.C., E.A. Miller, G.T. Gefalle, P.M. Wargo, W.H. Smith, and T.G. Siccama. 1984. *Armillaria mellea* and decline of red spruce. Plant Dis 68:794–795.

Carney, J.L., H.E. Garrett, and H.G. Hedrick. 1978. Influence of air pollutant gases on oxygen uptake of pine roots with selected ectomycorrhizae. Phytopathology 68:1160–1163.

Chang, T. and M. Alexander. 1983a. Effects of simulated acid precipitation on growth and nodulation of leguminous plants. Bull Environ Contam Toxicol 30:379–387.

Chang, T. and M. Alexander. 1983b. Effect of simulated acid precipitation on algal fixation of nitrogen and carbon dioxide in forest soils. Environ Sci Technol 17:11–13.

Chang, T. and M. Alexander. 1984. Effects of simulated acid precipitation on decomposition and leaching of organic carbon in forest soils. Soil Sci 138:226–234.

Cole, C.V. and J.W.B. Stewart. 1983. Impact of acid deposition on P cycling. Environ Exp Bot 23:235–241.

Coleman, D.C. 1983. The impacts of acid deposition on soil biota and C cycling. Environ Exp Bot 23:225–233.

Cook, R.B. 1983. The impact of acid deposition on the cycles of C, N, P, and S. *In*: The Major Biogeochemical Cycles and Their Interactions, B. Bolin and R.B. Cook (eds.), pp. 345–364. New York: Wiley.

Craft, C.B. and J.W. Webb. 1984. Effects of acidic and neutral sulfate salt solutions on forest floor arthropods. J Environ Qual 13:436–440.

Cronan, C.S. 1985. Comparative effects of precipitation acidity on three forest soils: Carbon cycling responses. Plant Soil 88:101–112.

Denison, R., B. Caldwell, B. Bormann, L. Eldred, C. Swanber, and S. Anderson. 1977. The effects of acid rain on nitrogen fixation in western Washington coniferous forests. Water Air Soil Pollut 8:21–34.

De Vries, W. and A. Breeuwsma. 1987. The relation between soil acidification and element cycling. Water Air Soil Pollut35:293–310.

Dighton, J. and R.A. Skeffington. 1987. Effects of artificial acid precipitation on the mycorrhizas of Scots pine seedlings. New Phytol 107:191–202.

Entry, J.A., K. Cromack, Jr., S.G. Stafford, and M.A. Castellano. 1987. The effect of pH and aluminum concentration on ectomycorrhizal formation in Abies balsamea. Can J For Res 17:865–871.

Evans, L.S. 1984. Botanical aspects of acidic precipitation. Bot Rev 50:449–490.

Fiedler, H.J. and S.D. Thakur. 1985. Effects of SO_2 and acid precipitation on soil fertility and plant nutrition in forest ecosystems. Beitr Forstwirtsch 19:25–34.

Firestone, M.K., R.B. Firestone, and J.M. Tiedje. 1980. Nitrous oxide from soil denitrification: Factors controlling its biological production. Science 208:749–750.

Firestone, M.K., J.G. McColl, K.S. Killham, and P.D. Brooks. 1984. Microbial response to acid deposition and effects on plant productivity. *In*: Direct and

Indirect Effects of Acidic Deposition on Vegetation, R.A. Lindhurst (ed.), pp. 51–63. Boston: Butterworths.
Fitzgerald, J.W., W.T. Swank, T.C. Strickland, J.T. Ash, D.D. Hale, T.L. Andrew, and M.E. Watwood. Sulfur pools and transformations in litter and surface soil of a hardwood forest. pp. 245–253. New York: Springer-Verlag.
Francis, A.J. 1982. Effects of acidic precipitation and acidity on soil microbial processes. Water Air Soil Pollut 18:375–394.
Francis, A.J., D. Olson, and R. Bernatsky. 1980. Effect of acidity on microbial processes in a forest soil. *In*: Ecological Impact of Acid Precipitation, Proceedings of an International Conference, D. Drabloes and A. Tollan (eds.), pp. 166–167. SNSF, Oslo, Norway.
Fritze, H. 1987. The influence of urban air pollution on soil respiration and fungal hyphal length. Ann Bot Fenn 24:251–256.
Garrett, H.E., J.L. Carney, and H.G. Hedrick. 1982. The effects of ozone and sulfur dioxide on respiration of ectomycorrhizal fungi. Can J For Res 12:141–145.
Grant, I.F., K. Bancroft, and M. Alexander. 1979. SO_2 and NO_2 effects on microbial activity in acid forest soil. Microb Ecol 5:85–89.
Gray, T.R.G. and P. Ineson. 1981. The effects of sulphur dioxide and acid rain on the decomposition of leaf litter. J Sci Food Agric 32:624–625.
Griffiths, A.P. and L. H. McCormick. 1984. Effects of soil acidity on nodulation of *Alnus glutinosa* and viability of *Frankia*. Plant Soil 79:429–434.
Hågvar, S. 1980. Effects of acid precipitation on soil and forest. 7. Soil animals. *In*: Ecological Impact of Acid Precipitation, Proceedings of an International Conference, D. Drabloes and A. Tollan (eds.), pp. 202–203. SNSF, Oslo, Norway.
Hågvar, S. 1984. Effects of liming and artificial acid rain on collembola and protura in coniferous forest. Pedobiologia 27:341–354.
Hågvar, S. 1986. Atmospheric deposition: Impact via soil biology. *In*: Acidification and its Policy Implications, T. Schneider (ed.), pp. 153–160. Amsterdam: Elsevier.
Hågvar, S. and T. Amundsen. 1981. Effects of liming and artificial acid rain in the mite (*Acari*) fauna in coniferous forest. Oikos 37:7–20.
Hågvar, S. and B.R. Kjøndal. 1981a. Decomposition of birch leaves: Dry weight loss, chemical changes, and effects of artificial acid rain. Pedobiologia 22:232–245.
Hågvar, S. and B.R. Kjøndal. 1981b. Effects of artificial acid rain on the microarthropod fauna in decomposing birch leaves. Pedobiologia 22:409–422.
Hay, G.W., J.H. James, and G.W. vanLoon. 1985. Solubilization effects of simulated acid rain on the organic matter of forest soil; preliminary results. Soil Sci 139:422–430.
Heagle, A.S., R.B. Philbeck, P.F. Brewer, and R.E. Ferrell. 1983. Response of soybeans to simulated acid rain in the field. J Environ Qual 12:538–543.
Hendrickson, O.Q. 1985. Variation in the C:N ratio of substrate mineralized during forest humus decomposition. Soil Biol Biochem 17:435–440.
Hern, J.A., G.K. Rutherford, and G.W. vanLoon. 1985. Chemical and pedogenetic effects of simulated acid precipitation on two eastern Canadian forest soils. I. Nonmetals. Can J For Res 15:839–847.
Ho, I. and J.M. Trappe. 1984. Effects of ozone exposure on mycorrhiza formation and growth of *Festuca arundinacea*. Environ Exp Bot 24:71–74.

Hovland, J. 1981. The effect of artificial acid rain on respiration and cellulase activity in Norway spruce needle litter. Soil Biol Biochem 13:23–26.

Hovland, J., G. Abrahamsen, and G. Ogner. 1980. Effects of artificial acid rain on decomposition of spruce needles and on mobilization and leaching of elements. Plant Soil 56:365–378.

Hung, L.L. and J.M. Trappe. 1983. Growth variation between and within species of ectomycorrhizal fungi in response to pH *in vitro*. Mycologia 75:234–241.

Johnson, A.H. and T.G. Siccama. 1983. Acid deposition and forest decline. Environ Sci Technol 17:294A–305A.

Johnson, D.W. 1984. Sulfur cycling in forests. Biogeochemistry 1:29–43.

Johnson, D.W. and D.E. Todd. 1984. Effects of acid irrigation on CO_2 evolution, extractable nitrogen, phosphorus, and aluminum in a deciduous forest soil. Soil Sci Soc Am J 48:664–666.

Johnson, D.W., J. Turner, and J.M. Kelly. 1982. The effects of acid rain on forest nutrient status. Water Resour Res 18:449–461.

Keane, K.D. and W.J. Manning. 1988. Effects of ozone and simulated acid rain on birch seedling growth and formation of ectomycorrhizae. Environ Pollut 52:55–65.

Kelly, J.M. and R.C. Strickland. 1984. CO_2 efflux from deciduous forest litter and soil in response to simulated acid rain treatment. Water Air Soil Pollut 23:431–440.

Kennedy, I.R. 1986. Acid Soil and Acid Rain: The Impact on the Environment of Nitrogen and Sulfur cycling. New York: Wiley.

Killham, K. and M.K. Firestone. 1982. Evaluation of accelerated H^+ applications in predicting soil chemical and microbial changes due to acid rain. Commun Soil Sci Plant Anal 13:995–1001.

Killham, K. and M.K. Firestone. 1983. Vesicular-arbuscular mycorrhizal mediation of grass response to acidic and heavy metal depositions. Plant Soil 72:39–48.

Killham, K. and M. Wainwright. 1981. Deciduous leaf litter and cellulose decomposition in soil exposed to heavy atmospheric pollution. Environ Pollut Ser A 26:79–85.

Killham, K., Firestone, M.K., and J.G. McColl. 1983. Acid rain and soil microbial activity: Effects and their mechanisms. J Environ Qual 12:133–137.

Klein, T.M. and M. Alexander. 1986. Effect of the quantity and duration of application of simulated acid precipitation on nitrogen mineralization and nitrification in a forest soil. Water Air Soil Pollut 28:309–318.

Klein, T.M., J.P. Kreitinger, and M. Alexander. 1983. Nitrate formation in acid forest soils from the Adirondacks. Soil Sci Soc Am J 47:506–508.

Klein, T.M., N.J. Novick, J.P. Kreitinger, and M. Alexander. 1984. Simultaneous inhibition of carbon and nitrogen mineralization in a forest soil by simulated acid precipitation. Bull Environ Contam Toxicol 32:698–703.

Kreutzer, K. and L. Zelles. 1986. The effect of acid irrigation and liming on the microbial activity in the soil. Forstwiss Centralbl (Hamburg) 105:314–317.

Labeda, D.P. and M. Alexander. 1978. Effects of SO_2 and NO_2 on nitrification in soil. J Environ Qual 7:523–526.

Larkin, R.P. and J.M. Kelly. 1987. Influence of elevated ecosystem S levels on litter decomposition and mineralization. Water Air Soil Pollut 34:415–428.

Lee, J.J. and D.E. Weber. 1983. Effects of sulfuric acid rain on decomposition rate and chemical element content of hardwood leaf litter. Can J Bot 61:872–879.

Lettl, A. 1984. The effect of atmospheric SO_2 pollution on the microflora of forest soils. Folia Microbiol 29:455–475.

Lettl, A. 1985. SO_2 pollution. II. Influence of inorganic sulphur compounds on bacterial communities of forest soils. Oekologia 4:121–133.

Ligocki, M.P., C. Leuenberger, and J.F. Pankow. 1985. Trace organic compounds in rain. II. Gas scavenging of neutral organic compounds. Atmos Environ 19:1609–1617.

Like, D.E. and R.M. Klein. 1985. The effect of simulated acid rain on nitrate and ammonium production in soils from three ecosystems of Camels Hump Mountain, Vermont. Soil Sci 140:352–355.

Ljungholm, K., B. Norén, and I. Wadsö. 1979. Microcalorimetric observations of microbial activity in normal and acidified soils. Oikos 33:24–30.

Lohm, U. 1980. Effects of experimental acidification on soil organism populations and decomposition. *In*: Ecological Impact of Acid Precipitation, Proceedings of an International Conference, D. Drabloes and A. Tollan (eds.) pp. 178–179. SNSF, Oslo, Norway.

Lohm, U., K. Larsson, and H. Nômmik. 1984. Acidification and liming of coniferous forest soil: Long-term effects on turnover rates of carbon and nitrogen during an incubation experiment. Soil Biol Biochem 16:343–346.

Mahoney, M.J., B.I. Chevone, J.M. Skelly, and L.D. Moore. 1985. Influence of mycorrhizae on the growth of loblolly pine seedings exposed to ozone and sulfur dioxide. Phytopathology 75:679–682.

Mancinelli, R.L. 1986. Alpine tundra soil bacterial responses to increased soil loading rates of acid precipitation, nitrate, and sulfate, Front Range, Colorado, U.S.A. Arct Alp Res 18:269–275.

Marx, D.H. and W.C. Bryan. 1975. The significance of mycorrhizae to forest trees. *In*: Forest Soils and Forest Land Management, B. Bernier and C.H. Winget (eds.) pp. 107–117. Laval University, Quebec.

McColl, J.G. and M.K. Firestone. 1987. Cumulative effects of simulated acid rain on soil chemical and microbial characteristics and conifer seedling growth. Soil Sci Soc Am J 51:794–800.

McCool, P.M., J.A. Menge, and O.C. Taylor. 1979. Effects of Ozone and HCl gas on the development of the mycorrhizal fungus *Glomus fasciculatus* and growth of 'Troyer' citrage. J Am Soc Hort Sci 104:151–154.

McFee, W.W. 1983. Sensitivity ratings of soils to acid deposition: A review. Environ Exp Bot 23:203–210.

McLaughlin, S.B. 1985. Effects of air pollution on forests—a critical review. J Air Pollut Control Assoc 35:512–534.

Meier, S., W.P. Robarge, R.I. Bruck, and L.F. Grand. 1989. Effects of simulated rain acidity on ectomycorrhizae of red spruce seedlings potted in natural soil. Environ Pollut 59:315–324.

Meyer, F.H. 1985. Effect of the nitrogen factor on the mycorrhizalcomplement of Norway spruce seedlings in humus from a damaged site. Allg Forstztg 9/10:208–219.

Miller, H.G. 1984. Deposition-plant-soil interactions. Philos Trans Royal Soc London B 305:339–351.

Mitchell, M.S., M.B. David, and R.B. Harrison. (in press). Sulfur dynamics of forest ecosystems. *In*: Sulfur Cycling in Terrestrial Systems and Wetlands, R.W. Howarth and S.W.B. Stewart (eds.). New York: Wiley.

Moloney, K.A., L.J. Stratton, and R.M. Klein. 1983. Effects of simulated acidic, metal-containing precipitation on coniferous litter decomposition. Can J Bot 61:3337–3342.

Morrison, I.K. 1984. Acid rain. For Abstr 45:483–506.

Nevell, W. and M. Wainwright. 1987. Nitrification and urea hydrolysis in deciduous woodland soils from a site exposed to heavy atmospheric pollution. Environ Pollut 45:49–59.

Nohrstedt, H. 1985. Studies of forest floor biological activities in an area previously damaged by sulphur dioxide emissions. Water Air Soil Pollut 25:301–311.

Nohrstedt, H.O. 1987. A field study on forest floor respiration response to artificial heavy metal-contaminated acid rain. Scand J For Res 2:13–19.

Novick, N.J., T.M. Klein, and M. Alexander. 1983. Effect of simulated acid precipitation on nitrogen mineralization and nitrification in forest soils. Water Air Soil Pollut 23:317–330.

Olson, R.A. 1983. The impacts of acid deposition on N and S cycling. Environ Exp Bot 23:211–223.

Padin, E. 1984. Adaptation of bacteria to external pH. *In*: Current Perspectives in Microbial Ecology, M.J. Klug and C.A. Reddy (eds.) pp.49–55. American Society for Microbiology, Washington, D.C.

Parkin, T.B., A.J. Sexstone, and J.M. Tiedje. 1985. Adaptation of denitrifying populations to low soil pH. Appl Environ Microbiol 49:1053–1056.

Persson, T., H. Lundkvist, A. Wirén, R. Hyvönen, and B. Wessén. 1989. Effects of acidification and liming on carbon and nitrogen mineralization and soil organisms in mor humus. Water Air Soil Pollut 45:77–96.

Popovic, B. 1984. Mineralization of carbon and nitrogen in humus from field acidification studies. Forest Ecol Manage 8:81–93.

Prescott, C.E. and D. Parkinson. 1985. Effects of sulphur pollution on rates of litter decomposition in a pine forest. Can J Bot 63:1436–1443.

Prescott, C.E., R.J.F. Bewley, and D. Parkinson. 1984. Litter decomposition and soil microbial activity in a forest receiving SO_2 pollution. *In*: Forest Soils and Treatment Impacts, E.L. Stone (ed.), pp. 448. Knoxville: University of Tennessee.

Press, M.C., J. Henderson, and J.A. Lee. 1985. Arylsulphatase activity in peat in relation to acidic deposition. Soil Biol Biochem 17:99–103.

Raynal, D.J., J.R. Roman, and W.M. Eichenlaub. 1982. Response of tree seedlings to acid precipitation. II. Effect of simulated acidified canopy throughfall on sugar maple seedling growth. Environ Exp Bot 22:385–392.

Rechcigl, J.E. and D.L. Sparks. 1985. Effect of acid rain on the soil environment: A review. Commun Soil Sci Plant Anal 16:653–680.

Reich, P.B., H.F. Stroo, A.W. Schoettle, and R.G. Amundson. 1986. Acid rain and ozone influence mycorrhizal infection in tree seedlings. J Air Pollut Control Assoc 36:724–726.

Reich, P.B., A.W. Schoettle, H.F. Stroo, J. Troiano, and R.G. Amundson. 1985. Effects of O_3, SO_2, and acidic rain on mycorrhizal infection in northern red oak seedlings. Can J Bot 63:2049–2055.

Reuss, J.O. and D.W. Johnson. 1986. Acid Deposition and the Acidification of Soils and Waters. Ecological Studies 59, New York: Springer-Verlag.

Ridout, L.M. 1982. Acid rain. Commonwealth Agricultural Bureaux No. F2, ii, 62 p. Farnum Royal, Slough, United Kingdom.

Roberts, T.M., T.A. Clarke, P. Ineson, and T.R. Gray. 1980. Effects of sulphur deposition on litter decomposition and nutrient leaching in coniferous soils. *In*: Effects of Acid Precipitation on Terrestrial Ecosystems, T.C. Hutchinson and M. Havas (eds.), pp. 381–393. New York: Plenum Press.

Shafer, S.R. 1988. Influence of ozone and simulated acidic rain on microorganisms in the rhizosphere of Sorghum. Environ Pollut 51:131–152.

Shafer, S.R., L.F. Grand, R.I. Bruck, and A.S. Heagle. 1985a. Formation of ectomycorrhizae on *Pinus taeda* seedlings exposed to simulated acidic rain. Can J For Res 15:66–71.

Shafer, S.R., R.I. Bruck, and A.S. Heagle. 1985b. Influence of simulated acidic rain on *Phytophthora cinnamomi* and phytophthora root rot of blue lupine. Phytopathology 75:996–1003.

Shriner, D.S. 1977. Effects of simulated rain acidified with sulfuric acid on host-parasite interactions. Water Air Soil Pollut 8:9–14.

Shriner, D.S., and E.B. Cowling. 1980. Effects of rainfall acidification on plant pathogens. *In*: Effects of Acid Precipitation on Terrestrial Ecosystems, T.C. Hutchinson and M. Haves (eds.), pp. 435–442. New York: Plenum Press.

Skelly, J.M. 1980. Photochemical oxidant impact on mediterranean and temperate forest ecosystems: Real and potential effects. *In*: Proceedings of Symposium on Effects of Air Pollutants on Mediterranean and Temperate Forest Ecosystems, P.R. Miller (ed.), pp. 38–50. USDA Forest Service General Technical Report No. PSW-43.

Smolander, A. and V. Sundman. 1987. *Frankia* in acid soils of forests devoid of actinorhizal plants. Physiol Plant 70:297–303.

Stopp, G.H., Jr. 1985. Acid rain—a bibliography of research annotated for easy access. Metuchen, New Jersey: Scarecrow Press.

Strayer, R.F. and M. Alexander. 1981. Effects of simulated acid rain on glucose mineralization and some physicochemical properties of forest soils. J Environ Qual 10:460–465.

Strayer, R.F., C.J. Lin, and M. Alexander. 1981. Effect of simulated acid rain on nitrification and nitrogen mineralization in forest soils. J Environ Qual 10:547–551.

Stroo, H.F. and M. Alexander. 1985. Effect of simulated acid rain on mycorrhizal infection of *Pinus strobus* L. Water Air Soil Pollut 25:107–114.

Stroo, H.F. and M. Alexander. 1986. Available nitrogen and nitrogen cycling in forest soils exposed to simulated acid rain. Soil Sci Soc Am J 50:110–114.

Stroo, H.F., P.B. Reich, A.W. Schoettle, and R.G. Amundson. 1988. Effects of ozone and acid rain on white pine (*Pinus strobus*) seedlings grown in five soils. II. Mycorrhizal infection. Can J Bot 66:1510–1516.

Tabatabai, M.A. 1985. Effect of acid rain on soils. CRC Crit Rev Environ Control 15:65–110.

Tamm, C.O. 1977. Acid precipitation and forest soils. Water Air Soil Pollut 7:367–369.

Timans, U. 1986. Effect of simulated acid rain and liming on nematodes. Forstwiss Centralbl (Hamburg) 105:335–337.

Turner, N.C., S. Rich, and P.E. Waggoner. 1973. Removal of ozone by soil. J Environ Qual 2:259–264.

Ulrich, B. 1981. An ecosystem hypothesis for the causes of silver fir (*Abies alba*) dieback. Forstwiss Centralbl (Hamburg)100:228–236.

van Breemen, N., C.T. Driscoll, and J. Mulder. 1984. Acidic deposition and internal proton sources in acidification of soils and waters. Nature 307:599–604.

vanLoon, G.W. 1984. Acid rain and soil. Can J Physiol Pharmacol 62:991–997.

Van Miegroet, H. and D.W. Cole. 1984. The impact of nitrification on soil acidification and cation leaching in a red alder ecosystem. J Environ Qual 13:586–590.

Wainwright, M. 1979. Microbial S-oxidation in soils exposed to heavy atmospheric pollution. Soil Biol Biochem 11:95–98.

Wainwright, M. 1980. Effect of exposure to atmospheric pollution on microbial activity in soil. Plant Soil 55:199–204.

Walker, N. and K.N. Wickramasinghe. 1979. Nitrification and autotrophic nitrifying bacteria in acid tea soils. Soil Biol Biochem 11:231–236.

Weber, D.E., R.A. Reinert, and K.R. Barker. 1979. Ozone and sulfur dioxide effects on reproduction and host-parasite relationships of selected plant-parasitic nematodes. Phytopathology 69:624–628.

Will, M.E., D.A. Graetz, and B.S. Roof. 1986. Effect of simulated acid precipitation on soil microbial activity in a Typic Quartzipsamment. J Environ Qual 15:399–403.

Wodzinski, R.S., D.P. Labeda, and M. Alexander. 1977. Toxicity of SO_2 and NO_x: Selective inhibition of blue-green algae by bisulfite and nitrite. J Air Pollut Control Assoc 27:891–893.

Wodzinski, R.S., D.P. Labeda, and M. Alexander. 1978. Effects of low concentrations of bisulfite-sulfite and nitrite on microorganisms. Appl Environ Microbiol 35:718–723.

Zelles, L., I. Scheunert, and K. Kreutzer. 1987. Effect of artificial irrigation, acid precipitation and liming on the microbial activity in soil of a spruce forest. Biol Fertil Soils 4:137–143.

6
The Atmosphere and the Rhizosphere: Linkages with Potential Significance for Forest Tree Health

WILLIAM H. SMITH

I. Introduction

In view of the obvious direct relationships between the deposition of atmospheric pollutants and aboveground portions of plants, it is not surprising that much of our research and appreciation of plant response to air pollutant stress is concentrated in aboveground vegetative parts (Treshow 1984, Weinstein and McCune 1979). Recent reviews (McLaughlin 1985, Smith 1990, 1985, Society of American Foresters 1984), however, have also emphasized the potential significance of air pollutant interaction with the belowground ecosystem. Special emphasis has been given to the possible influence of atmospheric contaminants on the ecology of the soil biota and on the processes of nutrient cycling in forest ecosystems. The direct and indirect consequences of atmospheric deposition on plant root physiology and on the ecology of the microbiota in the immediate vicinity of roots are two of the most important aspects of this more recent emphasis.

The root systems of forest trees are extremely impressive in function and structure. They provide trees with essentially all the moisture required for growth, the bulk of the inorganic nutrients essential for metabolism, massive storage of energy resources, and physical support and anchorage. Forest tree root systems are large. Stout (1956) exposed the root systems of 18 mature deciduous trees in Orange County, New York, and determined the average ratio between area of the root system and area of the crown to be 4.5:1. Leaphart and Grismer (1974) hydraulically excavated two 116-m^2 plots of mixed conifers in northern Idaho and measured total root length. They recorded 851 m of length per meter squared of soil to total root depth for a total living root length of 99 km per plot. Harris et al. (1973) have compared above- and belowground biomass for the Walker Branch Watershed Experimental Forest in eastern Tennessee. The largely deciduous forest types on Walker Branch had a mean rate of annual wood accumulation of 3250 kg ha^{-1} above ground. Total annual belowground biomass production of roots <0.5 cm in diameter was estimated at 9000 kg ha^{-1}.

In addition to emphasizing the significance of root function and mass, it is important to realize that metabolically active roots are concentrated in the upper portion of forest soil profiles. Kochenderfer (1973) has indicated that in excess of 80% of all root endings in West Virginia forests were located in the upper 60 cm of the soil profile. McMinn (1963) hydraulically excavated Douglas fir roots on Vancouver Island, British Columbia, and found that even large roots were usually located in the upper 25 cm of the soil profile. Wood (1979) conducted an intensive survey of tree root distribution in the northern hardwood forest in central New Hampshire. He observed that 50% of the fine roots ($\leq$0.6 mm diameter) were located in the upper 10 cm. That metabolically important root systems, which contain large fractions of tree biomass, are very close to soil surfaces emphasizes the critical need to more thoroughly understand the relationship between air pollutant deposition, root function and form, and tree health. This review concentrates on the interactions of atmospheric deposition and the rhizosphere.

The term "rhizosphere" was introduced by Hiltner (1904) to designate the zone of soil influenced by roots. Rovira et al. (1983) have refined the definition of rhizosphere to encompass that "zone of soil extending from the root-soil interface to the point in the soil where the microflora is unaffected by the root." The rhizosphere is a "concept," and in nature is discontinuous in time and space and variable in extension and character. Rovira et al. (1983) correctly emphasized that the rhizosphere is actually a gradient of numerous physical, chemical, and biological properties extending from just inside the root surface (endorhizosphere), through the root–soil interface (rhizoplane), through the inner rhizosphere to the outer rhizosphere, and eventually to the point beyond root influence on soil microbes (bulk soil).

II. The Rhizosphere

Since 1904, when the rhizosphere concept was introduced, there has been a gradual increase in our understanding of this part of the soil environment. While most of the early research was concentrated on agricultural or nonwoody species, some studies did focus on forest species including pine (Foster and Marks 1967, Tribunskaya 1955), fir (Maliszewska and Moreau 1960), yellow birch (Ivarson and Katznelson 1960), beech (Harley and Waid 1955), and poplar (De Leval and Remacle 1969). Early research efforts were characterized by experimental designs that limited extrapolation to natural soil environments. Many of the early studies employed seedling or young plants rooted in artificial soils that were grown under controlled environmental conditions. Over the past 20 years, however, rhizosphere investigations have been conducted in the field (Smith 1970a, 1970b, 1972, 1974), and observations have been made in natural soils.

The use of electron microscopy has greatly advanced our understanding of the structure and chemistry of the rhizosphere (Bae et al. 1972, Campbell and Rovira 1973, Foster 1981, Foster and Rovira 1976, Rovira and Campbell 1975).

The most useful general reviews of the rhizosphere concept include Baker and Cook (1974), Foster et al. (1983), Harley and Russell (1979), Rovira (1969), and Rovira et al. (1983).

A. Rhizosphere: General Character, Location, and Size

The rhizosphere is the very restricted zone of soil immediately surrounding plant roots. It is the zone of root exudation, sloughed root cells, and the root mucigel complex. Saprophytic, parasitic, and symbiotic microorganisms abound in the rhizosphere, and roots exchange ions, gases, and moisture here. The soil of the rhizosphere has greater bulk density, generally lower pH, and a more variable water potential gradient than bulk soil.

The rhizosphere concept is predicated on "root influence on soil microorganisms." It is generally agreed that this influence extends 1 to 2 mm from the root. Some fungi have been shown to be influenced by roots as much as 5 to 6 mm distant. Pathologists have quantified the influence of roots on soil pathogens and have used this information to model the rhizosphere "effect," to estimate the dimensions of the rhizosphere, and to differentiate the root surface from the soil immediately surrounding roots. Initial models were developed by Baker et al. (1967) and Baker (1971) in an effort to distinguish the rhizoplane from the rhizosphere. Mathematical models have been developed for the calculation of the rhizosphere volume (Gilligan 1979, Leonard 1980) and rhizosphere width (Drury et al. 1983, Gilligan 1979, Ferriss 1981). In a recent study, Reynolds et al. (1985) used *Phytophthora cinnamomi* infection of Fraser fir seedling radicals to estimate rhizosphere widths under varying soil moisture conditions. The model developed was

$$W = 4.01\ (1 - M)^{-0.72}$$

where W equals rhizosphere width in millimeters and M equals soil (moisture) matric potential in millibars. Mean rhizosphere width, under conditions of varying inoculum (chlamydospore) densities, varied from 5.18 mm at 0 mb to 0.39 mm at −100 mb. Polonenko and Mayfield (1979) employed a slide incubation chamber and a fluorescence stain to directly observe microbial influence in the rhizosphere and documented bacterial stimulation as much as 4 mm from pea roots.

The rhizosphere is a dynamic zone. The characteristics of the soil immediately surrounding roots change as the root matures and lengthens and eventually becomes senescent or woody (lignified). Foster et al. (1983)

have provided an excellent description of rhizosphere development and ultrastructure (Figure 6.1).

B. Rhizosphere Biology

The dominant feature and primary characteristic of rhizosphere soil relative to bulk soil is the stimulation of microbial numbers and activity in the former zone. Populations of viruses, bacteria, actinomycetes, yeasts, other fungi, insects, mites, other arthropods, nematodes, flagellates, and amoebas are all increased in rhizosphere ecosystems (Foster et al. 1983) (Table 6.1). Microbial stimulation in the soil close to roots is primarily caused by the eutrophic nature of the rhizosphere environment relative to the oligotrophic conditions characteristic of the bulk soil. Microbes enhanced in the rhizosphere encompass all ecological types, for example,

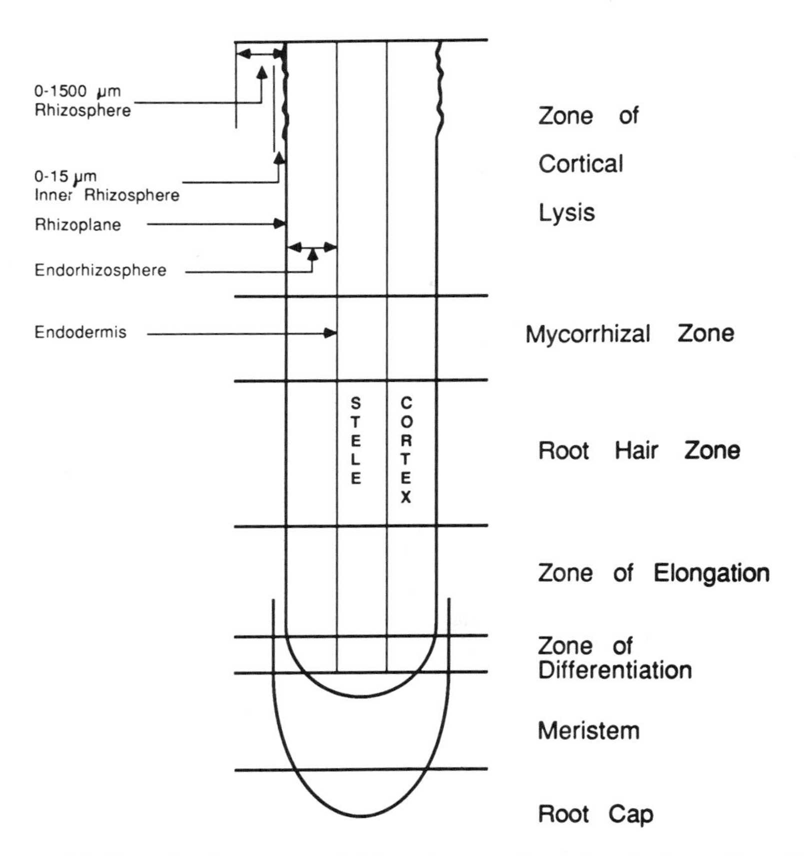

FIGURE 6.1. Root development and rhizosphere configuration (redrawn from Foster et al. 1983).

TABLE 6.1. Approximate maximum population densities for major microbial groups of the rhizosphere from Foster et al. (1983) and Foster (1985).[a]

Microbial group	Number per gram of rhizosphere soil
Bacteria	1×10^{9}
Actinomycetes	46×10^{6}
Fungi	$1 \times 10^{4-5}$
Algae	$1 \times 10^{3-5}$
Amoebas	18×10^{2}
Flagellates	50×10^{2}

[a]Bulk soil populations range from 10 to 100 times less.

saprophytes, parasites, and symbionts, and play extremely important and diverse roles in nutrient dynamics and root health. General reviews of rhizosphere biology include Bowen and Rovira (1973), Foster (1985), Foster et al. (1983), Newman and Watson (1977), Rovira et al. (1983), and Tinker and Sanders (1975).

1. Bacteria

In terms of numbers, bacteria exhibit the greatest stimulation in rhizosphere relative to bulk soil of all microbial groups. Bacterial cells are not randomly dispersed on the rhizoplane or in the rhizosphere, but tend to be located in colonies associated with sites having abundant nutrient resources. These sites typically include grooves between epidermal cells and wound sites from infection or feeding. Observations of rhizosphere bacteria with the electron microscope suggest that many are positioned within 10 μm of the root epidermis and commonly are not actually attached to the root surface (Foster et al. 1983).

The taxonomy of rhizosphere bacteria is incompletely understood. Current observations commonly do not allow assignment to known taxa. Members of the following genera have been frequently described in rhizosphere habitats: *Bacillus*, *Pseudomonas*, *Erwinia*, *Caulobacter*, *Arthrobacter*, *Micrococcus*, *Flavobacterium*, *Chromobacterium*, and *Hyphomycrobium* (Foster et al. 1983). Ecological groupings of rhizosphere bacteria are as diverse as taxonomic groupings and include saprophytes, root parasites, free-living nitrogen fixers, nodulating nitrogen fixers, and predatory bacteria (*Bdellovibrio*).

Alexander (1977) has suggested that rhizosphere stimulation may be most pronounced with short rods, including members of the genera *Pseudomonas*, *Flavobacterium*, and *Alcaligenes*. Several studies have indicated that rhizoplane and rhizosphere bacteria exhibit a higher incidence of phosphate solubilization (Azcon et al. 1976, Barea et al. 1976, 1978, Baya et al. 1981) and vitamin production (Baya et al. 1981, Cook and Lockhead 1959) than bacteria from bulk soil.

Rhizosphere processes and certain rhizosphere microorganisms are extremely important in the nitrogen economy of plants. Both free-living and nodulating nitrogen-fixing bacteria are significant components of the rhizosphere habitat. Nitrogen fixation by free-living bacteria in bulk soil is presumed limited by poor availability of appropriate carbon substrate (Postgate 1974). In the rhizosphere, on the other hand, carbon substrates are abundant and *Azotobacter* have been shown to form important rhizosphere associations with nonwoody species (Döbereiner et al. 1972). Barber and Lynch (1977) have emphasized the potential significance of free-living rhizosphere nitrogen fixers in the nitrogen dynamics of barley plants. Subsequently it was shown that both aerobic and anaerobic nitrogen-fixing bacteria develop in the rhizosphere of barley seedlings grown in nitrogen-deficient sand culture. The aerobic nitrogen-fixing bacterium was an endophyte that infected roots and formed "vesicle-like" structures (Pohlman and McColl 1982). Most research dealing with free-living nitrogen fixation, particularly in the rhizosphere, has been concentrated in nonwoody species, especially grasses (Döbereiner 1983). Woody plants may have less important relationships with free-living nitrogen fixers because their photosynthetic efficiencies may be insufficient to devote significant photosynthate to nitrogen fixation in the rhizosphere. In the temperate region, soil temperatures and moisture availabilities may also be marginal for free-living nitrogen fixers for much of the year.

Members of the *Rhizobium* genus can infect the root hairs of leguminous plants and initiate the formation of root nodules, within which they develop as intracellular symbionts and fix atmospheric nitrogen. The 12,000 leguminous species include trees, shrubs, and woody vines, of which less than 10% have been studied for nodulation (Postgate and Hill 1979). Globally important tree genera include *Dalbergia*, *Pterocarpus*, *Albizzia*, *Peltogyne*, *Acacia*, *Cassia*, *Castanospermum*, *Haemotoxylon*, and *Caragana*. *Gleditsia* and *Robinia* are the two most important native American genera, and, with 16 other arborescent genera, include 40 species.

Root nodule bacteria are considered facultative symbionts and can persist and compete in bulk soil. They clearly, however, are greatly stimulated in rhizosphere soil—especially rhizospheres of legume species. Parker et al. (1977) have comprehensively reviewed the rhizosphere environment relative to rhizobial species. Rhizosphere populations as high as 10^8 cells per gram of soil have been reported for *Rhizobium trifolii*.

2. Fungi

As in the case of bacteria, fungi of the rhizosphere are extremely abundant and extremely diversified. Ecological groups include saprophytes, facultative parasites, obligate parasites, mycoparasites, competitors, and symbionts. Fungal genera common to numerous rhizospheres include

Alternaria, *Aspergillus*, *Bipolaris*, *Ceratobasidium*, *Chaetomium*, *Cladosporium*, *Curvularia*, *Cylindrocarpon*, *Embellisia*, *Fusarium*, *Gaeumannomyces*, *Helminthosporium*, *Mortierella*, *Mucor*, *Penicillium*, *Pythium*, *Rhizoctonia*, *Syncephalis*, and *Verticillium* (Foster et al. 1983). Antagonistic genera (competitive with other fungal groups), including *Trichoderma* and *Gliocladium*, have recently been identified in rhizosphere habitats (Beagle-Ristaino and Papavizas 1985). In general, the eutrophic nature of the rhizosphere zone relative to bulk soil stimulates both spore and mycelial development. The greatest amount of literature regarding fungal inhabitants of the rhizosphere is associated with pathological and mycorrhizal species.

Most of the early, and much of the continuing, interest in the rhizosphere effect has been associated with studies of fungal pathogen ecology (Griffin 1985). The rhizosphere environment has been shown to be of critical importance for all types of fungal pathogens including unspecialized facultative parasites such as *Phytophthora cinnamomi*; more specialized facultative parasites such as *Macrophomina phaseoli*; and very specialized facultative parasites such as *Armillaria mellea*, *Phellinus weirii*, and *Heterobasidion annosum*.

A significant number of rhizosphere-pathogen studies have been done on forest species. Zoospores of *Phytophthora cinnamomi* were shown to be attracted to the roots of 23 species from Australian native forests when within 3–4 mm from roots (Hinch and Weste 1979). Similar results have been reported for *Eucalyptus* (Halsall 1978, Malajczuk et al. 1977, Tippett et al. 1976) and Fraser fir (Reynolds et al. 1985).

Spore germination, as well as spore attraction, has been shown to be stimulated in the rhizosphere. *Pythium ultimum* is a ubiquitous and virulent unspecialized parasite. Agnihotri and Vaartaja (1967) reported that most root-exudate amino acids and sugars from seedling red pine when supplied singly stimulated the germination of sporangia and growth of this pathogen. Certain mixtures of root-exudate components were particularly favorable for the development of germ tubes and mycelial growth.

Marx and Davey (1969a,b) studied the response of zoospores of *Phytophthora cinnamomi* to ectomycorrhizal and nonmycorrhizal roots of shortleaf and loblolly pine seedlings. The spores were not strongly attracted to either root type. Germination rate and vigor, however, were greatest near the tip and the region of cell elongation of nonmycorrhizal roots. This led the authors to conclude that ectomycorrhizal roots were less stimulating to *P. cinnamomi* than nonmycorrhizal roots. Marx (1972) has speculated that root exudates from mycorrhizal and nonmycorrhizal roots probably differ. Percentage of zoospore germination and mean length of germ tubes of *P. palmivora* have been shown to be greater when exposed to root exudates from susceptible *Theobroma cacao* (Turner 1963).

Macrophomina phaseoli is a facultative specialized parasite of considerable importance in tropical and warm temperate forest tree nurseries. This organism grows poorly in bulk soil (Smith 1969b) and persists by forming dormant, resistant sclerotia. Germination of these sclerotia in soil could be significantly increased by adding natural or synthetically prepared root exudate from sugar pine seedlings. The amino acid fraction of the artificial exudate was most effective in increasing germination (Smith 1969c).

Waitea circinata causes serious root disease in coniferous seedlings in Canadian nurseries. Ecologically it is similar to *M. phaseoli* as it develops poorly in the soil apart from a host and persists in the form of sclerotia (Agnihotri 1971). Although root-exudate components of seedling Mexican piñon pine were not required for sclerotial germination, carbohydrate and certain amino acid components did enhance vegetative growth from germinated sclerotia (Agnihotri and Vaartaja 1967).

Heterobasidion annosum is one of the most important specialized, root-inhabiting pathogens of the temperate forest regions of the world. Basidiospores are considered to be the main source of inoculum, but considerable speculation has been advanced concerning the role of conidia in the ecology of this pathogen (Hodges 1969). Conidia are known to be distributed and persistent in bulk soil, and Hüppel (1970) has provided evidence that they can infect unwounded intact roots of axenically grown Scotch pine and Norway spruce seedlings. In an effort to evaluate the ability of conidia to infect woody root tips of older trees, Smith (1974) collected exudates from mature red pine, tested their ability to enhance conidial germination in soil, and found that they provided only a very slight stimulation.

Rhizina undulata is an important root-inhibiting pathogen of British conifers, particularly in areas subjected to fire stress. Ascospores of this fungus are fairly persistent in the soil and may provide the pathogen with a mechanism for nonhost persistence. Jalaluddin (1967) has presented evidence that extracts of mature, heated roots of pine stimulated ascospore germination.

Fusarium oxysporum f. sp. *perniciosum* is a specialized parasite causing a wilt disease of mimosa. Redington and Peterson (1971) have suggested that the carbohydrate fraction of mimosa root exudate enhances spore germination and vegetative growth of this pathogen in the rhizosphere.

3. Mycorrhizas

The endorhizosphere, rhizoplane, and rhizosphere provide the habitat for symbiotic root-invading fungi. With the exception of a few families of higher plants, the roots of all terrestrial plants, including all forest trees, form mycorrhizal associations with soil fungi. Ectotrophic mycorrhizas invade the endorhizosphere (Figure 6.1), grow in the intercellular spaces of the root cortex and form a mycelial cover (mantle) over the root. The mantle actually consists of mycelium plus a mucilaginous gel and creates a mycorrhizosphere with a varied bacterial flora. Fine lateral roots are induced to branch dichotomously and assume a unique and characteristic appearance when colonized by ectotrophic mycorrhizal fungi. Endotrophic mycorrhizal fungi (vesicular-arbuscular [VA] mycorrhizae), on the other hand, do not cause dichotomous branching of roots and do not form a mycelial mantle on the root. These fungi are confined to the intercellular spaces of the endorhizosphere where they infect cortex cells, enter the cell lumens, and form specialized branching hyphae termed arbuscules. Endotrophic mycorrhizal fungi do not infect the cytoplasm of root cells because they do not penetrate the plasmalemma. They also do not penetrate the endodermis, so their habitat is confined to the endorhizosphere.

In both fungus–root associations, the symbioses involve fungal transfer of plant nutrients, especially phosphate, from soil to plant along with transfer of carbohydrates from the plant to the fungus (Foster et al. 1983, Rovira et al. 1983). Mycorrhizal associations are recognized to be especially important for forest tree species. In addition to nutrient dynamics, mycorrhizas may play important roles in the ecology of root-infecting pathogens (Marx 1972, Zak 1964).

It has been known for many years that the rhizosphere provides a favorable habitat for the very specialized fungi that form mycorrhizal associations. Pioneering work with forest trees demonstrated the importance of root-exudate components for mycorrhizal symbionts (Carrodus and Harley 1968, Melin 1955, 1963, Slankis 1958). Recent research has stressed the importance of root-exudate components for both ectotrophic and endotrophic mycorrhizal fungi (Azcon and Ocampo 1984, Fries et al. 1985, Graham 1982). The latter study emphasized the potential importance of root-exudate lipids in the stimulation of mycorrhizal fungi.

The mycorrhizal-rhizosphere relationship has been the topic of comprehensive and current reviews (Bowen and Theodorou 1973, Mosse et al. 1981, Rovira et al. 1983).

C. Rhizosphere Chemistry

The rhizosphere effect is based principally on microbial stimulation in the near-root environment. This stimulation is primarily the result of the abundance of nutritional substrates in the vicinity of roots relative to bulk soil. Recent evidence has demonstrated the significant amount of carbon, fixed by photosynthesis, that is provided to heterotrophs in the rhizosphere environment (Newman 1978). Controlled environment and field studies with wheat plants have indicated that more than 15% of the net carbon fixed by photosynthesis is lost via the roots (Martin 1977a, Martin and Puckridge 1982). This amounts to approximately 1000 kg carbon ha^{-1} annually input to wheat rhizospheres (Martin and Foster 1985). Tree root studies employing $^{14}CO_2$ have also revealed impressive carbon inputs to the rhizosphere. Reid (1974) reported that ponderosa pine roots may release as much as 5% of the total activity (on a dry weight basis) while lodgepole pine roots may release up to 7% of the total root activity (Reid and Mexal 1977).

1. Sources of Organic Compounds in the Rhizosphere

The sources of chemicals of the rhizosphere are varied and diverse (Rovira et al. 1983). Among the most important are plant mucilages, mucigel, lysates, secretions, and root exudates.

Mucilage is the gel layer on the surface of active roots. It is typically less than 1 μm thick and consists of mucilaginous pectins and hemicelluloses. Four primary sources of plant mucilages are recognized: (1) mucilage associated with the root cap and produced by Golgi vesicles, (2) mucilage consisting of hydrolysates of polysaccharides of the primary cell walls between epidermal cells and sloughed root cap cells, (3) mucilage released by epidermal cells that lack secondary walls, including root hairs, and (4) mucilage produced by bacterial decomposition of primary cell walls of nonliving epidermal cells (Rovira et al. 1983).

Mucigel has been defined as the gelatinous material at the surface of plant roots grown in natural soil (Jenny and Grossenbacher 1963). This material is extremely diverse and includes intact and modified plant mucilages, bacterial cells and their metabolic products including capsules and slimes, and colloidal mineral and organic matter from the soil. Mucilages are distinguished from mucigels by designating the former as entirely of plant origin whereas the latter are produced by the full root-soil-microbe complex (Rovira et al. 1983).

Lysates are compounds released from autolysis of older epidermal cells when the plasmalemma ceases to function. With time autolyzed cells become decomposed by microbes, and this decomposition process also releases by-products of microbial activity into the rhizosphere.

The most intensively and longest studied, and perhaps the most important, sources of rhizosphere organic materials are root secretion and root exudation. Root secretion is the metabolic (active) release of low or high molecular weight compounds from root systems. Root exudation is the nonmetabolic (passive) release of low molecular weight chemicals into the rhizosphere. Root exudates leak from root cells to soil either into the intercellular spaces of the cortex and then to soil through cell junctions or directly through epidermal cell walls (Rovira et al. 1983). In practice, it is impossible to distinguish exuded and secreted material. This review will assume that exudation is the dominant process because there is no evidence that plants invest significant amounts of energy in active secretion.

2. Root Exudates

The process of root exudation has been studied intensively for more than 25 years, and we have substantial appreciation of the sites of exudation and the qualitative and quantitative nature of root-exudate compounds (Rovira 1965a, 1965b, 1969, Rovira and McDougall 1967). Root exudation can occur from all root types over their entire age and length. Considerable evidence, however, recognizes the area of major release as the zone of elongation (Figure 1) proximal to the root tip. The compounds released, while generally simple compounds of low molecular weight, are very diverse and encompass sugars, amino acids, proteins, lipids, and organic acids. This review concentrates primarily on root exudation from forest trees (Smith 1977).

Bowen and Theodorou (1973) have reviewed studies on root exudates of forest trees. In general, woody plants have received less attention than herbaceous species. Studies on 10 species of forest tree seedlings have reported 21 amino acids/amides, 10 carbohydrates, and 14 organic acids. The qualitative characteristics of the exudates are generally similar to those of nonwoody plant species. To enable very approximate comparison of quantitative relationships, the data and recalculations of Hale et al. (1971) for crop plants are compared with the data for some forest tree species (Table 6.2). Because of the variation in techniques, and ages and sizes of seedlings, employed by the numerous investigators whose data are incorporated into Table 6.2, comparisons must be considered as *suggestive* only. With this in mind, tree species appear to release more materials in the amino acid/amide group and comparable amounts in the carbohydrate group. The quantitative aspects of organic acid exudation by herbaceous seedlings have not received intensive consideration, and no comparison can be provided in Table 6.2. Tree seedlings release greater amounts of organic acids than seedlings of the two other major root-exudate groups.

Woody plants, particularly trees, are long lived relative to herbaceous plants, and age becomes a particularly important variable with regard to root exudation. Most of the tree root-exudate literature discusses the exudation of seedlings or very young plants. There is evidence, however, that exudation, similar to that occurring from seedling roots, also may occur from unsuberized second and higher order nonwoody roots and tips of woody roots (cf. terminology of Lyford and Wilson 1964) of mature trees. To examine the exudation from the roots of mature trees, Smith (1970a) developed a technique for collection based on a modified air-layering procedure developed by Lyford and Wilson (1966). This system has been successfully employed for the past 20 years in New England forests on both coniferous and deciduous species. One of the first applications was to compare the exudates from sugar maple seedling roots with those from the unsuberized tips of new woody roots of a mature

TABLE 6.2. Amounts of amino acids/amides, carbohydrates, and organic acids contained in seedling root exudates from various crop and forest tree species.[a]

Species	Exudate group	Exudate amount (recalculated) μg plant^{-1} week^{-1}	Reference
	Amino acids/Amides		
Crop species			
Wheat		0.81	Hale et al. 1971
Peas		3.30	Hale et al. 1971
Alfalfa		0.37	Hale et al. 1971
Sorghum		0.22	Hale et al. 1971
Sunnhemp		0.28	Hale et al. 1971
Ragi		0.15	Hale et al. 1971
Tomato		0.05	Hale et al. 1971
Tree species			
Pinus radiata		0.73	Hale et al. 1971
Pinus radiata		104.3	Smith 1969a
Pinus lambertiana		210.7	Smith 1969a
Pinus banksiana		44.8	Smith 1969a
Pinus rigida		20.3	Smith 1969a
Robinia pseudoacacia		31.0	Smith 1969a
Acer saccharum		1.15	Smith 1969a
	Carbohydrates		
Crop species			
Sorghum		38.3	Hale et al. 1971
Sunnhemp		31.7	Hale et al. 1971
Ragi		19.6	Hale et al. 1971
Tomato		7.46	Hale et al. 1971
Peanut		3.80	Hale et al. 1971
Tree species			
Pinus radiata		9.1	Smith 1969a
Pinus lambertiana		23.7	Smith 1969a
Pinus banksiana		0.07	Smith 1969a
Pinus rigida		0.70	Smith 1969a
Robinia pseudoacacia		6.3	Smith 1969a
Acer saccharum		13.32	Smith 1969a
	Organic acids		
Tree species			
Pinus radiata		224.5	Smith 1969a
Pinus lambertiana		382.2	Smith 1969a
Pinus banksiana		189.0	Smith 1969a
Pinus rigida		46.2	Smith 1969a
Robinia pseudoacacia		53.9	Smith 1969a
Acer saccharum		32.6	Smith 1969a

[a]From Smith 1977.

sugar maple growing in southern Connecticut as a codominant member of a natural forest community (Smith 1970b). Carbohydrates released from seedlings were more diverse and were released in greater amounts than those from roots of the mature tree. Three amino acids contained in the mature tree exudate were absent from the seedling exudate. Amino compounds released commonly by both root types were exuded in greater quantities by mature tree roots, with the exception of glutamine, which was equal in the two exudates. The release of comparatively large amounts of organic acids was consistent with previous observations of tree seedling exudation patterns (Smith 1969a). Roots of mature trees produced six times more organic acids per milligram dry weight than did seedling roots. Extrapolation of root-exudate information from studies employing seedling or young plants to phenomena involving root exudates from unsuberized root tips of mature trees may only be made with considerable risk.

The technique for collecting mature tree root exudates has been used for 10 years in the second-growth northern hardwoods of the Hubbard Brook Experimental Forest (Bormann et al. 1970) in New Hampshire. Information has been collected from mature individuals of the three major species *Fagus grandifolia*, *Betula alleghaniensis*, and *Acer saccharum* of the northern deciduous forest (Table 6.3). These data suggest that while the qualitative patterns of root exudation are similar for all three species, the quantitative differences are considerable. The release of total organic materials (per milligram dry root) by yellow birch is approximately two times the release by American beech and approximately three times that of sugar maple. For all three species, organic acids constitute the largest organic exudate fraction, while amino acids/amides comprise the smallest fraction (Smith 1976).

The inorganic fraction of both tree and herbaceous plant root exudates has received little research consideration. Bowen (1968) has measured the loss of chloride along roots of *Pinus radiata* seedlings labeled with ^{36}Cl. By employing atomic absorption spectrophotometry and selected methods of anion analysis, the mature tree root exudates from Hubbard Brook have been examined for several inorganic ions (Table 6.4).

While the procedure permitting examination of root exudates from mature trees (Smith 1970a) has been extremely useful, it is not without disadvantages. Among these are (1) the procedure (to date) permits collection only from the unsuberized tips of woody roots and does not sample the exudation of nonwoody or mycorrhizal roots, (2) the time required for root development before collection mandates that exudates have only been sampled late in the growing season, and (3) the number of trees sampled has not been large, because a considerable investment of time and labor is required by the method. The first two disadvantages can probably be addressed by alteration and refinement of the procedure.

The actual amount of organic and inorganic compounds introduced into forest soils by roots depends on an estimation of the quantity and distribution of roots releasing exudates. Because the quantitative estimation of root exudation provided by the Smith technique is probably conservative (does not include fine root or injured root release) relative

TABLE 6.3. Organic root exudation patterns of mature trees in northern hardwood forest ($\mu g \times 10^{-1}$ of each compound released during 14 days per mg of oven-dried root).[a]

Exudates	Tree species		
	Betula alleghaniensis	*Fagus grandifolia*	*Acer saccharum*
Carbohydrates			
Glucose	12.36 ± 5.21[b]	6.32 ± 3.36	0.21[c]
Fructose	16.59 ± 2.54[b]	11.08 ± 6.40	0.83 ± 0.62
Ribose	3.35 ± 0.74	ND	ND
Sucrose	21.17[c]	ND	13.14 ± 5.26
Amino acids/amides			
Alanine	1.42[c]	0.04[c]	1.21 ± 0.47
Arginine	0.60[c]	0.12 ± 0.08	ND
Asparagine	0.31 ± 0.13	0.07 ± 0.03	ND
Aspartic acid	0.52 ± 0.28	0.11	ND
Cystine	0.02 ± 0.01[b]	0.02[c]	0.03[c]
Glutamine	0.78 ± 0.68	0.13[c]	0.75[c]
Glycine	0.45[c]	0.06 ± 0.02	0.05 ± 0.01
Homoserine	0.15 ± 0.01[b]	0.23 ± 0.06	0.22 ± 0.03[b]
Leucine/Isoleucine	ND	ND	ND
Lysine	1.07 ± 0.57	0.59 ± 0.16	0.18 ± 0.14[b]
Methionine	0.08 ± 0.03[b]	0.01[c]	ND
Phenylalanine	ND	ND	0.13 ± 0.08[b]
Proline	1.69[c]	2.36 ± 0.64	ND
Serine	1.19[c]	0.51[c]	0.32[c]
Threonine	0.03 ± 0.003	0.08 ± 0.03	0.29 ± 0.24
Tyrosine	ND	ND	0.03[c]
Valine	0.27[c]	0.07 ± 0.05	0.37[c]
Organic acids			
Acetic	15.38 ± 6.99	5.92[c]	21.42 ± 10.35[b]
Aconitic	ND	17.47[c]	ND
Citric	10.06[c]	ND	10.93 ± 4.29[b]
Fumaric	20.02[c]	5.62 ± 5.54[b]	ND
Malic	20.74 ± 8.86[b]	ND	ND
Malonic	ND	ND	1.85[c]
Oxalic	10.28 ± 2.26	24.68 ± 0.10[b]	ND
Succinic	4.03[c]	3.74 ± 1.98	ND
Total organic material	142.56	79.23	50.75

[a]From dominant trees growing in undisturbed area of Hubbard Brook Experimental Forest, New Hampshire; data are mean and standard error for three trees unless otherwise indicated.
[b]Detected in exudate from two trees.
[c]Detected in exudate from one tree.
ND, not detected.

to the release under natural conditions, and because the number of exudation regions of roots is large, the biological and chemical significance of root exudation in forest soils is considerable. When short-term exu-

TABLE 6.4. Inorganic root exudation patterns of mature members of the northern hardwood forest (μg of each ion released during 14 days per mg of oven-dry root).[a,b]

	Tree species		
Root exudate ion	*Betula alleghaniensis*	*Fagus grandifolia*	*Acer saccharum*
I. Cations			
1. Na^+	55.58 ± 6.62	18.52 ± 1.93	17.43 ± 2.79
2. K^+	12.80 ± 1.33	4.23 ± 0.49	5.09 ± 0.84
3. Ca^{2+}	4.81 ± 0.47	2.00 ± 0.22	3.15 ± 0.36
4. NH_4^+	1.59 ± 0.12	0.53 ± 0.08	0.59 ± 0.07
5. Mg^{2+}	0.25 ± 0.04	0.07 ± 0.01	0.36 ± 0.09
II. Anions			
1. SO_4^{2-}	11.19 ± 0.82	2.43 ± 0.14	2.90 ± 0.27
2. Cl^-	5.10 ± 0.96	0.24 ± 0.03	0.43 ± 0.06
3. PO_4^{2-}	1.60 ± 0.10	0.23 ± 0.03	0.28 ± 0.07
4. NO_3^-	0.12 ± 0.02	0.06 ± 0.01	0.05 ± 0.01
Total inorganic material	93.05	28.31	30.28

[a]From Smith 1976.
[b]Composition of inorganic root exudates from mature, dominant trees growing in undisturbed area of Hubbard Brook Experimental Forest, New Hampshire. Data are mean and standard error for 10 roots sampled from one tree for all anions and NH_4^+ and from two trees for all other cations.

dation patterns (see Tables 6.3 and 6.4) are expanded to the entire growing season (Table 6.5), the kinds and amounts of materials released are very impressive.

3. Organic Acids

Organic acids are quantitatively significant and qualitatively diverse in the root exudation of forest trees (see Tables 6.2, 6.3, and 6.5). Organic acids are also very important in complexing trace metals in soil ecosystems. It is important, therefore, to review organic acid differences in bulk soil relative to rhizosphere soil.

Soil organic acids may be conveniently divided into two broad groups, nonhumified and humified (Figure 6.2). Nonhumified acids are diverse and include simple aliphatic acids (e.g., acetic and oxalic acids) and complex aromatic and heterocyclic acids (e.g., vanillic, tannic, and benzoic acids). These acids are important intermediate components of plant metabolism, and some represent very important components of forest tree root exudation (Tables 6.2, 6.3, and 6.5). They are also introduced to soil via organic matter decomposition. In bulk soil the concentration of these acids is relatively low and may approximate 0.5 to 0.9 mmol 100 g^{-1} of soil (Tan 1985). The concentration of nonhumified organic acids in rhizosphere soil, however, may be elevated manyfold, especially in regions of unsuberized roots exhibiting high metabolic activity.

TABLE 6.5. Root-exudate materials released (by calculation) during growing season by three principal tree species in northern hardwood forest (Hubbard Brook Experimental Forest, New Hampshire) (from Smith 1976).

Exudates	Tree species (kg ha^{-1} yr^{-1})			
	Betula alleghaniensis	*Fagus grandifolia*	*Acer saccharum*	Total
Carbohydrates				
Glucose	0.38	0.46	0.003	0.84
Fructose	0.52	0.81	0.01	1.34
Ribose	0.10	ND	ND	0.10
Sucrose	0.66	ND	0.24	0.90
Total	1.66	1.27	0.25	3.18
Amino acids/amides				
Alanine	0.04		0.02	0.06
Arginine	0.01		ND	0.018
Asparagine	0.009	0.002	ND	0.01
Aspartic acid	0.01	0.008	ND	0.018
Cystine	0.005	0.005	0.0004	0.002
Glutamine	0.02	0.008	0.01	0.04
Glycine	0.01	0.001	0.0008	0.01
Homoserine	0.004	0.009	0.004	0.02
Lysine	0.03	0.004	0.003	0.07
Methionine	0.002	0.0007	ND	0.003
Phenylalanine	ND	ND	0.002	0.002
Proline	0.05	0.17	ND	0.22
Serine	0.03	0.03	0.005	0.06
Threonine	0.0008	0.005	0.005	0.01
Tyrosine	ND	ND	0.0004	0.0004
Valine	0.008	0.005	0.006	0.019
Total	0.22	0.29	0.05	0.56
Organic acids				
Acetic	0.48	0.43	0.39	1.30
Aconitic	ND	1.27	ND	1.27
Citric	0.31	ND	0.20	0.51
Fumaric	0.62	0.41	ND	1.03
Malic	0.65	ND	ND	0.65
Malonic	ND	ND	0.03	0.03
Oxalic	0.32	1.80	ND	2.12
Succinic	0.12	0.27	ND	0.39
Total	2.50	4.18	0.62	7.30
Cations				
Na^+	17.47	13.54	3.22	34.23
K^+	4.02	3.09	0.94	8.05
Ca^{2+}	1.51	1.46	0.58	3.55
NH_4^+	0.49	0.38	0.10	0.97
Mg^{2+}	0.07	0.05	0.06	0.18
Total	23.56	18.52	4.90	46.98
Anions				
SO_4^{2-}	3.51	1.77	0.53	5.81
Cl^-	1.60	0.17	0.07	1.84
PO_4^{2-}	0.50	0.16	0.05	0.71
NO_3^-	0.03	0.04	0.009	0.07
Total	5.64	2.14	0.65	8.43
Grand total	33.58	26.40	6.47	66.45

ND, not detected.

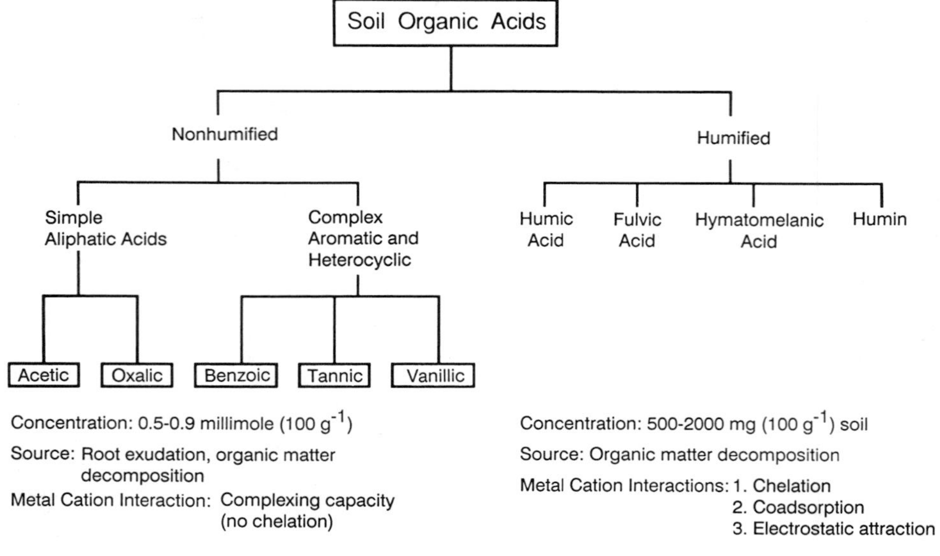

FIGURE 6.2. Organic acids of soil (from Tan 1985).

Humified acids are also diverse, and when classified by their differences in solubility in alkali, acid, and alcohol they are conveniently separated into fulvic, humic, and hymatomelanic acids and humin. The source of humic substances is plant residue conversion by microorganisms. Amino compounds synthesized by microbes react with modified lignins, quinones, and reducing sugars to form complex polymers (Stevenson 1982). The concentration of humified acids (fulvic and humic) in bulk soil is highly variable and ranges from 500 to 2000 mg 100 g^{-1} soil (Tan 1985).

Organic acid chemistry is probably very different in rhizosphere soil relative to bulk soil because of inputs of low molecular weight organic acids in root exudates to the rhizosphere. The significance of this difference to trace metal dynamics may be great (Section IV. C.).

4. Hydrogen Ion Concentration

In general, the pH of rhizosphere soil is thought to be lower than the pH of surrounding bulk soil. While the processes regulating soil pH are numerous and often competing, the net of rhizosphere processes (e.g., respiration) increases the hydrogen ion concentration in the near-root environment. Acidification of the rhizosphere may be one of the important mechanisms used by plants to make macronutrients, for example, phosphate, and micronutrients, notably iron and zinc, available (Olsen et al. 1981). The pH at the root surface may differ from the pH a few millimeters away by 1 or 2 units (Nye 1981).

The form in which nitrogen is taken up by plant roots may be extremely important in regulating rhizosphere pH (Nye 1981, Smiley 1974). This may be particularly important with reference to nitrogen input to forest systems from the atmosphere. No net charge can cross the root symplast-free space boundary. The rules of electrical neutrality require exchange of cations and anions in the nutrient uptake process. Plants that absorb nitrogen as NO_3^- tend to raise the pH in the rhizosphere. Uptake of the NO_3^- anion must be neutralized either by transfer of negative charges to metabolites, primarily organic anions within the plant, or into the rhizosphere as OH^- and HCO_3^-. Plants that absorb nitrogen as NH_4^+ lower the pH of the rhizosphere via the release of H^+. In plants fixing nitrogen via symbiosis, the uncharged N_2 molecule crosses the nodule-soil boundary. If this is the major source of nitrogen, cation intake will probably exceed anion intake and the root will lower rhizosphere pH.

Rhizosphere acidification by the metabolism of rhizosphere bacteria is also of substantial importance. For example, nitrification of ammonia, released by ammonification of organic nitrogen during decay of root parts, leads to hydrogen ion production (Dixon and Wheeler 1983):

$$2NH_4^+ + 30_2 \rightarrow 2NO_2^- + 2H_20 + 4H^+$$

5. Other Chemical Differences

In addition to differences in organic matter and pH, numerous other chemical characteristics differ in rhizosphere soil relative to bulk soil. Root surfaces have a large negative water potential. This results from the evapotranspiration demand of the aboveground plant parts. During the day, and with appropriate meteorological conditions, available moisture in the rhizosphere may become quite deficient.

Rhizosphere soil may have a lower osmotic potential than bulk soil. Salts moved through bulk soil by mass flow may be rapidly adsorbed or precipitated in the rhizosphere. Low osmotic potential decreases total water potential in the rhizosphere, thus promoting movement of water from the bulk soil to the root surface.

The root and microbial respiration in the rhizosphere tends to reduce the oxygen-to-carbon-dioxide ratio in the rhizosphere relative to bulk soil. This causes the rhizosphere to have a generally lower redox potential compared to bulk soil.

D. Rhizosphere Physical Characteristics

Foster et al. (1983) have emphasized that roots must constantly penetrate "new soil" to encounter nutrients, especially those of reduced solubility. Plants must continually produce new roots and extend old roots. Foster et al. estimated that in 1 day, each root tip may produce 18,000 new cells and extend 5 cm. Our experience with woody root tips of tree roots is that they extend at less than 5 cm per day, but still at an impressive rate of growth (Smith 1970b). As roots progress through the soil, compressive forces at root tips and along root surfaces reduce the spaces between soil particles. Pore spaces are smaller in the rhizosphere than in bulk soil. As a result, the bulk density of rhizosphere soil is higher than in nonrhizosphere soil.

E. Rhizosphere Summary

Research over the past eight decades has allowed us to partition the soil ecosystem into that part influenced by plant roots and that part devoid of root influence. The former is designated the rhizosphere, and extends approximately 1–5 mm around the roots of all plants as they grow through soil. The rhizosphere is characterized by intense biological activity because it provides abundant nutrition for microbial growth. The complex interaction between plant roots and microorganisms in the rhizosphere produces a biologically, chemically, and physically unique soil environment (Table 6.6) of enormous consequence to plant health. It is presently believed that the rhizosphere region of soil exerts dominant regulation over nutrient uptake by roots, root disease from biotic infection, and microbial saprophyte and symbiont ecology. In view of this, it is critically important to evaluate the potential interactions between atmospheric deposition and rhizosphere structure and function. Section IV presents several hypotheses for alteration of rhizosphere dynamics by atmospheric deposition.

TABLE 6.6. Characteristics of rhizosphere soil and bulk soil compared.

Characteristic	Rhizosphere soil	Bulk soil
Biological		
Microorganisms	Diverse and abundant	Limited and few
Respiration	High	Low
Symbiont habitat	Excellent	Poor
Mycorrhizas	Abundant	Nonexistent, dormant
Root pathogens	Active	Dormant
Chemical		
low molecular weight organic compounds	Diverse and abundant	Restricted
Organic acids	Nonhumified	Humified
pH	Lower	Higher
Water potential	Lower (more negative)	Higher (less negative)
O_2:CO_2 ratio	Lower	Higher
Redox potential	Lower	Higher
Osmotic potential	Lower	Higher
Physical		
Pore space	Less	More
Bulk density	Higher	Lower

IV. Hypotheses for Linkages Between Atmospheric Deposition and Rhizosphere Dynamics

The complexity and importance of rhizosphere processes allows formulation of multiple hypotheses describing potential impacts of atmospheric deposition on the nature and functions of soil near roots. Six hypotheses, however, appear to describe the most important potentials from current literature: (a) alteration of rhizosphere chemistry by aboveground stress; and alterations to rhizosphere regulation of (b) nutrient uptake, (c) heavy metal uptake, (d) aluminum uptake, (e) habitat of microbes controlling nutrient uptake, and (f) pathogen and saprophyte ecology.

A. Abnormal Physiology in Aboveground Tree Parts Causes Alteration of Rhizosphere Effects by Changing the Chemistry and Biology of the Rhizosphere

It is recognized that root growth and root exudation are regulated by numerous plant, environmental, and cultural variables. Root growth and exudation are primary determinants of rhizosphere populations and functions. Thus, it is important to understand how these root processes are regulated. It is especially critical to appreciate how atmospheric deposition may influence root growth and exudation. Very few studies have examined the former and none have explored the latter. Over the past few decades, however, numerous investigations have increased our understanding of the factors regulating root exudation. Review of these studies suggests some potential air pollution effects.

Two excellent reviews have inventoried the factors affecting root exudation (Hale et al. 1971, Hale and Moore 1979). In general, four types of influencing variables are recognized: plant factors (species, age, mechanical damage); environmental factors (temperature, light, nutrients, moisture, pH); rhizosphere biota (saprophytes, parasites, symbionts); and chemicals applied to foliage (pesticides). Some of these (for example, mechanical damage, moisture stress, pH, and pesticides) appear to have relevance to air pollution effects because they may stress plants in a manner similar to air contaminant stress. Unfortunately, only a few studies of the relevant factors have been carried out with woody plants.

Mechanical damage to roots and aboveground plant parts appears to stimulate root exudation and cause qualitative changes in root-exudate patterns. Ayers and Thornton (1968) experimentally damaged wheat and pea roots by swirling and rinsing them in sand and found increases in root exudate nitrogen compounds. Hamlen et al. (1972) presented evidence indicating that severe clipping of alfalfa plants increased the amount and altered the quality of sugars released in root exudate. These studies emphasize the need to appreciate the influence of forest tree root injury and defoliation on exudation and the rhizosphere processes.

Moisture availability is a dominant regulator of root physiology. Vancura and Garcia (1969) suggested that moisture stress increased root exudation of millet. Martin (1977b) provided evidence that reduced moisture availability increased both root autolysis and root secretion. Some work has been conducted on young forest trees. Reid (1974) evaluated the influence of water stress on ponderosa pine seedlings. In general, higher negative water potentials increased the loss of ^{14}C from roots. In an investigation of 7-year-old lodgepole pine, roots subjected to a water potential of -400 kPa exuded more ^{14}C material than roots at either 0 or -200 kPa, although the authors noted that the former roots also received the greatest amount of translocated label (Reid and Mexal 1977). This limited evidence suggests that drought conditions may generally increase release of organic materials into the rhizosphere.

The influence of pH on root exudation is particularly important with reference to the potential for forest soil acidification. Unfortunately the influence of hydrogen ion concentration on root exudation has not received detailed study. Studies with nonwoody species grown in solution culture have revealed conflicting evidence. McDougall (1970) recorded a decrease in exudation of ^{14}C from wheat seedlings when the pH of the collecting solution was increased from pH 5.9 to pH 7.0. Bonish (1973), on the other hand, found that a certain fraction of red clover root exudation increased as the pH was increased.

A large number of chemicals that are applied to foliage have been evaluated with respect to potential to alter root exudation and/or rhizosphere microbes. Most of the chemicals screened include those otherwise employed in plant management practices, including fertilizers (Agnihotri 1964, Balasubramanian and Rangaswami 1969, Vrany 1972, Vrany and Macura 1971, Vrany et al. 1962), growth regulators (Gupta 1971, Sethunathan 1970a, b, Sullia 1968), and pesticides (Balakrishnan and Raj 1970, Lai and Semeniuk 1970, Smith 1979, Vrany 1965). With the exception of Smith (1979) in which we examined the exudation patterns in American elm trees after injection of carbendazim, none of these studies was conducted with forest trees. In addition, while almost all of the studies mentioned previously recorded alteration of root exudates and/or changes in rhizosphere microflora following foliar treatment, the range and variability of responses literally defy generalization. It is extremely important to realize, however, that substantial evidence has been accumulated indicating that changes in rhizosphere chemistry and biology can be mediated by materials applied to the foliage. This has considerable implication for pollutants deposited on foliar surfaces and their potential to influence root processes.

Ozone, for example, is an important regional-scale air pollutant with substantial potential to influence directly and indirectly forest tree root metabolism. Ozone is known to cause reduction in apparent photosynthesis via direct impairment of chloroplast function, reduced CO_2 uptake resulting from ozone-induced stomatal closure, or both (Smith 1990). Regardless of mechanism, a sustained reduction in photosynthesis will ultimately influence root physiology, alter root exudation, and change the microflora of the rhizosphere. In addition to depressing photosynthesis in the foliage of numerous species, ozone has been shown to inhibit the allocation and translocation of photosynthate (e.g. sucrose) from the shoots to the roots (Jacobson 1982, Tingey 1974). Tingey et al. (1971) found exposure of radish to ozone (0.05 ppm for 8 h, 5 days wk^{-1} for 5 weeks) inhibited hypocotyl growth by 50%. Ponderosa pine exposed to 0.10 ppm ozone for 6 h day^{-1} for 20 weeks stored significantly less sugar and starch in their roots than did control plants (Tingey et al. 1976).

Numerous other papers reviewed in U.S. Environmental Protection Agency (1987) indicate carbohydrate allocation to roots is reduced following ozone exposure. Clearly this altered allocation will alter root exudation and rhizosphere processes. Manning et al. (1971) for example, have reported that pinto bean plants exposed to 0.1–0.15 $\mu l\ l^{-1}$ of ozone for 8 h day^{-1} for 28 days had poor root growth, and *Rhizobium* nodules developed only on unfumigated bean plants. Tingey and Blum (1973) recorded similar results when soybean plants were exposed to 0.75 ppm ozone for 1 h. In addition to these influences, severe (high-dose) ozone, or other gaseous pollutant exposure, may result in acute foliar morbidity and mortality. This could represent the equivalent of defoliation. It is known that defoliation can influence both the quantity and quality of root exudation. Late-season defoliation of sugar maple has been shown to alter patterns of carbohydrate, amino acid, and organic acid exudation (Smith 1972). Defoliated maples released greater quantities of fructose, cystine, glutamine, lysine, phenylalanine, and tyrosine, whereas foliated maples exuded greater amounts of sucrose, glycine, homoserine, methionine, threonine, and acetic acid (Table 6.7).

Despite the fact that extremely limited evidence is available from woody plant studies, a large number of investigations have indicated the ability of environmental variables and foliar chemical applications to alter the chemistry and biology of the rhizosphere. To date, these studies have not included air pollutants. Some air pollutant studies, however, have indicated the ability of regionally important pollutants such as ozone to have the capability to influence carbon allocation to roots and root physiology. Effects of this kind will certainly be ultimately translated into rhizosphere impacts. It is imperative to design and implement studies to detail this important possibility in woody plants.

TABLE 6.7. Influence of defoliation on quantities of compounds exuded by roots of sugar maples. [Data are micrograms $\times$ 10^{-1} of each material released during 14 days per milligram of oven dry root].[a]

	1969[b]		1970[c]	
Compound	Control	Defoliated	Control	Defoliated
Carbohydrates				
Fructose	2.6 ± 0.2	4.3 ± 0.1[d]	4.3 ± 1.2	6.1 ± 0.9
Glucose	0	0	Trace	Trace
Sucrose	7.3 ± 0.1	2.9 ± 0.2[d]	7.9 ± 0.9	6.0 ± 0.3
Amino acids/amides				
Alanine	0	0	1.8 ± 0.4	1.7 ± 0.1
Cystine	0.2 ± 0.1	0.5 ± 0.1	0	0
Glutamine	2.3 ± 0.3	3.4 ± 0.5	3.6 ± 0.7	4.3 ± 1.1
Glycine	0.5 ± 0.1	0.3 ± 0.1	1.9 ± 0.3	1.1 ± 0.3
Homoserine	1.1 ± 0.2	0.3 ± 0.1[d]	2.4 ± 0.6	1.8 ± 0.3
Lysine	0.8 ± 0.1	1.8 ± 0.2[d]	0	0.5 ± 0.2[d]
Methionine	0	0	1.5 ± 0.1	1.4 ± 0.4
Phenylalanine	0	0	2.7 ± 0.6	3.1 ± 0.5
Threonine	Trace	Trace	3.4 ± 0.3	0[d]
Tyrosine	0	0	0.9 ± 0.7	1.1 ± 0.2
Organic acids				
Acetic	49.7 ± 10.1	24.3 ± 9.4	63.2 ± 11.1	58.1 ± 13.3
Malonic	0	0	Trace	0

[a]From Smith 1972.
[b]Mean and standard error of three replicate determinations using one composite exudate sample from 19 and 20 roots of control and defoliated trees, respectively.
[c]Mean and standard error of three replicate determinations using one composite exudate sample from 17 and 23 roots of control and defoliated trees, respectively.
[d]Control and defoliated data are significantly different at 95% level.

B. Atmospheric Deposition Alters Rhizosphere Regulation of Nutrient Uptake

Normal plant growth and development requires roots to grow through the soil and acquire adequate amounts of essential nutrients—nitrogen, phosphorus, potassium, calcium, magnesium, sulfur, boron, copper, manganese, iron, zinc, and molybdenum. These nutrient elements are commonly present in bulk soil in amounts far in excess of vegetative needs. To be taken up by plants, however, nutrients must be "available" and ultimately present as free ions.

A discussion of the details of nutrient uptake by plants is beyond the scope of this review (see Haynes 1980 and Figure 6.3). The important point here is that the rhizosphere mediates the transformation of nutrients from unavailable to available forms. Rhizosphere processes that help provide free ions for symplast uptake include pH regulation, protonation, reduction, and complexation. Hydrogen ions and organic acids produced by both the root and microorganisms of the rhizosphere play central roles

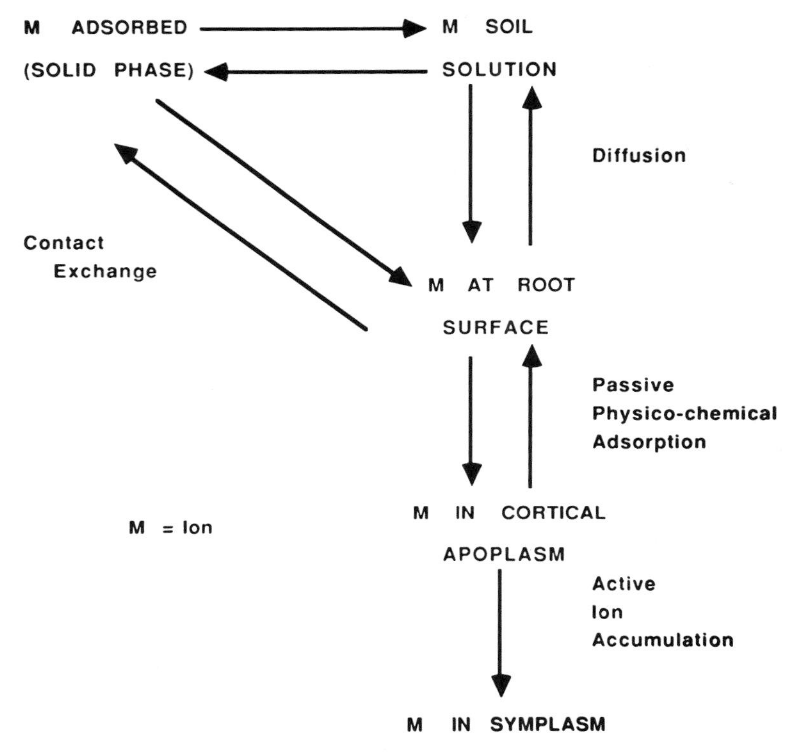

FIGURE 6.3. Ion uptake by plant roots (redrawn from Haynes 1980) (M, ion).

in these processes. Nonhumified low molecular weight organic acids in root exudates may be particularly important because they form soluble complexes with nutrient ions. Nutrients in these complexes are thought to be more available than nutrients in strong chelates formed by the humified organic acids prevalent in bulk soil (Stevenson 1982, Tan 1985).

The following paragraphs review studies that clearly demonstrate the importance of rhizosphere processes in tree uptake of selected nutrients. There are, however, no studies demonstrating effects of atmospheric deposition on rhizosphere levels of hydrogen ions and organic acids. Mechanisms for such effects *could* involve pollutant stress in aboveground tree parts (Section IV. A.) and/or pollutant-induced changes in soil chemistry (Smith 1990).

1. Phosphorus

The major forms of phosphorus in bulk soil are only very slightly soluble and generally "unavailable" for plant uptake. Rhizosphere processes allow plants to absorb phosphorus. The most important of these include phosphorus solubilization by root exudates, rhizosphere bacteria, or mycorrhizal fungi. Moghimi et al. (1978a) studied the uptake of phosphate from ^{32}P-labeled synthetic hydroxyapatite by wheat, corn, and pea plants. Wheat plants produced a water-soluble rhizosphere product, either a root-exudate compound or rhizosphere microbe product, capable of releasing phosphate from calcium phosphate. In a subsequent investigation, Moghimi et al. (1978b) provided evidence that the phosphate-solubilizing rhizosphere product was 2-ketogluconic acid. This acid is one of the strongest monobasic carboxylic acids and serves as a readily available source of hydrogen ions for the dissolution of hydroxyapatite (Moghimi and Tate 1978). It was not established whether 2-ketogluconic acid was produced by rhizosphere bacteria or released directly by root exudation. Bowen and Rovira (1966) indicated that the presence of bacteria increased uptake and translocation of phosphate in plants. Numerous bacterial groups capable of solubilizing phosphorus (termed phosphobacteria) are abundant in plant rhizospheres, for example, *Bacillus* sp., *Pseudomonas* sp., and *Agrobacterium* sp. (Ocampo et al. 1978). Mycorrhizal roots have been shown to be more efficient in phosphate uptake than nonmycorrhizal roots (Gerdemann 1974, Sanders et al. 1975). Cromack et al. (1976) have presented an integrated hypothesis for the evolution of the mycorrhizal condition of forest tree roots. Initially tree roots released organic acids (oxalic, malic, and others) into the rhizosphere to extract phosphorus. Because of the abundant supply of phosphorus and organic acid metabolites, fungi colonized the rhizoplane region of these roots. Eventually, fungal mycelium produced mantles around roots and hyphae penetrated the root cortex (endorhizosphere). Root release of oxalate and acetate were gradually reduced while the supply of malate and carbohydrates to the fungi was increased. Fungi assumed the role of oxalic acid secretion and ensured a continual availability of phosphorus. Whether or not this hypothesis is correct, it is clear that mycorrhizal and other rhizosphere processes regulate phosphorus availability and uptake.

2. Potassium

Potassium release from silicate minerals in bulk forest soils is necessary for normal tree nutrition. Boyle et al. (1967) have emphasized the potential of low molecular weight organic acids, characteristic of the rhizosphere, to weather biotite. In experiments with seedling Monterey pine, the complexing ability (ability to form soluble compounds) of oxalic, malonic, citric, malic, and lactic acids from the rhizosphere was shown to be effective in releasing potassium from biotite (Boyle 1967).

3. Iron

Iron is abundant in bulk soil but commonly not available to plants. Evidence has been accumulated that certain plants are especially efficient in solubilizing iron. When certain of these plants are grown in solution culture they exhibit an "iron-stress response mechanism" that can decrease ambient solution pH from above 7.0 to 3.7 within a few hours. It can be calculated that a decrease in three pH units can cause an increase in Fe^{3+} concentration by a factor of 1 billion (Olsen et al. 1981). These authors speculated that plants stressed by iron deficiency release reductants into the rhizosphere and increase the solubility of iron. Tomato plants, for example, convert p-coumaric acid to caffeic acid, which is able to reduce Fe^{3+} to Fe^{2+}. Iron is absorbed by plant roots primarily in the uncomplexed Fe^{2+} form (Olsen et al. 1981).

4. Other Micronutrients

In addition to iron, other heavy metal micronutrients (zinc, copper, manganese) are common in soils but may be unavailable for root uptake. Rhizosphere processes are again considered essential for normal root uptake. Rhizosphere processes may involve solubilization, reduction, and complexation.

Sarkar and Wyn Jones (1982) have studied the effect of varying rhizosphere pH on micronutrient uptake of bean plants in pot culture. Variations in rhizosphere pH were induced by applying different sources of nitrogen to greenhouse soil that had an initial pH of 7 or 8. Strong inverse relationships were shown between shoot and root zinc and manganese and pH of the rhizosphere. The relationship was linear with zinc and curvilinear with manganese. It is clear that reduced rhizosphere pH makes these elements available by preventing formation of insoluble hydroxides. Linehan et al. (1985) have studied the bulk soil and rhizosphere soil solution concentrations of copper, manganese, and zinc in barley fields over the course of a growing season. Evidence for rhizosphere-to-bulk soil solution concentration ratios of 3 for copper and zinc and 15 for manganese were presented. The authors speculated that acidification of the rhizosphere and complexation with rhizosphere organic and amino acids may have contributed to elevated solution concentrations in the rhizosphere.

Formation of organic–metal complexes has been demonstrated for zinc by Hodgson et al. (1965) and for manganese by Olomu et al. (1973). Elgawhary et al. (1970) employed ^{65}Zn to demonstrate that complexing agents can increase the transport and diffusion of zinc to roots.

As in the case of iron, a significant proportion of Mn^{3+} is reduced to Mn^{2+} before root uptake (Sarkar and Wyn Jones 1982). All the processes leading to enhanced micronutrient availability (i.e., solubilization, complexation, and reduction) occur in the rhizosphere.

Microorganisms of the rhizosphere may alter availability of micronutrients. Zinc and copper uptake by roots may be increased by mycorrhizal fungi (Tinker 1984). Certain rhizosphere bacteria, on the other hand, may reduce the availability of manganese by oxidizing divalent manganese to tetravalent manganese (Douka 1977).

5. Nitrogen

Nitrogen deficiency is probably the most widespread nutritional disorder in plants. Unlike the elements previously discussed, however, nitrogen deficiency largely results from biological competition for this scarce resource rather than mineral insolubility. The biological intensity of rhizosphere soil may, in fact, enhance nitrogen stress by increasing the competition for the uptake of this element. For those species with symbiotic associations with nitrogen-fixing microbes, either free-living or nodulated, the rhizosphere plays an important positive role in nitrogen nutrition by providing a habitat for *Azotobacter*, *Azospirillum*, *Frankia*, *Rhizobium*, and others.

C. Atmospheric Deposition Alters Rhizosphere Regulation of Heavy Metal Uptake

Approximately 90 elements constitute the inorganic component of forest soils. Of these elements, 89% are present in concentrations less than 1000 $\mu g\ g^{-1}$ (1000 ppm) or less than 0.1% and are appropriately designated trace elements. Trace metals with a density greater than 6 $g\ cm^{-3}$ are designated "heavy metals". Heavy metals are input to forest soils via both natural and anthropogenic processes. Natural sources include in situ soil weathering and atmospheric deposition of metals derived from soil processes, oceanic processes, and volcanic eruptions. Anthropogenic sources include fossil fuel combustion for electricity generation, industrial processes, motor vehicle use, and agricultural activities (National Research Council 1980, Parekh and Hussain 1981).

Some heavy metals are biologically essential and are required for normal plant growth and development (see Section IV. B). Other heavy metals have not been proven to be required for plant growth and development and are appropriately designated biologically nonessential. At sufficient concentration or dose, *all* heavy metals are toxic to plants (Table 6.8).

Heavy metal particles are deposited to forests by both wet and dry processes. Dry deposition is presumed more effective for large particles and elements such as cadmium and lead (Galloway et al. 1982). These latter authors have provided an excellent inventory of heavy metal deposition rates ($kg\ ha^{-1}\ yr^{-1}$) for urban, rural, and remote environments. For certain heavy metals, for example, cadmium, lead, manganese, and zinc, there is evidence from studies in the eastern United States that wet deposition of the metals is highest during the growing season (warmest months) (Lindberg 1982).

Extensive evidence is available to support the suggestion that heavy metals deposited from the atmosphere to forest systems are accumulated in the upper soil horizons or forest floors (Andresen et al. 1980, Friedland et al. 1984a,b, Miller and McFee 1983, Page and Chang 1979, Siccama

TABLE 6.8. Heavy metals that are essential and nonessential for plant health have natural and pollution sources.[a]

Essential	Nonessential	
COPPER (Cu)	antimony (Sb)	SILVER (Ag)
IRON (Fe)	bismuth (Bi)	THALLIUM (Tl)
MANGANESE (Mn)	CADMIUM (Cd)	tellurium (Te)
molybdenum (Mo)	CHROMIUM (Cr)	thorium (Th)
ZINC (Zn)	cobalt (Co)	tin (Sn)
	LEAD (Pb)	uranium (U)
	MERCURY (Hg)	VANADIUM (V)
	NICKEL (Ni)	

[a]Metals designated by all capital letters are thought to be especially significant for considerations of forest health risk.

et al. 1980, Smith and Siccama 1981). Because this accumulation is commonly in the soil horizons with maximum root activity and maximum activity of the soil micro- and macrobiota, it is appropriate to consider rhizosphere regulation of heavy metal availability and/or toxicity to tree roots and other biotic components of the soil ecosystem. A very large literature provides detail on the biochemical and physiological toxicity of heavy metals to higher plants (e.g., Antonovics et al. 1971, Bowen 1966, Energy Research Development Administration 1975, Foy et al. 1978, Foy 1983) and to microorganisms (e.g., Jernelov and Martin 1975, Somers 1961, Summers and Silver 1978). It is not necessary to review this information here, but it is appropriate to summarize selected aspects of our understanding of rhizosphere processes associated with heavy metal availability.

It is generally assumed that the heavy metals accumulated from the atmosphere in the forest floor are retained in bulk soil in largely insoluble form as hydroxide, sulfide, or other low-solubility precipitates or strongly chelated by humic acids. In this form the heavy metals are not available for root or microbial uptake, and the potential for toxicity is presumed low. The rhizosphere processes of pH regulation, protonation, solubilization, complexation, and reduction, however, may transform unavailable metals to available metals, as previously discussed for heavy metal micronutrients.

1. Lead

Lead deposited from the atmosphere may have a residence time approximating 5000 years in the surface organic soil horizons (Benninger et al. 1975), and long-term concentration increase can be predicted so long as inputs to forest soils exceed outputs (Smith 1990). Rhizosphere characteristics that may be especially important in mobilizing lead in the forest floor include pH, hydrogen ion availability, organic acid availability, and phosphate level (Zimdahl and Koeppe 1979). Rolfe (1973) studied the uptake of lead by 2-year-old seedlings of eight forest tree species. He concluded that the only soil factor important in lead uptake by the trees, other than lead concentration in the soil, was phosphorus levels. Lead uptake was reduced by half when high levels of phosphorus were present. Rhizosphere availability of phosphorus may critically regulate lead uptake in forest soils.

Reddy and Patrick (1977) have stressed the importance of oxidation-reduction potential (redox potential) and pH in the uptake of lead by rice. More lead was taken up at lower pH. Similar importance of lower pH on increased lead uptake was emphasized by Arvik and Zimdahl (1974). In addition, mycorrhizal roots may be especially efficient for heavy metal uptake, including lead (Killham and Firestone 1983). Some evidence has been presented indicating that lead taken up by roots of some plants is distributed to aboveground tissues while in other plants it is retained in the endorhizosphere as an insoluble precipitate (e.g., lead phosphate) (Smith 1990).

2. Manganese, Zinc, and Copper

Micronutrients present in excess can be phytotoxic if available for root uptake. As previously discussed in Section IV.B., rhizosphere processes generally enhance availability of these elements.

Microsites in the rhizosphere with low oxygen availability may reduce manganese to the divalent (available) form (Foy et al. 1978). Evidence for complexation of manganese with low molecular weight compounds is conflicting and inconclusive. Hofner (1970) has suggested that manganese may complex with an amino acid. Bremner and Knight (1970), on the other hand, have indicated that manganese is generally present in the uncomplexed form.

Substantial evidence has been presented stressing the importance of low molecular weight complex formations in the case of zinc and copper (Bremner and Knight 1970, Bremner 1974). Organic and amino acid complexes with these elements may substantially increase availability in rhizosphere soil relative to bulk soil.

3. Other Heavy Metals

Other heavy metals that may accumulate in bulk soil of the forest floor and have potential for toxicity to tree or soil organisms include cadmium, chromium, mercury, nickel, silver, thallium, and vanadium. With the exception of cadmium and mercury, there is only limited information on the deposition, mobilization, and rhizosphere dynamics of these elements.

Cadmium may be present in bulk soil adsorbed onto humic acids (Riffaldi and Levi-Minzi 1975). Available cadmium is well known to be readily taken up by plant roots (Führer 1982, Haghiri 1973, John et al. 1972a). It is reasonable to propose that rhizosphere processes could mediate the conversion of bound cadmium to ionic cadmium. Cadmium chemistry in soil is similar to zinc chemistry, and complex formation with low molecular weight rhizosphere compounds may be important in the former as well as the latter. Reduced soil pH has been shown to increase cadmium uptake by radish (Lagerwerff 1971) and lettuce (John et al. 1972b). High cation exchange capacity, high organic matter content, and low pH, characteristic of rhizosphere soil relative to bulk soil, all may contribute to increased cadmium availability in the rhizosphere.

D. Atmospheric Deposition Alters Rhizosphere Regulation of Aluminum Uptake

Aluminum is the third most abundant element on Earth and a major structural component of soils. At bulk soil pH in excess of 5.5, aluminum is generally tightly bound in clay mineral particles and is typically unavailable for root or microbial uptake. When bulk soil pH falls below the range of 5.0 to 5.5, however, aluminosilicates become unstable and some aluminum bound within the clay crystals moves to exchangeable positions on clay surfaces. In strongly acid soils, for example, most forest soils, clay surfaces are saturated primarily with exchangeable aluminum ions and secondarily with hydrogen ions (Foy 1985). Because many of the exchangeable cations in acid forest soils are aluminum and hydrogen, and since a cation–anion balance is required in soil solutions, the input of anions such as sulfate or nitrate (whether as neutral salts or acids) via atmospheric deposition may cause increased soil solution concentrations of aluminum and hydrogen via exchange processes (Johnson et al. 1981, Wolt, this volume).

Atmospheric input of sulfuric acid and nitric acid to noncalcareous higher elevation watersheds in the White Mountain and Adirondack regions have lead to relatively high concentrations of dissolved aluminum in surface and groundwaters (Cronan and Schofield 1979) and in bulk soil leachates (Cronan 1980). If ionic aluminum (Al^{3+}) is available for root or microbial uptake, it is toxic at sufficient concentration. Aluminum is phytotoxic because it interferes with cell division in root tips and lateral roots, increases cell wall rigidity by cross-linking pectins, reduces DNA replication by increasing the rigidity of the double helix, reduces availability of phosphorus at the rhizoplane or in the endorhizosphere, decreases root respiration, interferes with enzymes governing sugar phosphorylation and deposition of cell wall polysaccharides, and interferes with the uptake, transport, and use of several nutrients including calcium, magnesium, phosphorus, and iron (Foy 1971, 1983, Foy et al. 1978). Unfortunately, only limited research has been conducted regarding the influence of soluble aluminum on the nutrient uptake processes of forest trees. Soil solution levels of aluminum may be especially high in northeastern forests, especially in spring after acidic snowmelt (Table 6.9). Cumming et al. (1985) have examined the influence of 0, 1, and 4 mg l^{-1} of aluminum on potassium uptake of red spruce seedlings grown in solution culture. Aluminum was seen to increase seedling uptake of potassium under most treatment regimes.

Aluminum is a component of microbial cells (Brock 1966). At sufficient concentration, however, aluminum is toxic to microorganisms as well as to plants. Wood et al. (1984a,b) and Wood and Cooper (1984) have demonstrated the toxicity of aluminum to rhizobium in the rhizosphere of white clover.

In addition to pH, the availability of soluble aluminum in bulk soil is influenced by the kind of clay minerals present, the amount of organic matter present, and the concentrations of other cations, anions, and total salts present.

There is substantial evidence that rhizosphere processes have the ability to regulate aluminum availability. The single most important regulator is rhizosphere pH. Changing the pH of the rhizosphere by only 0.1 unit can change the theoretical aluminum solubility twofold (Foy and Fleming 1976; Foy et al. 1978). Foy (1971), in fact, emphasizes that aluminum-tolerant cultivars of wheat, barley, rice, peas, and hybrid corn are tolerant because they can increase the pH of the rhizosphere and thereby reduce the availability of aluminum for root uptake. The mechanism for pH regulation in this case may be differential anion/cation uptake, especially nitrate nitrogen vs. ammonium nitrogen uptake (Bartlett and Riego 1972, Dodge and Hiatt 1972, Mugwira and Patel 1977). The uptake of nitrate nitrogen raises the pH of the rhizosphere. As a result, nitrate nutrition in acid rhizosphere soil could serve to reduce aluminum uptake by reducing aluminum availability.

Other potential mechanisms of rhizosphere regulation of aluminum availability involve phosphorus precipitation and complexation by low molecular weight organic compounds. Above pH 5, phosphorus precip-

TABLE 6.9. Forest soil solution and forest stream concentrations of aluminum from some northeastern locations.

Location	Medium	Aluminum concentration (mg l^{-1})	Reference
Adirondack Mountains	Conifer stands (2) soil leachate	3.57, 2.74	Mollitor and Raynal 1982
White Mountains	Forest soil leachate	0.48	Cronan 1980
Hubbard Brook Experimental Forest, New Hampshire	First-order stream	0.71	Johnson et al 1981

itation of aluminum as insoluble aluminum phosphate is thought to be important (Bartlett and Riego 1972). The importance of organic matter complexation of Al in the rhizosphere or endorhizosphere is incompletely understood. Mollitor and Raynal (1982) have compared bulk soil chemistry in a hardwood and coniferous forest in the central Adirondacks. These investigators found higher aluminum concentrations in conifer site leachates and suggested that the higher organic anion concentrations, characteristic of the conifer soils sampled, may contribute to the elevated aluminum concentrations. We have recently provided evidence for a strongly descending gradient of aluminum concentration from bulk soil through the rhizospheres of mature red spruce trees in New Hampshire (Smith 1989).

E. Atmospheric Deposition Alters Rhizosphere Regulation of Microbes Mediating Uptake of Plant Nutrients from Soil

Microorganisms that mediate and facilitate element uptake by tree roots have roles of very great importance in nutrient cycling in forest ecosystems. Forests frequently flourish in regions of low, marginal, or poor soil nutrient status. In addition to nutrient conservation and tight control over nutrient cycling, trees have evolved critically significant relationships with free-living and symbiotic bacteria and fungi that enhance nutrient supply and uptake. In Section III, we emphasized the critical significance of the rhizosphere habitat for free-living and symbiotic microbes. The bulk soil environment is a very harsh habitat for most microorganisms. It is characterized by a variety of chemicals that exert bacteriostatic and fungistatic influence on spore germination and vegetative development. Substrates available for microbial nutrition are generally extremely deficient for most microbes in bulk soil. The supply of readily utilizable carbon is especially limiting. The rhizosphere habitat provides amelioration of these limitations.

Alteration of the relationship between tree roots and nutrient supplying microbes caused directly or indirectly by atmospheric deposition could have extremely important consequences for tree health and growth. A most important alteration would be any significant and prolonged change in the quality or quantity of low molecular weight carbon compounds supplied to the microbes by the roots. Another important alteration would be significant change in the chemistry of the rhizosphere that would adversely influence microbial metabolism and growth. Heavy metal toxicity is a special concern (Kelly and Reanney 1984, Wood and Wang 1983).

1. Free-Living Bacteria

The relative importance of nitrogen-fixing free-living bacteria and algae in the rhizospheres of forest trees is very poorly appreciated. The influence of atmospheric deposition on these potentially significant organisms has not been studied.

2. Symbiotic Bacteria

The environmental factors that influence members of the *Rhizobium* genus have been reviewed by Postgate and Hill (1979). Evidence has been provided to add air pollution to this list (Smith 1990). Treatment of clover plants with ozone has been shown to reduce the growth and nodulation of treated plants (Letchworth and Blum 1977). Presumably ozone stress imposed on clover leaves altered root/rhizosphere physiology in some manner capable of restricting root infection, nodule formation, or bacteroid movement.

Heavy metals have been implicated as capable of altering nitrogen fixation in legumes (Döbereiner 1966, Huang et al. 1974). Vesper and Weidensaul (1978) subjected soybeans to cadmium and nickel at 1, 2.5, and 5 ppm and copper and zinc at 1, 5, and 10 ppm in sand culture. The degree of toxicity was cadmium > nickel > copper > zinc. Cadmium dramatically reduced nodule number, dry weight, and nitrogen fixation. Fixation by nickel treated plants was very low. Copper suppressed nodulation, but inhibited nitrogen fixation directly only at 5 and 10 ppm. Zinc reduced nodulation, but only slightly inhibited fixation.

Recently, studies have focused on potential effects of pH and aluminum changes in the environment on *Rhizobium* symbioses. Wood et al. (1984a) evaluated the influence of manganese and aluminum at various pH levels on the white clover–*Rhizobium trifolii* symbiosis. This work was conducted under laboratory conditions using axenic culture. Manganese at 200 μM increased root elongation in the pH range 4.3–5.5, but did not affect root hair formation, numbers of *Rhizobium* in the rhizosphere, or nodule formation. Aluminum at 50 μM inhibited root elongation and root hair formation at pH 4.3 and 4.7. This level of aluminum also inhibited *Rhizobium* multiplication in the rhizosphere and reduced nodule formation at pH 5.5. A subsequent study by these same authors used the same laboratory design to examine aluminum–phosphate–calcium interactions with white clover symbiosis (Wood et al. 1984b). In the presence of 10 μM phosphate, 50 μM aluminum reduced or inhibited root elongation and root hair formation at pH <5.0, and *Rhizobium* multiplication in the rhizosphere and nodule formation at pH <6.0. In the absence of aluminum, root elongation and root hair formation were reduced at pH <4.3, and *Rhizobium* multiplication and nodule formation were inhibited at pH <5.0. Root hair formation was more sensitive to aluminum at pH <5.0 than was root elongation. These studies suggest that aluminum and low pH are more limiting to white clover–*Rhizobium* symbioses than calcium deficiency or high manganese concentration under laboratory conditions. The inhibitory effect of aluminum was judged to derive from the polymeric rather than the monomeric form (Wood and Cooper 1984).

3. Symbiotic Actinomycetes

The endophyte of *Alnus* and other actinorhizal plant nodules has been identified as *Frankia*, an actinomycete. The ecology and physiology of the endophytes associated with woody plants with alder-type nodules is only now being studied. The few investigators working in this area have been hampered by serious limitations in culturing the microsymbiont. Some level of success, however, has been achieved in establishing and characterizing pure cultures of Frankia during the past decade (Baker 1989). At present, *Frankia* can be cultivated rather simply and used to inoculate numerous actinorhizal genera for plantation purposes. It is important to continue studies directed to improving our understanding of nutrient and cultural requirements of alder-type endophytes, because in vitro cultivation experiments have shown actinomycete growth to be influenced by heavy metals in growth media. Cadmium has been shown to be toxic over a wide range of media concentrations. Aluminum and nickel are generally toxic above 10 ppm, while lead and vanadium appear to be relatively nontoxic (Waksman 1967).

4. Mycorrhizal Fungi

The uptake of nutrients from the forest floor and bulk soil throughout the relatively large interroot distances of forest vegetation is achieved by the longevity of ectomycorrhizal and endomycorrhizal roots. These fungal symbioses are critically important to the development of forest vegetation because of the role these specialized roots play in the uptake of relatively immobile ions such as phosphate, zinc, copper, and molybdenum. Unfortunately, this capacity appears to place mycorrhizal roots at risk if heavy metal micronutrients or other heavy metals are taken up in amounts in excess of some toxicity threshold for either the fungus or the root.

Limited evidence for increased heavy metal uptake by mycorrhizal roots has been presented for zinc, copper, and manganese (Bowen 1973). Unfortunately, little is known concerning the relative tolerance of symbiotic fungi to heavy metal contamination. Fungi are known to accumulate metal ions, and it has been hypothesized that they may combine these with oxalic acid and then dispose of them as insoluble oxalate salts (Cromack et al. 1975).

Under laboratory conditions, but with field-collected root material, Bowen et al. (1974) observed enhancement of zinc uptake from solution with ectomycorrhizas of Monterey pine and with mycorrhizas of hoop pine compared with uninfected short roots. Zinc amendments of 45 and 135 $\mu g\ g^{-1}$ of soil decreased both nodulation and mycorrhizae of pinto bean when compared to amounts in nonamended soil (McIlveen et al. 1975). Recently Killham and Firestone (1983) provided laboratory evidence in support of the potential of mycorrhizal roots for excess heavy metal uptake. These investigators grew perennial bunchgrass with and without the endomycorrhizal fungus, *Glomus fasciculatum*. Plants were exposed to copper, nickel, lead, zinc, iron, and cobalt in rain simulants of pH 3.0, 4.0, and 5.6. Metal concentrations in the roots and shoots of mycorrhizal plants were greater than those of nonmycorrhizal plants. Mycorrhizal enhancement of plant metal uptake increased with greater acidity and higher heavy metal content of treatments.

With current information we cannot even approximate the threshold levels of any heavy metal that might exert an adverse influence on a mycorrhizal fungus in soil. In view of the extraordinarily large number of fungi capable of forming mycorrhizal associations, estimated to be 2000 species for Douglas-fir alone (Trappe 1977), and general variation of fungi in response to heavy metal exposure, for example the decrease in formation of ectomycorrhizae by some and the increase by others following application of copper fungicide to seedlings (Göbb and Pümpel 1973), it is assumed that the threshold range would be very broad, variable, and species specific.

The influence of simulated acid rain on nitrogen uptake by the fungus *Glomus mosseae*, which is endomycorrhizal with the roots of sweetgum seedlings, has been studied by Haines and Best (1976). Natural forest soil profiles collected in North Carolina and supporting the growth of sweetgum seedlings (originating from planted seeds) were treated with artificial acidic rain. Treatments included sterilized soil and soil inoculated with a mycorrhizal fungus. When test soil columns received NH_4-N in acid, concentrations of NH_4-N in soil solutions were not significantly depleted by mycorrhizal roots but were significantly depleted by nonmycorrhizal roots when compared to the NH_4-N concentrations in columns containing soil alone. When soil solution acidity was adjusted to pH 2.0 by rain simulants, NO_3-N concentrations were unchanged by mycorrhizal columns. The authors judged that if mycorrhizal roots are more subject to competitive inhibition of cation uptake by H^+ ions than nonmycorrhizal roots, then acidification of rhizosphere solutions may result in decreased cation uptake by mycorrhizal roots.

It is difficult, at the present time, to generalize the influence of dry or wet deposition influence on mycorrhizal formations. The experimental evidence available is inconclusive or in conflict. Ho and Trappe (1984) have indicated that prolonged ozone exposure reduced the intensity of mycorrhiza formation in fescue. Reich et al. (1985), on the other hand, recorded that mycorrhizal infection of northern red oak seedlings was significantly increased by laboratory or field exposure to ozone. No influence of ozone or sulfur dioxide treatment was observed on mycorrhizal intensity of loblolly pine seedlings (Mahoney et al. 1985). Application of rain simulants of pH 3.5 at ambient rates had no influence on mycorrhizal intensity of eastern white pine seedlings in six of nine soils tested by Stroo and Alexander (1985). In contrast, Shafer et al. (1985) concluded that rain simulants of pH 3.2 and 4.0 inhibited mycorrhizal development in loblolly pine, but that a simulant of pH 2.4 enhanced mycorrhizal root formation.

F. Atmospheric Deposition Alters Rhizosphere Regulation of Root Pathogen or Soil Saprophyte Ecology

The rhizosphere is of special significance to root-infecting fungi, because root exudates provide the nutrition for fungal propagules to germinate and to support vegetative growth in the rhizosphere. Exudates may also provide energy for infection and may attract motile spores of Phycomycetes (Lockwood and Filonow 1981). As in the case of symbionts, any significant change in the quantity or quality of root exudation or rhizosphere chemistry could have profound influence on pathogen spore germination, infection, and growth in the rhizosphere. Unfortunately this review did not uncover any air pollution studies directly addressing this important topic for forest trees.

More information is available on atmospheric deposition effects on saprophytic microbes that participate in important soil functions (Myrold, this volume). Several of these studies have investigated pollutant effects on bulk soil processes such as total microbial respiration rather than effects on individual microbial species. Bewley and Stotzky (1983d) employed a continuous perfusion technique to investigate influences of simulated acid rain and cadmium, alone and in combination, on ammonification and nitrification in soil returned to the laboratory from Ossining, New York. Ammonification was relatively insensitive to both cadmium and acidity and occurred even in soils treated with pH 2.0 or 1000 ppm cadmium. Nitrification, on the other hand, was more sensitive and was retarded in ammonium nitrogen-supplemented soils exposed to pH 2.5 and inhibited in soil exposed to pH 2.0.

In an effort to evaluate the effects of cadmium and zinc on carbon mineralization and the mycoflora, Bewley and Stotzky (1983a) applied these heavy metals to "stored" (8 years) glucose-amended soils. The test soils were also amended with 9% kaolinite or montmorillorite to evaluate mitigation of heavy metal toxicity by clay minerals. Results indicated that the threshold of important inhibitory effect on microbial activity in soils by zinc may be at approximately 10,000 ppm, and the threshold for cadmium approximately 5,000 ppm. The clays did not reduce the toxicity of the heavy metals for the microbes. In a subsequent study (Bewley and Stotzky 1983b), both metals were applied simultaneously to test soils. In spite of a gradual lengthening in the time before initiation of glucose degradation as the metal concentration increased, the combination of metals resulted in an additive, rather than a synergistic, response. The growth of *Aspergillus niger* in soil acidified to pH levels of 3.6 to 4.2 was reduced by the addition of either 100 or 250 ppm cadmium or 1000 ppm zinc (Bewley and Stotzky 1983c).

The studies of Bewley and Stotzky (1983a,b,c,d) and others emphasize the importance of redundancy in microbial ecosystems. After a sufficient length of time, a microbial population tolerant of zinc and cadmium will be selected. Babich, Bewley, and Stotzky (1983) have proposed the use of an "ecological dose 50% (EcD50)" concept to quantify the inhibition of microbe-mediated soil processes by pollutants. Application of this strategy or similar procedure is desperately needed for rhizosphere processes!

V. Conclusions

The influence of atmospheric deposition on the belowground ecosystem is poorly characterized. More information on the direct and indirect consequences of air pollution on tree root physiology and on the ecology of microbiota in the vicinity of tree roots is essential for a comprehensive understanding of air contaminant impacts on forest health and growth. Root biomass of forest trees is extremely large and primarily located in the uppermost portion of the soil profile. The focus of primary interest with respect to potential interaction of the atmosphere with tree roots is the rhizosphere. The rhizosphere "connects" tree roots to the soil ecosystem and is defined as that zone of soil where the microflora is affected by the root.

The rhizosphere is actually quite restricted, generally extending 1 to 2 mm beyond the root surface, but has unique and important biological, chemical, and physical characteristics relative to soil devoid of roots (bulk soil). Populations of numerous and varied microbes are stimulated in the rhizosphere. Bacteria and fungi exhibit the greatest rhizosphere stimulation of all taxonomic groups. In addition, numerous ecological groups, including saprophytes, facultative parasites, obligate parasites, mycoparasites, and symbionts, are enhanced in rhizosphere soil relative to bulk soil. The endorhizosphere, rhizoplane, and rhizosphere provide the habitat for endrotrophic and ectotrophic fungi that form mycorrhizal symbioses with forest trees.

Microorganisms are stimulated in the rhizosphere because it is a very eutrophic environment. Organic compounds are supplied to the rhizosphere by plant mucilages, mucigel, lysates, secretions, and root exudates. The latter are particularly important as they supply microbes with an abundant and diversified supply of low molecular weight, soluble carbon compounds. The organic fraction of forest tree root exudates are dominated by organic acids. Other chemical characteristics that distinguish rhizosphere soil from bulk soil include lower pH, lower water potential, lower osmotic potential, lower redox potential, and higher bulk density in the former relative to the latter.

It is concluded that the rhizosphere region of soil exerts dominant regulation over nutrient uptake by roots, root disease from biotic infection, and microbial saprophyte and symbiont ecology. It is, therefore, critically essential to evaluate the potential interactions between pollutants deposited to forests from the atmosphere and rhizosphere structure and function.

Following an intensive review of current literature, it is concluded that six hypotheses appear to describe the most important potential atmosphere-rhizosphere interactions. These are summarized in the following sections.

A. Abnormal Physiology in Aboveground Tree Parts might Cause Alteration of Rhizosphere Effects by Changing the Chemistry and Biology of the Rhizosphere

Studies evaluating air pollutant stress applied to aboveground tree parts and subsequent alteration of rhizosphere chemistry or biology are not available. Substantial evidence was uncovered, however, indicating that changes in rhizosphere chemistry and biology can be mediated by numerous environmental variables acting above ground and that several chemicals (not pollutants) applied to foliage are capable of changing rhizosphere chemistry. Ozone exposure has been shown to alter carbon allocation to roots and to alter root physiology.

B. Atmospheric Deposition might Alter Rhizosphere Regulation of Nutrient Uptake

Numerous nutrients, essential for tree growth, are not readily available for root uptake as they have a tendency in bulk soil to form insoluble precipitates or strong chelates with organic material. Rhizosphere processes that help provide free ions for root uptake include pH regulation, protonation, solubilization, reduction, and complexation. Hydrogen ions and organic acids, produced both by tree roots and by microbes of the rhizosphere, play central roles in these processes. Phosphorus is solubilized in the rhizosphere by root exudates, bacteria, and/or mycorrhizal fungi. Organic acids play a key role in this solubilization. Potassium is released from silicate minerals by organic acids of the rhizosphere. Hydrogen ions generated in the rhizosphere maintain low pH and reduce iron making this important micronutrient available for root uptake. Rhizosphere solubilzation and complexation are critical for the uptake of zinc and copper, while reduction is important in the uptake of manganese. For those trees dependent on microbial fixation for nitrogen supply, the rhizosphere provides habitat for free-living or symbiotic microbes. It is obvious that alterations of hydrogen ion and organic acid quantities in the rhizosphere would have substantial potential to alter the uptake of these essential macro- and micronutrients and cause nutrient stress.

C. Atmospheric Deposition might Alter Rhizosphere Regulation of Heavy Metal Uptake

It is generally assumed that heavy metals accumulated from the atmosphere in forest floors are retained in bulk soil in largely insoluble forms, for example, as hydroxide, sulfide, or other low-solubility precipitates or strong humic acid chelates. In this form the heavy metals are not available for root or microbial uptake, and the potential for toxicity is presumed low. The rhizosphere processes of pH regulation, protonation, solubililization, reduction, and complexation, however, may transform unavailable metals to available metals, as was previously discussed for heavy metal micronutrients. Rhizosphere characteristics that may be particularly important in lead uptake include pH, hydrogen ion availability, organic acid availability, and phosphate availability. As previously indicated, organic acids capable of complexing copper and zinc in the rhizosphere may be important in root uptake of these elements. Manganese reduction to the divalent form may allow excess uptake of this heavy metal. Complexing agents and reduced rhizosphere pH may also facilitate cadmium uptake. Again, any alteration of rhizosphere processes caused by chemical changes induced by air pollution could easily alter availability and uptake of these potentially toxic elements.

D. Atmospheric Deposition might Alter Rhizosphere Regulation of Aluminum Uptake

In many forest soils, clay surfaces are saturated primarily with exchangeable aluminum ions and secondarily with hydrogen ions. Because most of the exchangeable cations in acid forest soils are aluminum and hydrogen, and because a cation-anion balance is required in soil solutions, the input of anions such as sulfate or nitrate via atmospheric deposition may cause increased soil solution concentrations of aluminum and hydrogen via exchange processes. The single most important regulator of aluminum availability to tree roots is rhizosphere pH. Other important rhizosphere regulators of aluminum availability are phosphorus availability and complexing low molecular weight organic compounds. Again, small changes in rhizosphere chemistry could cause big changes in root uptake of aluminum.

E. Atmospheric Deposition might Alter Rhizosphere Regulation of Microbes Mediating Uptake of Plant Nutrients from Soil

In addition to nutrient conservation and tight control over nutrient cycling, trees have evolved critically significant relationships with free-living and symbiotic bacteria and fungi that enhance nutrient supply and uptake. The rhizosphere provides a high-quality habitat for these beneficial microbes. Evidence is being accumulated that altered pH, aluminum, or available heavy metal concentrations in the rhizosphere have the ability to adversely influence microbes essential to normal nutrient absorption.

F. Atmospheric Deposition might Alter Rhizosphere Regulation of Root Pathogen and Soil Saprophyte Ecology

The rhizosphere provides chemical orientation and nutrition for spore germination, infection, and vegetative development to root pathogens. Comparable benefits are provided to saprophytes in the soil around roots. Change in the quality or quantity of organic compounds in the rhizosphere or other chemical characteristics of the rhizosphere has the potential to increase or decrease root disease or increase or decrease critically important soil processes such as ammonification, nitrification, and carbon mineralization.

The rhizosphere is defined as that zone of soil influenced by plant roots. It is appropriate, therefore, that the hypotheses proposed here emphasize potential rhizosphere changes resulting from atmospheric deposition impacts on vegetation. Direct impacts of atmospheric deposition on bulk soil properties and subsequent alteration of rhizosphere processes are probably less likely. Forest soils at particular risk of acidification and aluminum mobilization from atmospheric deposition may, however, have rhizosphere perturbations mediated via bulk soil dynamics rather than tree dynamics. Pulses of acidity to bulk soil, for example, those that may be associated with spring snowmelt conditions, may have the potential to influence rhizosphere processes via direct soil effect.

This review has clearly established the extreme importance of the rhizosphere to forest tree health. It has also clearly defined important hypotheses for potential linkages between atmospheric deposition of pollutants and rhizosphere functions (Figure 6.4). Our understanding of the importance and detail of rhizosphere function has increased dramatically in the last 20 years. Most of this research, however, has been conducted with nonwoody species using solution or artificial soil substrates, and has been carried out in controlled environment facilities.

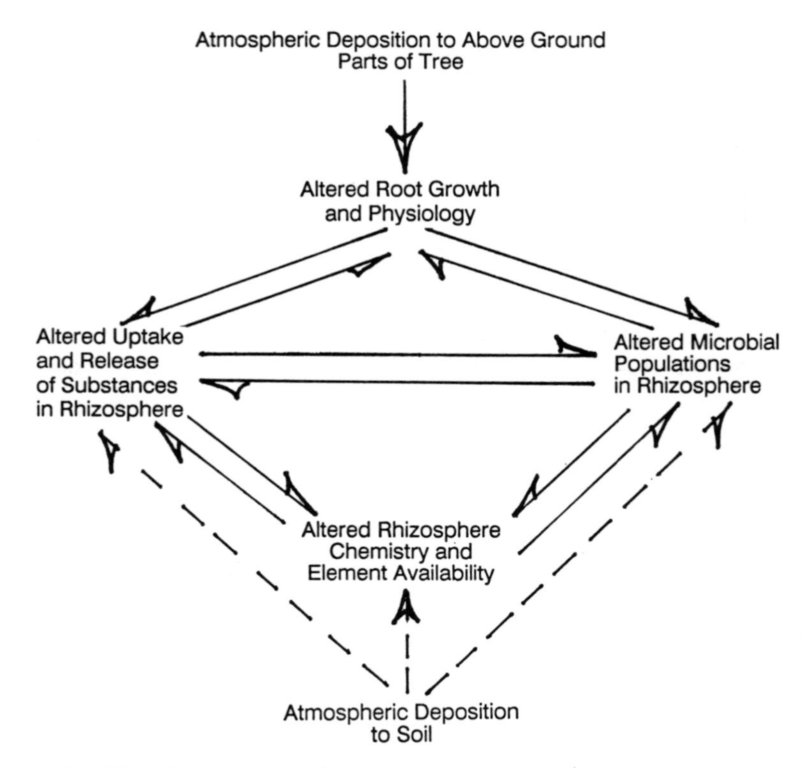

FIGURE 6.4. Hypotheses summarizing potential interactions between atmospheric deposition and rhizosphere dynamics. Hypotheses emphasize rhizosphere alterations mediated by atmospheric deposition influence on forest trees. Effects on rhizosphere dynamics mediated by direct atmospheric deposition influence on bulk soil properties are of uncertain significance.

There is a critical need to develop more investigations employing forest tree species, to use natural forest soils, and to perform experiments and observations in field environments. Over the next 20 years we can advance our understanding of rhizosphere function and structure as much, or more, than we have in the past 20 years if we combine field research and natural soil studies with controlled environment and artificial soil investigations.

VI. Research Priorities

These priority research areas follow directly from the hypotheses that have been developed. Future research efforts with rhizosphere topics must strive to combine both laboratory and field studies. Research under controlled environmental conditions will allow understanding of rhizosphere mechanisms. Research under field conditions will allow mechanisms and hypotheses developed under controlled conditions to be evaluated in actual soil ecosystems. Rhizosphere research should be conducted on all age classes of trees: seedlings, saplings, and mature trees. More use of scanning and transmission electron microscopes and microprobe x-ray analysis should be used in rhizosphere studies of woody plants. Their application to rhizosphere investigations of grass and agricultural crops has proven very valuable.

General research areas, not in priority order, are as follows:

1. Determine how the deposition of air pollutants to aboveground tree parts influences root physiology. Determine how alterations in root physiology change the input of organic and inorganic compounds to the rhizosphere. Place special emphasis on organic acid dynamics.
2. Determine how the deposition of air pollutants to aboveground tree parts *or* to the soil may alter the rhizosphere availability of essential nutrients, especially, phosphorus, potassium, iron, zinc, copper, manganese, and nitrogen.
3. Determine how the deposition of air pollutants to aboveground tree parts *or* to the soil may alter the rhizosphere availability of heavy metals, especially lead, copper, zinc, manganese, cadmium, and nickel.
4. Determine how the deposition of air pollutants to aboveground tree parts *or* to the soil may alter the rhizosphere availability of aluminum.
5. Determine how the deposition of air pollutants to aboveground tree parts *or* to the soil may alter the rhizosphere as a habitat for microbes that mediate the uptake of tree nutrients *and* how the deposition may influence the organisms themselves. Place special emphasis on mycorrhizal fungi.
6. Determine how the deposition of air pollutants to aboveground tree parts *or* to the soil may alter the rhizosphere as a habitat for tree root pathogens or saprophytic microbes that participate in root disease or important soil processes and how the deposition may influence the organisms themselves.

Rhizosphere research, especially that conducted on large trees under field conditions, is especially difficult to perform. Tree roots are difficult to culture and manipulate. Root excavation is labor intensive and root growth discontinuous during the growing season. Long-term effects of relatively low-level atmospheric deposition are difficult to investigate in short-term observations or experiments. The potential importance of rhizosphere processes, however, in regional-scale air pollutant effects fully justifies expanded research investment in this important area.

Acknowledgements. The author wishes to acknowledge primary support of this work from the National Council of the Paper Industry for Air and

Stream Improvement and contributing support from USDA Cooperative Agreement No. 23-993 and USDA Competitive Grant No. 85-CRCR-1-1807.

References

Agnihotri, V.P. 1964. Studies on aspergilli. XIV. Effect of foliar spray of urea on the aspergilli of the rhizosphere of *Triticum vulgare* L. Plant Soil 20:364–370.

Agnihotri, V.P. 1971. Effects of certain fungitoxicants on the viability and pathogenicity of sclerotia of *Waitea circinata*. Phytopathol Z 50:71–80.

Agnihotri, V.P. and O. Vaartaja. 1967. Root exudates from red pine seedlings and their effects on *Pythium ultimum*. Can J Bot 45:1031–1040.

Alexander, M. 1977. Introduction to Soil Microbiology. New York: Wiley.

Andresen, A.M., A.H. Johnson, and T.G. Siccama. 1980. Levels of lead, copper, and zinc in the forest floor in the northeastern United States. J Environ Qual 9:293–296.

Antonovics, J., A.D. Bradshaw, and R.G. Turner. 1971. Heavy metal tolerance in plants. Adv Ecol Res 7:1–85.

Arvik, J.H. and R.L. Zimdahl. 1974. The influence of temperature, pH, and metabolic inhibitors on uptake of lead by plant roots. J Environ Qual 3:374–376.

Ayers, W.A. and R.H. Thorton. 1968. Exudation of amino acids by intact and damaged roots of wheat and peas. Plant Soil 28:193–207.

Azcòn, R. and J.A. Ocampo. 1984. Effect of root exudation on VA mycorrhizal infection at early stages of plant growth. Plant Soil 82:133–138.

Azcòn, R., J.M. Barea, and D.S. Hayman. 1976. Utilization of rock phosphate in alkaline soils by plants inoculated with mycorrhizal fungi and phosphate-solubilizing bacteria. Soil Biol Biochem 8:135–138.

Babich, H., R.J.F. Bewley, and G. Stotzky. 1983. Application of the "ecological dose" concept to the impact of heavy metals on some microbe-mediated ecologic processes in soil. Arch Environ Contam Toxicol 12:421–426.

Bae, H.C., E.H. Bota-Robles, and L.E. Casida. 1972. The microflora of soil as viewed by transmission electron microscopy. Appl Microbiol 23:637–641.

Baker, D.D. 1989. Methods for the isolation, culture and characterization of the Frankiaceae, soil actinomycetes and symbionts of actinorhizal plants. *In*: Isolation of Biotechnological Organisms from Nature, D. Labeda (ed.), pp. 213–236. New York: McGraw Hill.

Baker, K.F. and R.J. Cook. 1974. Biological Control of Plant Pathogens. St. Paul: American Phytopathological Society.

Baker, R. 1971. Analyses involving inoculum density of soilborne plant pathogens in epidemiology. Phytopathology 61:1280–1292.

Baker, R., C.L. Maurer, and R.A. Maurer. 1967. Ecology of plant pathogens in soil. VIII. Mathematical models and inoculum density. Phytopathology 57:662–666.

Balakrishnan, S. and J. Lam Raj. 1970. Effect of TMV and certain chemicals on the rhizosphere microflora of tobacco. Agric Res J Kerala 8:114–120.

Balasubramanian, A. and G. Rangaswami. 1969. Studies on the influence of foliar nutrient sprays on the root exudation pattern in four crop plants. Plant Soil 30:210–220.

Barber, D.A. and J.M. Lynch. 1977. Microbial growth in the rhizosphere. Soil Biol Biochem 9:305–308.

Barea, J.M., E. Navarro, and E. Montoya. 1976. Production of plant growth regulators by rhizosphere phosphate-solubilizing bacteria. J Appl Bacteriol 40:129–134.

Barea, J.M., J.A. Ocampo, R. Azcon, J. Olivares, and E. Montoya. 1978. Effects of ecological factors on the establishment of Azotobacter in the rhizosphere. Ecol Bull (Stockholm) 26:325–330.

Bartlett, R.J. and D.C. Riego. 1972. Toxicity of hydroxy aluminum in relation to pH and phosphorus. Soil Sci 114:194–200.

Baya, A.M., R.S. Boethling, and A. Ramos-Cormenzana. 1981. Vitamin production in relation to phosphate solubilization by soil bacteria. Soil Biol Biochem 13:527–531.

Beagle-Ristaino, J.E. and G.C. Papavizas. 1985. Survival and proliferation of propagules of *Trichoderma* spp. and *Gliocladium virens* in soil and in plant rhizosphere. Phytopathology 75:729–732.

Benninger, L.K., D.M. Lewis, and K.K. Turekian. 1975. The use of natural ^{210}Pb as a heavy metal tracer in the river-estuarine system. *In*: Marine Chemistry and Coastal Environment, T.M. Church (ed.), pp. 201–210. Am. Chem. Soc. Symp. Ser. 18. Washington, D.C.: American Chemical Society.

Bewley, R.J.F. and G. Stotzky. 1983a. Effects of cadmium and zinc on microbial activity in soil: Influence of clay minerals. Part I: Metals added individually. Sci Total Environ 31:41–55.

Bewley, R.J.F. and G. Stotzky. 1983b. Effects of cadmium and zinc on microbial activity in soil: Influence of clay minerals. Part II. Metals added simultaneously. Sci Total Environ 31:57–69.

Bewley, R.J.F. and G. Stotzky. 1983c. Effects of combinations of simulated acid rain and cadmium or zinc on microbial activity in soil. Environ Res 31:332–339.

Bewley, R.J.F. and G. Stotzky. 1983d. Effects of cadmium and simulated acid rain on ammonification and nitrification in soil. Arch Environ Contam Toxicol 12:285–291.

Bonish, P.M. 1973. Cellulase in red clover exudates. Plant Soil 38:307–314.

Bormann, F.H., T.G. Siccama, G.E. Likens, and R.H. Whittaker. 1970. The Hubbard Brook ecosystem study: Composition and dynamics of the tree stratum. Ecol Monogr 40:373–388.

Bowen, G.D. 1968. Chloride efflux along *Pinus radiata* roots. Nature 218:686–687.

Bowen, G.D. 1973. Mineral nutrition in ectomycorrhizae. *In*: Ectomycorrhizae, Their Ecology and Physiology, G.C. Marks and T.T. Kozlowski (eds.), pp. 151–205. New York: Academic Press.

Bowen, G.D. and A.D. Rovira. 1966. Microbial factor in short-term phosphate uptake studies with plant roots. Nature 211:665–666.

Bowen, G.D. and A.D. Rovira. 1973. Are modelling approaches useful in rhizosphere biology. Bull Ecol Res Comm (Stockholm) 17:433–450.

Bowen, G.D. and C. Theodorou. 1973. Growth of ectomycorrhizal fungi around seeds and roots. *In*: Ectomycorrhizae, Their Physiology and Ecology, G.C. Marks and T.T. Kozlowski (eds.), pp. 107–149. New York: Academic Press.

Bowen, G.D., M.F. Skinner, and D.I. Bevege. 1974. Zinc uptake by mycorrhizal and uninfected roots of *Pinus radiata* and *Araucaria cunninghamii*. Soil Biol Biochem 6:141–144.

Bowen, H.J.M. 1966. Trace Elements in Biochemistry. New York: Academic Press.

Boyle, J.R. 1967. Biological Weathering of Micas in Rhizospheres of Forest Trees. Unpublished Ph.D. Thesis, School of Forestry, Yale University, New Haven, Connecticut.

Boyle, J.R., G.K. Voigt, and B.L. Sawhney. 1967. Biotite flakes: Alteration by chemical and biological treatment. Science155:193–195.

Bremner, I. 1974. Heavy-metal toxicities. Q Rev Biophys 7:75–124.

Bremner, I. and A.H. Knight. 1970. The complexes of zinc, copper and manganese present in ryegrass. Br J Nutr 24:279–289.

Brock, T.D. 1966. Principles of Microbial Ecology. Englewood Cliffs, New Jersey: Prentice-Hall.

Campbell, R. and A.D. Rovira. 1973. The study of the rhizosphere by scanning electron microscopy. Soil Biol Biochem 6:747–752.

Carrodus, B.B. and J.L. Harley. 1968. Note on the incorporation of acetate and the TCA cycle in mycorrhizal roots of beech. New Phytol 67:557–560.

Cook, F.D. and A.G. Lockhead. 1959. Growth factor relationships of soil microorganisms as affected by proximity to the plant root. Can J Microbiol 5:323–334.

Cromack, K., R.L. Todd, and C.D. Monk. 1975. Patterns of Basidiomycete nutrient accumulation in conifer and deciduous forest litter. Soil Biol Biochem 7:265–268.

Cromack, K., P. Sollins, R.L. Todd, R. Fogel, A.W. Todd,W. Fender, M. Crossley, and D. Crossley. 1976. The role of oxalic acid and bicarbonate in calcium cycling by fungi and bacteria: Some possible implications for soil animals. Ecol Bull (Stockholm) 25:246–252.

Cronan, C.S. 1980. Solution chemistry of a New Hampshire subalpine ecosystem: A biogeochemical analysis. Oikos 34:272–281.

Cronan, C.S. and C.L. Schofield. 1979. Aluminum leaching response to acid precipitation: Effects on high-elevation watersheds in the northeast. Science 204:304–305.

Cumming, J.R., R.T. Eckert, and L.S. Evans. 1985. Effect of aluminum on potassium uptake by red spruce seedings. Can J Bot 63:1099–1103.

De Leval, J. and J. Remacle. 1969. A microbiological study of the rhizosphere of poplar. Plant Soil 31:31–47.

Dixon, R.O.D. and C.T. Wheeler. 1983. Biochemical, physiological and environmental aspects of symbiotic nitrogen fixation. *In*: Biological Nitrogen Fixation in Forest Ecosystems: Foundations and Applications, J.C. Gordon and C. T. Wheeler (eds.), pp. 107–171. Boston: Dr. W. Junk Publishers.

Döbereiner, J. 1966. Manganese toxicity effects on nodulation and nitrogen fixation of beans (*Phaseolus vulgaris* L.) in acid soils. Plant Soil 24:153–166.

Döbereiner, J. 1983. Dinitrogen fixation in rhizosphere and phyllosphere associations. *In*: Inorganic Plant Nutrition, Encyclopedia of Plant Physiology (New Series) Vol. 15, A. Lauch Lü and R.L. Bieleski (eds.), pp. 330–350. New York: Springer-Verlag.

Döbereiner, J., J.M. Day, and P.J. Dart. 1972. Nitrogenase activity and oxygen sensitivity of the *Paspalum notatum–Azotobacter paspali* association. J Gen Microbiol 71:103–116.

Dodge, C.S. and A.J. Hiatt. 1972. Relationship of pH to ion uptake imbalance by varieties of wheat (*Triticum vulgare*). Agron J 64:476–477.

Douka, C.E. 1977. Study of bacteria from manganese concentrations. Precipitation of manganese by whole cells and cell-free extracts of isolated bacteria. Soil Biol Biochem 9:89–97.

Drury, R.E., R. Baker, and G.J. Griffin. 1983. Calculating the dimensions of the rhizosphere. Phytopathology 73:1351–1354.

Elgawhary, S.M., W.L. Lindsay, and W.D. Kemper. 1970. Effect of complexing agents and acids on the diffusion of zinc to a simulated root. Soil Sci Soc Am Proc 34:211–214.

Energy Research and Development Administration. 1975. Biological Implications of Metals in the Environment. ERDA Symp Ser. 42. Washington, D.C.: Energy Research and Development Administration.

Ferriss, R.S. 1981. Calculating rhizosphere size. Phytopathology 71:1229–1231.

Foster, R.C. 1981. The ultrastructure and histochemistry of the rhizosphere. New Phytol 89:263–273.

Foster, R.C. 1985. The biology of the rhizosphere. *In*: Ecology and Management of Soilborne Plant Pathogens, C.A.Parker, A.D. Rovira, K.J. Moor, P.T.W. Wong, and J.F. Kollinorgen (eds.), pp. 75–79. St. Paul: American Phytopathology Society.

Foster, R.C. and G.C. Marks. 1967. Observations on the mycorrhizas of forest trees. II. The rhizosphere of *Pinus radiata* D. Don. Aust J Biol Sci 20:915–926.

Foster, R.C. and A.D. Rovira. 1976. The ultrastructure of the wheat rhizosphere. New Phytol 76:343–352.

Foster, R.C., A.D. Rovira, and T.W. Cock. 1983. Ultrastructure of the Root-Soil Interface. St. Paul: American Phytopathology Society.

Foy, C.D. 1971. Effects of aluminum on plant growth. *In*: The Plant Root and Its Environment, E.W. Carson (ed.), pp. 601–642. Charlottesville: University of Virginia Press.

Foy, C.D. 1983. The physiology of plant adaptation to mineral stress. Iowa State J Res 57:355–391.

Foy, C.D. 1985. Personal communication, Dr. Charles D. Foy, Soil Scientist, Plant Stress Laboratory, Plant Physiology Institute, ARS-USDA, Beltsville, Maryland 20705.

Foy, C.D. and A.L. Fleming. 1976. The physiology of plant tolerance to aluminum and manganese in acid soils. *In*: Symposium on Crop Tolerance to Suboptimal Land Conditions, Annual Meeting American Society Agronomy, Houston, Texas, November 28-December 3, 1976.

Foy, C.D., R.L. Chaney, and M.C. White. 1978. The physiology of metal toxicity in plants. Annu Rev Plant Physiol 29:511–566.

Friedland, A.J., A.H. Johnson, T.G. Siccama, and D.L. Mader. 1984a. Trace metal profiles in the forest floor of New England. Soil Sci Soc Am J 48:422–425.

Friedland, A.J., A.H. Johnson, and T.G. Siccama. 1984b. Trace metal content of the forest floor in the Green Mountains of Vermont: Spatial and temporal patterns. Water Air Soil Pollut 21:161–170.

Fries, N., M. Bardet, and K. Serck-Hanssen. 1985. Growth of ectomycorrhizal fungi stimulated by lipids from a pine root exudate. Plant Soil 86:287–290.

Führer, J. 1982. Early effects of excess cadmium uptake in *Phaseolus vulgaris*. Plant Cell Environ 5:263–270.

Galloway, J.N., J.D. Thornton, S.A. Norton, H.L. Volchok, and R.A.N. McLean. 1982. Trace metals in atmospheric deposition: A review and assessment. Atmos Environ 16:1677–1700.

Gerdemann, J.W. 1974. Mycorrhizae. *In*: The Plant Root and Its Environment, E.W. Carson (ed.), pp. 205–217. Charlottesville: University Press of Virginia.

Gilligan, C.A. 1979. Modeling rhizosphere infection. Phytopathology 69:782–784.

Göbb, F. und B. Pümpel. 1973. Einfluss von "Grünkupfer Linz" auf Pflanzenausbildung, Mykorrhizabesatz sowie Frosthauarte von Zirbenjungpflanzen. Eur J For Pathol 3:242–245.

Graham, J.H. 1982. Effect of citrus root exudates on germination of chlamydospores of the vesicular-arbuscular mycorrhizal fungus, *Glomus epigaeum*. Phytopathology 72:951.

Griffin, D.M. 1985. Soil as an environment for the growth of root pathogens. *In*: Ecology and Management of Soilborne Plant Pathogens, C.A. Parker, A.D. Rovira, K.J. Moor, P.T.W. Wong, and J.F. Kollmorgen (eds.), pp. 187–190. St. Paul: American Phytopathology Society.

Gupta, P.C. 1971. Foliar spray of gibberellic acid and its influence on the rhizosphere and rhizoplane mycoflora. Plant Soil 34:233–236.

Haghiri, F. 1973. Cadmium uptake by plants. J Environ Qual 2:93–96.

Haines, B. and G.R. Best. 1976. The influence of an endomycorrhizal symbiosis on nitrogen movement through soil columns under regimes of artificial throughfall and artificial acid rain. *In*: Proceedings, 1st International Symposium Acid Precipitation and the Forest Ecosystem, L.S. Dochinger and T.A. Seliga (eds.), pp. 951–961. Gen. Tech. Rep. No NE–23, USDA, Forest Service, Washington, D.C.

Hale, M.G. and L.D. Moore. 1979. Factors affecting root exudation. Adv Agron 31:93–124.

Hale, M.G., C.L. Foy and F.J. Shay. 1971. Factors affecting root exudation. Adv Agron 23:89–109.

Halsall, D.M. 1978. A comparison of *Phytophthora cinnamomi* infection in *Eucalyptus sieberi*, a susceptible species, and *E. maculata*, a field resistant species. Aust J Bot 26:643–655.

Hamlen, R.A., F.L. Lukezic, and J.R. Bloom. 1972. Influence of clipping height on the neutral carbohydrate levels of root exudates of alfalfa plants grown under gnotobiotic conditions. Can J Plant Sci 52:643–649.

Harley, J.L. and R.S. Russell. 1979. The Soil Root Interface. Academic Press, New York. 395 p.

Harley, J.L. and J.S. Waid. 1955. The effect of light upon the roots of beech and its surface population. Plant Soil 7:96–112.

Harris, W.F., R.A. Goldstein, and G.S. Henderson. 1973. Analysis of forest biomass pools, annual primary production, and turnover of biomass for a mixed deciduous forest watershed. *In*: Proceedings of the Working Party on Forest Biomass of IUFRO, H. Young (ed.), pp. 41–64. Orono: University of Maine Press.

Haynes, R.J. 1980. Ion exchange properties of roots and ionic interactions within the root apoplasm: Their role in ion accumulation by plants. Bot Rev 46:75–99.

Hiltner, L. 1904. Über neuere Erfahrungen und Probleme auf dem Gebiet der Bodenbakteriologie und unter besonderer Berücksich- tigung der Gründüngung und Brache. Arb Dtsch Landwirtsch Ber 98:59–78.

Hinch, J. and G. Weste. 1979. Behavior of *Phytophthora cinnamomi* zoospores on roots of Australia forest species. Aust J Bot 27:679–691.

Ho, I. and J.M. Trappe. 1984. Effects of ozone exposure on mycorrhiza formation and growth of *Festuca arundinacea*. Environ Exp Bot 24: 71–74.

Hodges, C.S. 1969. Modes of infection and spread of *Fomes annosus*. Annu Rev Phytopathol 7:247–266.

Hodgson, J.F., H.R. Geering, and W.A. Norvell. 1965. Micronutrient cation complexes in soil solutions. Partition between complexed and uncomplexed forms with solvent extraction. Soil Sci Soc Am Proc 29:665–669.

Hofner, W. 1970. Elsen und manganhatlige verbindungen in blutungssaft von *Helianthus annuus*. Physiol Plant 23:673–677.

Huang, C., F.A. Bazzaz, and L.N. Vanderhoef. 1974. The inhibition of soybean metabolism by cadmium and lead. Plant Physiol 54:122–124.

Hüppel, A. 1970. Inoculation of pine and spruce seedlings with conidia of Fomes annosus. *In*: *Fomes annosus*: Proceedings of 3rd International Conference on *Fomes annosus*, C.S. Hodges, J. Rishbeth, and A. Yde-Andersen (eds.), pp. 54–56. USDA Forest Service, Washington, D.C.

Ivarson, K.G. and H. Katznelson. 1960. Studies on the rhizosphere microflora of yellow birch seedlings. Plant Soil 12:30–40.

Jacobson, J.S. 1982. Ozone and the growth and productivity of agricultural crops. *In*: Effects of Gaseous Air Pollution in Agriculture and Horticulture, M.H. Unsworth and D.P. Ormrod (eds.), pp. 293–304. London: Butterworth.

Jalaluddin, M. 1967. Studies on *Rhizina undulata*. Mycelial growth and ascospore germination. Trans Br Mycol Soc 50:449–459.

Jenny, H. and K.A. Grossenbacher. 1963. Root soil boundary zone as seen in the electron microscope. Soil Sci Soc Am Proc 27:273–277.

Jernelov, A. and A.L. Martin. 1975. Ecological implications of metal metabolism by microorganisms. Annu Rev Microbiol 29:61–77.

John, M.K., H.H. Chuah, and C.J. Van Laerhoven. 1972a. Cadmium contamination of soil and its uptake by oats. Environ Sci Technol 6:555–557.

John, M.K., C.J. Van Laerhoven, and H.H. Chuah. 1972b. Factors affecting plant uptake and phytotoxicity of cadmium added to soils. Environ Sci Technol 6:1005–1009.

Johnson, N.M., C.T. Driscoll, J.S. Eaton, G.E. Likens, and W.H. McDowell. 1981. "Acid rain," dissolved aluminum, and chemical weathering at the Hubbard Brook Experimental Forest, New Hampshire. Geochim Cosmochim Acta 45:1421–1437.

Kelly, W.J. and D.C. Reanney. 1984. Mercury resistance among soil bacteria: Ecology and transferability of genes encoding resistance. Soil Biol Biochem 16:1–8.

Killham, K. and M.K. Firestone, 1983. Vesicular arbuscular mycorrhizal mediation of grass response to acidic and heavy metal depositions. Plant Soil 72:39–48.

Kochenderfer, J.N. 1973. Root distribution under some forest types native to West Virginia. Ecology 54:445–448.

Lagerwerff, J.V. 1971. Uptake of cadmium, lead, and zinc by radish from soil and air. Soil Sci 111:129–133.

Lai, M. and G. Semeniuk. 1970. Picloram-induced increase of carbohydrate exudation from corn seedlings. Phytopathology 60:563–564.

Leaphart, C.D. and M.A. Grismer. 1974. Extent of roots in the forest soil mantle. J For 72:358–359.

Leonard, K.J. 1980. A reinterpretation of the mathematical analysis of rhizoplane and rhizosphere effects. Phytopathology 70:695–696.

Letchworth, M.B. and V. Blum. 1977. Effects of acute ozone exposure on growth, nodulation and nitrogen content of ladino clover. Environ Pollut 14:303–312.

Lindberg, S.E. 1982. Factors influencing trace metal, sulfate, and hydrogen ion concentrations in rain. Atmos Environ 16:1701–1709.

Linehan, D.J., A.H. Sinclair, and M.C. Mitchell. 1985. Mobilization of Cu, Mn, and Zn in the soil solutions of barley rhizospheres. Plant Soil 86:147–149.

Lockwood, J.L. and A.B. Filonow. 1981. Responses of fungi to nutrient-limiting conditions and to inhibitory substances in natural habitats. Adv Microb Ecol 5:1–61.

Lyford, W.H. and B.F. Wilson. 1964. Development of the root system of *Acer rubrum* L. Harvard Forest Paper No. 10, Harvard Forest, Petersham, Massachusetts.

Lyford, W.H. and B.F. Wilson. 1966. Controlled growth of forest tree roots: Technique and application. Harvard Forest Paper No. 16, Harvard Forest, Petersham, Massachusetts.

Mahoney, M.J., B.I. Chevone, J.M. Skelly, and L.D. Moore. 1985. Influence of mycorrhizae on the growth of loblolly pine seedlings exposed to ozone and sulfur dioxide. Phytopathology 75:679–682.

Malajczuk, N., A.J. McComb, and C.A. Parker. 1977. Infection by *Phytophthora cinnamomi* Rands of roots of *Eucalyptus calophylla* R. Br. and *Eucalyptus marginata* Donner. Sm. Aust J Bot 25:483–500.

Maliszewska, W. and R. Moreau. 1960. A study of the fungal microflora in the rhizosphere of fir (*Albies alba* Mill). *In*: The Ecology of Soil Fungi, D. Parkinson and J. Waid (eds.), pp. 209–220. Liverpool, United Kingdom: Liverpool University Press.

Manning, W.J., W. Feder, P.M. Papia, and I. Perkins. 1971. Influence of foliar ozone injury on root development and root surface fungi on pinto bean plants. Environ Pollut 1:305–312.

Martin, J.K. 1977a. Factors influencing the loss of organic carbon from wheat roots. Soil Biol Biochem 9:1–7.

Martin, J.K. 1977b. Effect of soil moisture on the release of organic carbon from wheat roots. Soil Biol Biochem 9:303–304.

Martin, J.K. and R.C. Foster. 1985. A model system for studying the biochemistry and biology of the root-soil interface. Soil Biol Biochem 17:261–269.

Martin, J.K. and D.W. Puckridge. 1982. Carbon flow through the rhizosphere of wheat crops in South Australia. *In*: The Cycling of Carbon, Nitrogen, Sulfur, and Phosphorus in Terrestrial and Aquatic Ecosystems, I.E. Galbally and J.R. Freney (eds.) pp.77–82. Canberra, Australia: Australian Academy of Science.

Marx, D.H. 1972. Ectomycorrhizae as biological deterrents to pathogenic root infections. Annu Rev Phytopathol 10:429–454.

Marx, D.H. and C.B. Davey. 1969a. The influence of ectotrophic mycorrhizal fungi on the resistance of pine roots to pathogenic infections. III. Resistance of aseptically formed mycorrhizae to infection by *Phytophthora cinnamomi*. Phytopathology 59:549–558.

Marx, D.H. and C.B. Davey. 1969b. The influence of ectotrophic mycorrhizal fungi on the resistance of pine roots to pathogenic infection. IV. Resistance of naturally occurring mycorrhizae to infections by *Phytophthora cinnamomi*. Phytopathology 59:559–565.

McDougall, B.M. 1970. Movement of ^{14}C-photosynthate into the roots of wheat seedlings and exudation of ^{14}C from intact roots. New Phytol 69:37–46.

McIlveen, W.D., R.A. Spotts, and D.D. Davis. 1975. The influence of soil zinc on nodulation, mycorrhizae, and ozone sensitivity to pinto bean. Phytopathology 65:645–647.

McLaughlin, S.B. 1985. Effect of air pollution on forests. A critical review. J Air Pollut Control Assoc 35:512–534.

McMinn, R.G. 1963. Characteristics of Douglas-fir root systems. Can J Bot 41:105–122.

Melin, E. 1955. Nyare undersökningar över skogstrauadens mykorrhizasvampar och det pysiologiska vauarelspelet mellan dem och trauadens rötter. Uppsala Univ Arsskr 3:1–29.

Melin, E. 1963. Some effects of forest tree roots on mycorrhizal Basidiomycetes. *In*: 13th Symposium of Society of General Microbiology, London, pp. 125–145. London: Cambridge University Press.

Miller, W.P. and W.W. McFee. 1983. Distribution of cadmium, zinc,copper, and lead in soils of industrialized northwestern Indiana. J Environ Qual 12:29–33.

Moghimi, A. and M.E. Tate. 1978. Does 2-ketogluconate chelate calcium in the pH range 2.4 to 6.4? Soil Biol Biochem 10:289–292.

Moghimi, A., D.G. Lewis, and J.M. Oades. 1978a. Release of phosphate from calcium phosphates by rhizosphere products. Soil Biol Biochem 10:277–281.

Moghimi, A., M.E. Tate, and J.M. Oades. 1978b. Characterization of rhizosphere products especially 2-ketogluconic acid. Soil Biol Biochem 10:283–287.

Mollitor, A.V. and D.J. Raynal. 1982. Acid precipitation and ionic movements in Adirondack forest soils. Soil Sci Soc Am J 46:137–141.

Mosse, B., D.P. Stribley, and F. LeTacon. 1981. Ecology of mycorrhizae and mycorrhizal fungi. Adv Microb Ecol 5:137–169.

Mugwira, L.M. and S.V. Patel. 1977. Aluminum tolerance in triticale. II. Differential pH changes in ion uptake imbalance by triticale, wheat, and rye. Agron J 69:407–412.

National Research Council. 1980. Trace Element Geochemistry of Coal Resource Development Related to Environmental Quality and Health. Washington, D.C.: National Academy Press.

Newman, E.I. 1978. Root micro-organisms: Their significance in the ecosystem. Biol Rev 53:511–554.

Newman, E.I. and A. Watson. 1977. Microbial abundance in the rhizosphere: A computer model. Plant Soil 48:17–56.

Nye, P.H. 1981. Changes of pH across the rhizosphere induced by roots. Plant Soil 61:7–25.

Ocampo, J.A., J.M. Barea, and E. Montoya. 1978. Bacteriostasis and the inoculation of phosphate-solubilizing bacteria in the rhizosphere. Soil Biol Biochem 10:439–440.

Olomu, M.O., G.J. Raca, and C.M. Cho. 1973. Effect of flooding on the Eh, pH and concentration of Fe and Mn in several Manitoba soils. Soil Sci Soc Am Proc 37:220–224.

Olsen, R.A., R.B. Clark, and J.H. Bennett. 1981. The enhancement of soil fertility by plant roots. Am Sci 69:378–384.

Page, A.L. and A.C. Chang. 1979. Contamination of soil and vegetation by atmospheric deposition of trace elements. Phytopathology 69:1007–1011.

Parekh, P.P. and L. Husain. 1981. Trace element concentrations in summer aerosols at rural sites in New York State and their possible sources. Atmos Environ 15:1717–1725.

Parker, C.A., M.J. Trinick, and D.L. Chatel. 1977. Rhizobia as soil and rhizosphere inhabitants. *In*: A Treatise on Dinitrogen Fixation, Vol. IV, Agronomy and Ecology, pp. 311–352. New York: Wiley.

Pohlman, A.A. and J.G. McColl. 1982. Nitrogen fixation in the rhizosphere and rhizoplane of barley. Plant Soil 69:341–352.

Polonenko, D.R. and C.I. Mayfield. 1979. A direct observation technique for studies on rhizoplane and rhizosphere colonization. Plant Soil 51:405–420.

Postgate, J.R. 1974. New Advances and future potential in biological nitrogen fixation. J Appl Bacteriol 37:185–202.

Reddy, C.N. and W.H. Patrick Jr. 1977. Effect of redox potential and pH on the uptake of cadmium and lead by rice plants. J Environ Qual 6:259–262.

Redington, C.B. and J.L. Peterson. 1971. Influence of environment on *Albizzia julibrissin* root exudation and exudate effect on *Fusarium oxysporium* f. sp. *perniciosum* in soil. Phytopathology 61:812–815.

Reich, P.B., A.W. Schoettle, H.F. Stroo, J. Troiano, and R.G. Amundson. 1985. Effects of O_3, SO_2, and acidic rain on mycorrhizal infection in northern red oak seedlings. Can J Bot 63:2049–2055.

Reid, C.P.P. 1974. Assimilation, distribution and root exudation of ^{14}C by ponderosa pine seedlings under induced water stress. Plant Physiol 54:44–49.

Reid, C.P.P. and J.G. Mexal. 1977. Water stress effects on root exudation by lodgepole pine. Soil Biol Biochem 9:417–421.

Reynolds, K.M., D.M. Benson, and R.I. Bruck. 1985. Epidemiology of Phytophthora root rot of Fraser fir: Estimates of rhizosphere width and inoculum efficiency. Phytopathology 75:1010–1014.

Riffaldi, R. and R. Levi-Minzi. 1975. Adsorption and desorption of Cd on humic acid fraction of soils. Water Soil Air Pollut 5:179–184.

Rolfe, G.L. 1973. Lead uptake by selected tree seedlings. J Environ Qual 2:153–157.

Rovira, A.D. 1965a. Plant root exudates and their influence upon soil microorganisms. *In*: Ecology of Soil-Borne Plant Pathogens, K.F. Baker and W.G. Snyder (eds.), pp. 170–186. Berkeley: University of California Press.

Rovira, A.D. 1965b. Interactions between plant roots and soil microorganisms. Annu Rev Microbiol 19:241–266.

Rovira, A.D. 1969. Plant root exudates. Bot Rev 35:35–57.

Rovira, A.D. and R. Campbell. 1975. A scanning electron microscope study of interactions between microorganisms and *Galumannomyces graminis* (Syn. *Ophiobolus graminis*) on wheat roots. Microb Ecol 2:177–185.

Rovira, A.D. and B.M. McDougall. 1967. Microbiological and biochemical aspects of the rhizosphere. *In*: Soil Biochemistry, A.D. McLaren and G.H. Peterson (eds.), pp. 417–463. New York: Marcel Dekker.

Rovira, A.D., G.D. Bowen, and R.G. Foster. 1983. The significance of rhizosphere microflora and mycorrhizas in plant nutrition. *In*: Inorganic Plant Nutrition, Encyclopedia of Plant Physiology, New Series, Vol. 15, A. Lauchli and R.L. Bieleski (eds.), pp. 61–93. New York: Springer-Verlag.

Sanders, F.E., B. Mosse, and P.B. Tinker. 1975. Endomycorrhizas. New York: Academic Press.

Sarkar, A.N. and R.G. Wyn Jones. 1982. Effect of rhizosphere pH on the availability and uptake of Fe, Mn, and Zn. Plant Soil 66:361–372.

Sethunathan, N. 1970a. Foliar sprays of growth regulators and rhizosphere effect in *Cajanus cajan* Millsp. I. Quantitative changes. Plant Soil 33:62–65.

Sethunathan, N. 1970b. Foliar sprays of growth regulators and rhizosphere effect in *Cajanus cajan* Millsp. II. Qualitative changes in the rhizosphere and certain metabolic changes in the plant. Plant Soil 33:71–74.

Shafer, S.R., L.F. Grand, R.I. Bruck, and A.S. Heagle. 1985. Formation of ectomycorrhizae on *Pinus taeda* seedlings exposed to simulated acidic rain. Can J For Res 15:66–71.

Siccama, T.G., W.H. Smith, and D.L. Mader. 1980. Changes in lead, zinc, copper, dry weight, and organic matter content of the forest floor of white pine stands in central Massachusetts over 16 years. Environ Sci Technol 14:54–56.

Slankis, V. 1958. The role of auxin and other exudates in mycorrhizal symbiosis of forest trees. *In*: The Physiology of Forest Trees, K.V. Thimann (ed.), pp. 427–443. New York: Ronald Press.

Smiley, R.W. 1974. Rhizosphere pH as influenced by plants, soils, and nitrogen fertilizers. Soil Sci Soc Am Proc 38:795–799.

Smith, W.H. 1969a. Release of organic materials from the roots of tree seedlings. For Sci 15:138–143.

Smith, W.H. 1969b. Comparison of mycelial and sclerotial inoculum of *Macrophomina phaseoli* in the mortality of pine seedlings under varying soil conditions. Phytopathology 59:379–382.

Smith, W.H. 1969c. Germination of *Macrophomina phaseoli* sclerotia as affected by *Pinus lambertiana* root exudate. Can J Microbiol 15:1387–1391.

Smith, W.H. 1970a. Technique for collection of root exudates from mature trees. Plant Soil 32:238–241.

Smith, W.H. 1970b. Root exudates of seedling and mature sugar maple. Phytopathology 60:701–703.

Smith, W.H. 1972. Influence of artificial defoliation on exudates of sugar maple. Soil Biol Biochem 4:111–113.

Smith, W.H. 1974. Influence of root exudates of red pine on the germination of conidia of *Fomes annosus*. *In*: *Fomes annosus*, E.G. Kuhlman (ed.), pp. 231–237. Washington, D.C.: USDA Forest Service.

Smith, W.H. 1976. Character and significance of forest tree root exudates. Ecology 57:324–331.

Smith, W.H. 1977. Tree root exudates and the forest soil ecosystem: Exudate chemistry, biological significance, and alteration by stress. *In*: The Belowground Ecosystem: A Synthesis of Plant-Associated Processes, J.K. Marshall (ed.), pp. 289–301. Range Sci. Dept. Sci. Ser. No. 26, Ft. Collins: Colorado State University.

Smith, W.H. 1979. Carbendazim exudation from roots of American elm. Soil Biol Biochem 11:687–688.

Smith, W.H. 1985. Forest quality and air quality. J For 83:82–92.

Smith, W.H. 1990. Air Pollution and Forests, Second Edition. New York: Springer-Verlag. 618 pp.

Smith, W.H. and T.G. Siccama. 1981. The Hubbard Brook Ecosystem Study: Biogeochemistry of lead in the northern hardwood forest. J Environ Qual 10:323–333.

Smith, W.H. and Pooley, A.S. 1989. Red spruce rhizosphere dynamics: Spatial distribution of aluminum and zinc in the near-root soil zone. For Sci 35:1114–1124.

Society of American Foresters. 1984. Acidic Deposition and Forests. Society of American Foresters, Bethesda, Maryland.

Somers, E. 1961. The fungitoxicity of metal ions. Ann Appl Biol 49:246–253.

Stevenson, F.J. 1982. Humus Chemistry. New York: Wiley.

Stout, B.B. 1956. Studies of the Root Systems of Deciduous Trees. Black Rock Forest Bull No 15. Harvard Forest, Petersham, Massachusetts.

Stroo, H.F. and M. Alexander. 1985. Effect of simulated acid rain on mycorrhizal infection of *Pinus strobus* L. Water Air Soil Pollut 25:107–114.

Sullia, S.B. 1968. Effect of foliar spray of hormones on the rhizosphere of leguminous weeds. Plant Soil 29:292–295.

Summers, A.O. and S. Silver. 1978. Microbial transformations of metals. Annu Rev Microbiol 32:637–672.

Tan, K.H. 1985. Decomposition by organic acids. *In*: 1986 McGraw-Hill Yearbook of Science and Technology, S.P. Parker (ed.), pp. 406–408. New York: McGraw Hill.

Tingey, D.T. 1974. Ozone-induced alterations in the metabolite pools and enzyme activities of plants. *In*: Air Pollution Effects on Plant Growth, M. Dugger (ed.), pp. 40–57, Am. Chem. Soc. Symp. Ser. No. 3. Washington, D.C.: American Chemical Society.

Tingey, D.T. and V. Blum. 1973. Effects of ozone on soybean nodules. J Environ Qual 2:341–342.

Tingey, D.T., W.W. Heck, and R.A. Reinert. 1971. Effect of low concentrations of ozone and sulfur dioxide on foliage, growth, and yield of radish. J Am Soc Hortic Sci 96:369–371.

Tingey, D.T., R.G. Wilhour, and E. Standley. 1976. The effect of chronic ozone exposures on the metabolite content of ponderosa pine seedlings. For Sci 22:234–241.

Tinker, P.B.H. 1984. The role of microorganisms in mediating and facilitating the uptake of plant nutrients from soil. Plant Soil 76:77–91.

Tinker, P.B.H. and F.E. Sanders. 1975. Rhizosphere microorganisms and plant nutrition. Soil Sci 119:363–368.

Tippett, J.T., A.A. Holland, G.C. Marks, and T.P. O'Brien. 1976. Penetration of *Phytophthora cinnamomi* into disease tolerant and susceptible eucalyptus. Arch Microbiol 108:231–242.

Trappe, J.M. 1977. Selection of fungi for ectomycorrhizal inoculation in nurseries. Annu Rev Phytopathol 15:203–222.

Treshow, M. (ed). 1984. Air Pollution and Plant Life. New York: Wiley.

Tribunskaya, A.J. 1955. Investigation of the microflora of the rhizosphere of pine seedlings. Mikrobiologiya (Moscow)24:188–192.

Turner, P.D. 1963. Influence of root exudates of cacao and other plants on spore development of *Phytophthora palmivora*. Phytopathology 53:1337–1339.

U.S. Environmental Protection Agency. 1987. Effects of ozone and other photochemical oxidants on vegetation. *In*: Criteria Document for Ozone and Other

Photochemical Oxidants. Washington, D.C.: U.S. Environmental Protection Agency.

Vancura, V. and J.L. Garcia. 1969. Root exudates of reversibly wilted millet plants (*Panicum miliaceum* L.). Ecol Plant (Czechoslovakia) 4:93–98.

Vesper, S.J. and T.C. Weidensaul. 1978. Effects of cadmium, nickel, copper, and zinc on nitrogen fixation by soybeans. Water Air Soil Pollut 9:413–422.

Vrany, J. 1965. Effect of foliar application on the rhizosphere microflora. *In*: Proceedings, Symposium Relationships Between Soil Microorganisms and Plant Roots, pp. 84–90. Prague: Czechoslovakia Academy of Science.

Vrany, J. 1972. The effect of foliar application of urea on the root fungi of wheat growing in soil artificially contaminated with *Fusarium* spp. Folia Microbiol 17:500–504.

Vrany, J. and J. Macura. 1971. Changes in bacterial population during the colonization of wheat rhizosphere, following aseptic cultivation and foliar application of urea. Zentralbl Bakteriol Parasitenkd Infektonskr Hyg Abt 2 126:399–408.

Vrany, J., V. Vancura, and J. Macura. 1962. The effect of foliar application of some readily metabolized substances, growth regulators, and antibiotics on rhizosphere microflora. Folia Microbiol 7:61–70.

Waksman, S.A. 1967. The Actinomycetes. A Summary of Current Knowledge. New York: Ronald Press.

Weinstein, L.H. and D.C. McCune. 1979. Air pollution stress. *In*: Stress Physiology in Crop Plants, H. Mussell and R. Staples (eds.), pp. 328–342. New York: Wiley.

Wood, J.M. and H.K. Wang. 1983. Microbial resistance to heavy metals. Environ Sci Technol 17:582A–590A.

Wood, M. and J.E. Cooper. 1984. Aluminum toxicity and multiplication of *Rhizobium trifolii* in a defined growth medium. Soil Biol Biochem 16:571–576.

Wood, M., J.E. Cooper, and A.J. Holding. 1984a. Soil acidity factors and nodulation of *Trifolium repens*. Plant Soil 78:367–379.

Wood, M., J.E. Cooper, and A.J. Holding. 1984b. Aluminum toxicity and nodulation of *Trifolium repens*. Plant Soil 78:381–391.

Wood, T. 1979. Biological and Chemical Control of Phosphorus Cycling in a Northern Hardwood Forest Ecosystem. Ph.D. thesis. School of Forestry and Environmental Studies, Yale University, New Haven, Connecticut.

Zak, B. 1964. Role of mycorrhizae in root disease. Annu Rev Phytopathol 2:377–392.

Zimdahl, R.L. and D.E. Koeppe. 1979. Uptake by plants. *In*: Lead in the Environment, W.R. Boggess (ed.), pp. 99–104. Washington, D.C.: National Science Foundation.

Index